PHYSICS FROM FISHER INFORMATION
A Unification

This book defines and develops a unifying principle of physics, that of 'extreme physical information'. The information in question is, perhaps surprisingly, not Shannon or Boltzmann entropy but, rather, Fisher information, a simple concept little known to physicists.

Both statistical and physical properties of Fisher information are developed. This information is shown to be a physical measure of disorder, sharing with entropy the property of monotonic change with time. The information concept is applied 'phenomenally' to derive most known physics, from statistical mechanics and thermodynamics to quantum mechanics, the Einstein field equations, and quantum gravity. Many new physical relations and concepts are developed including new definitions of disorder, time and temperature. The information principle is based upon a new theory of measurement, one which incorporates the observer into the phenomenon that he/she observes. The 'request' for data creates the law that, ultimately, gives rise to the data. The observer creates his or her local reality.

This fascinating work will be of great interest to students and researchers from all areas of physics with an interest in new ways of looking at the subject.

PHYSICS FROM FISHER INFORMATION

A Unification

B. ROY FRIEDEN

Optical Sciences Center, The University of Arizona

CAMBRIDGE
UNIVERSITY PRESS

PUBLISHED BY THE PRESS SYNDICATE OF THE UNIVERSITY OF CAMBRIDGE
The Pitt Building, Trumpington Street, Cambridge CB2 1RP, United Kingdom

CAMBRIDGE UNIVERSITY PRESS
The Edinburgh Building, Cambridge CB2 2RU, UK http://www.cup.cam.ac.uk
40 West 20th Street, New York, NY 10011-4211, USA http://www.cup.org
10 Stamford Road, Oakleigh, Melbourne 3166, Australia

First published 1998

Printed in the United Kingdom at the University Press, Cambridge

Typeset in Times 11/14pt [KT]

A catalogue record for this book is available from the British Library

Library of Congress Cataloguing in Publication data

Frieden, B. Roy, 1936–
    Physics from Fisher information : a unification / B. Roy Frieden.
        p.   cm.
    Includes index.
    ISBN 0-521-63167-X (hardbound)
    1. Physical measurements.   2. Information theory.   3. Physics-
-Methodology.   I. Title.
    QC39.F75   1998
    530.8–dc21   98–20461   CIP

ISBN 0 521 63167 X hardback

To my wife Sarah
and to my children Mark, Amy and Miriam

Contents

0

Introduction

0.1 Aim of the book

The overall aim of this book is to develop a theory of measurement that incorporates the observer into the phenomenon under measurement. By this theory, the observer becomes both a collector of data and an activator of the physical phenomenon that gives rise to the data. These ideas have probably been best stated by J. A. Wheeler (1990), (1994):

All things physical are information-theoretic in origin and this is a participatory universe ... Observer participancy gives rise to information; and information gives rise to physics.

The measurement theory that will be presented is largely, in fact, a quantification of these ideas. However, the reader might be surprised to find that the 'information' that is used is not the usual Shannon or Boltzmann entropy measures, but one that is relatively unknown to physicists, that of R. A. Fisher.

During the same years that quantum mechanics was being developed by Schroedinger (1926) and others, the field of classical measurement theory was being developed by R. A. Fisher (1922) and co-workers (see Fisher Box, 1978, for a personal view of his professional life). According to classical measurement theory, the quality of any measurement(s) may be specified by a form of information that has come to be called Fisher information. Since these formative years, the two fields – quantum mechanics and classical measurement theory – have enjoyed huge success in their respective domains of application. And until recent times it has been presumed that the two fields are distinct and independent.

However, the two fields actually have strong overlap. The thesis of this book is that all physical law, from the Dirac equation to the Maxwell–Boltzmann velocity dispersion law, may be unified under the umbrella of classical meas-

John A. Wheeler, from a photograph taken *c.* 1970 at Princeton University. Sketch by the author.

urement theory. In particular, the information aspect of measurement theory – Fisher information – is the key to the unification.

Fisher information is part of an overall theory of physical law called the principle of extreme physical information (EPI). The unifying aspect of this principle will be shown by example, i.e., by application to the major fields of physics: quantum mechanics, classical electromagnetic theory, statistical mechanics, gravitational theory, etc. The defining paradigm of each such discipline is either a wave equation, a field equation or a distribution function of some sort. These will be derived by use of the EPI principle. A separate chapter is devoted to each such derivation. New effects are found, as well, by the information approach.

Such a unification is, perhaps, long overdue. Physics is, after all, the science of measurement. That is, physics is a quantification of *observed* phenomena.

And observed phenomena contain noise, or fluctuations. The physical paradigm equations (defined above) define the fluctuations or errors from ideal values that occur in such observations. That is, *the physics lies in the fluctuations*. On the other hand, classical Fisher information is a scalar measure of these very physical fluctuations. In this way, Fisher information is intrinsically tied into the laws of fluctuation that define theoretical physics.

EPI theory proposes that all physical theory results from observation: in particular, *imperfect* observation. Thus, EPI is an observer-based theory of physics. We are used to the concept of an imperfect observer in addressing quantum theory. But the imperfect observer does not seem to be terribly important to classical electromagnetic theory, for example, where it is assumed (wrongly) that fields are known exactly. The same comment can be made about the gravitational field of general relativity. What we will show is that, by admitting that any observation is imperfect, one can derive both the Maxwell equations of electromagnetic theory and the Einstein field equations of gravitational theory. The EPI view of these equations is that they are expressions of fluctuation in the values of measured field positions. Hence, the four-positions (r, t) in Maxwell's equations represent, in the EPI interpretation, random excursions from an ideal, or mean, four-position over the field.

Dispensing with the artificiality of an 'ideal' observer reaps many benefits for purposes of *understanding* physics. EPI is, more precisely, an expression of the 'inability to know' a measured quantity. For example, quantum mechanics is derived from the viewpoint of the inability to know an ideal position. We have found, from teaching the material in this book, that students more easily understand quantum mechanics from this viewpoint than from the conventional viewpoint of derivative operators that somehow represent energy or momentum. Furthermore, that *the same* inability to know also leads to the Maxwell equations when applied to that scenario is even more satisfying. It is, after all, a human desire to find common cause in the phenomena we see.

Unification is also, of course, the major aim of physics, although EPI is probably not the ultimate unification that many physicists seek. Our aim is to propose a *comprehensive* approach to deriving physical laws, based upon a new theory of measurement. Currently, the approach presumes the existence of sources and particles. EPI derives major classes of particles, but not all of them, and does not derive the sources. We believe, however, that EPI is a large step in the right direction. Given its successes so far, the sources and remaining particles should eventually follow from these considerations as well.

At this point we want to emphasize *what this book is not about*. This is not a book whose primary emphasis is upon the *ad hoc* construction of Lagrangians and their extremization. That is a well-plowed field. Although we often derive a physical law via the extremization of a Lagrangian integral, the information viewpoint we take leads to other types of solutions as well. Some solutions arise, for example, out of *zeroing* the integral. (See the derivation of the Dirac equation in Chap. 4.) Other laws arise out of a combination of both zeroing and extremizing the integral. Similar remarks may be made about the process by which the Lagrangians are *formed*. The zeroing and extremizing operations actually allow us to *solve for* the Lagrangians of the scenarios (see Chaps. 4–9, and 11). In this way we avoid, to a large degree, the *ad hoc* approach to Lagrange construction that is conventionally taken. This subject is discussed further in Secs. 1.1 and 1.8.8. The rationale for both zeroing and extremizing the integral is developed in Chap. 3. It is one of *information transfer* from phenomenon to data.

The layout of the book is, very briefly, as follows. The current chapter is intended to derive and exemplify mathematical techniques that the reader might not be familiar with. Chap. 1 is an introduction to the concept of Fisher information. This is for single-parameter estimation problems. Chap. 2 generalizes the concept to multidimensional estimation problems, ending with the scalar information form I that will be used thereafter in the applications Chaps. 4–11. Chap. 3 introduces the concept of the 'bound information' J, leading to the principle of extreme physical information (EPI). This is derived from various points of view. Chaps. 4–11 apply EPI to various measurement scenarios, in this way deriving the fundamental wave equations and distribution functions of physics. Chap. 12 is a chapter-by-chapter summary of the key points made in the development. The reader in a hurry might choose to read this first, to get an idea of the scope of the approach and the phenomena covered.

0.2 Level of approach

The level of physics and mathematics that the reader is presumed to have is that of a senior undergraduate in physics. Calculus, through partial differential equations, and introductory matrix theory, are presumed parts of his/her background. Some notions from elementary probability theory are also used. But since these are intuitive in nature, the appropriate formula is usually just given, with reference to a suitable text as needed.

A cursory scan through the chapters will show that a minimal amount of prior knowledge of physical theory is actually used or needed. In fact, *this is the nature of the information approach taken* and is one of its strengths. The main physical input to each application of the approach is a simple law of invariance that is obeyed by the given phenomenon.

The overall mathematical notation that is used is that of conventional calculus, with additional matrix and vector notation as needed. Tensor notation is only used where it is a 'must' – in Chaps. 6 and 11 on classical and quantum relativity, respectively. No extensive operator notation is used; this author believes that specialized notation often hinders comprehension more than it helps the student to understand theory. Sophistication *without* comprehension is definitely not our aim.

A major step of the information principle is the extremization and/or zeroing of a scalar integral. The integral has the form

$$K \equiv \int d\mathbf{x} \mathscr{L}[\mathbf{q}, \mathbf{q}', \mathbf{x}], \ \mathbf{x} \equiv (x_1, \ldots, x_M), \ d\mathbf{x} \equiv dx_1 \cdots dx_M, \ \mathbf{q}, \ \mathbf{x} \ \text{real},$$

$$\mathbf{q} \equiv (q_1, \ldots, q_N), \ q_n \equiv q_n(\mathbf{x}), \ \mathbf{q}'(\mathbf{x}) \equiv \partial q_1/\partial x_1, \partial q_1/\partial x_2, \ldots, \partial q_N/\partial x_M.$$
$$(0.1)$$

Mathematically, $K \equiv K[\mathbf{q}(\mathbf{x})]$ is a 'functional', i.e., a single number that depends upon the values of one or more functions $\mathbf{q}(\mathbf{x})$ continuously over the domain of $\mathbf{x}$. Physically, K has the form of an 'action' integral, whose extremization has conventionally been used to derive fundamental laws of physics (Morse and Feshbach, 1953). Statistically, we will find that K is the 'physical information' of an overall system consisting of a measurer and a measured quantity. The limits of the integral are fixed and, usually, infinite. The dimension M of $\mathbf{x}$-space is usually 4 (space-time). The functions q_n of $\mathbf{x}$ are probability amplitudes, i.e., whose squares are probability densities. The q_n are to be found. They specify the physics of a measurement scenario. Quantity $\mathscr{L}$ is a known function of the q_n, their derivatives with respect to all the x_m, and $\mathbf{x}$. $\mathscr{L}$ is called the 'Lagrangian' density (Lagrange, 1788). It also takes on the role of an information density, by our statistical interpretation.

The solution to the problem of extremizing the information K is provided by a mathematical approach called the 'calculus of variations'. Since the book makes extensive use of this approach, we derive it in the following.

0.3 Calculus of variations

0.3.1 Derivation of Euler–Lagrange equation

We find the answer to the lowest-dimension version $M = N = 1$ of the problem, and then generalize the answer as needed. Consider the problem of finding the single function $q(x)$ that satisfies

$$K = \int_a^b dx \mathscr{L}[x, q(x), q'(x)] = extrem., \quad q'(x) \equiv dq(x)/dx. \qquad (0.2)$$

A well-known example is the case $\mathscr{L} = \frac{1}{2}mq'^2 - V(q)$ of a particle of mass m moving with displacement amplitude q at time $x \equiv t$ in a known field of potential $V(q)$. We will return to this problem below.

Suppose that the solution to the given problem is the function $q_0(x)$ as shown in Fig. 0.1. Of course at the endpoints (a, b) the function has the values $q_0(a)$, $q_0(b)$, respectively. Consider any finite departure $q(x, \varepsilon)$ from $q_0(x)$,

$$q(x, \varepsilon) = q_0(x) + \varepsilon \eta(x), \qquad (0.3)$$

with ε a finite number and $\eta(x)$ any perturbing function. Any function $q(x, \varepsilon)$ must pass through the endpoints so that, from Eq. (0.3),

$$\eta(a) = \eta(b) = 0. \qquad (0.4)$$

Eq. (0.2) is, with this representation $q(x, \varepsilon)$ for $q(x)$,

$$K = \int_a^b dx \mathscr{L}[x, q(x, \varepsilon), q'(x, \varepsilon)] \equiv K(\varepsilon), \qquad (0.5)$$

a function of the small parameter ε. (Once x is integrated out, only the ε-dependence remains.)

We use ordinary calculus to find the solution. By the construction (0.3), $K(\varepsilon)$ attains the extremum value when $\varepsilon = 0$. Since an extremum value is attained there, $K(\varepsilon)$ must have zero slope at $\varepsilon = 0$ as well. That is,

$$\frac{\partial K}{\partial \varepsilon}\bigg|_{\varepsilon=0} = 0. \qquad (0.6)$$

The situation is sketched in Fig. 0.2.

We may evaluate the left-hand side of Eq. (0.6). By Eq. (0.5), $\mathscr{L}$ depends upon ε only through quantities q and q'. Therefore, differentiating Eq. (0.5) gives

$$\frac{\partial K}{\partial \varepsilon} = \int_a^b dx \left[\frac{\partial \mathscr{L}}{\partial q} \frac{\partial q}{\partial \varepsilon} + \frac{\partial \mathscr{L}}{\partial q'} \frac{\partial q'}{\partial \varepsilon} \right]. \qquad (0.7)$$

The second integral is

$$\int_a^b dx \frac{\partial \mathscr{L}}{\partial q'} \frac{\partial^2 q}{\partial x \partial \varepsilon} = \frac{\partial \mathscr{L}}{\partial q'} \frac{\partial q}{\partial \varepsilon}\bigg|_a^b - \int_a^b \frac{\partial q}{\partial \varepsilon} \frac{d}{dx}\left(\frac{\partial \mathscr{L}}{\partial q'} \right) dx \qquad (0.8)$$

after an integration by parts. (In the usual notation, setting $u = \partial \mathscr{L}/\partial q'$ and $dv = \partial^2 q/\partial x \partial \varepsilon$.)

We now show that the first right-hand term in Eq. (0.8) is zero. By Eq. (0.3),

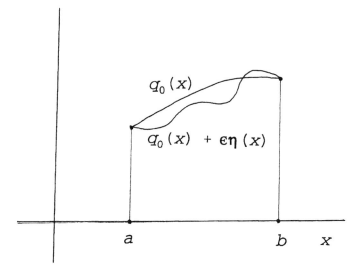

Fig. 0.1. Both the solution $q_0(x)$ and any perturbation $q_0(x) + \varepsilon\eta(x)$ from it must pass through the endpoints $x = a$ and $x = b$.

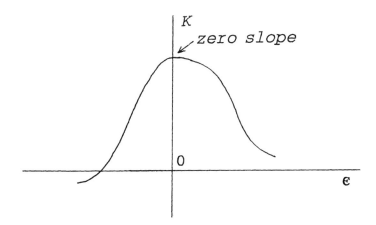

Fig. 0.2. K as a function of perturbation size parameter ε.

$$\frac{\partial q}{\partial \varepsilon} = \eta(x), \tag{0.9}$$

so that by Eq. (0.4)

$$\left.\frac{\partial q}{\partial \varepsilon}\right|_b = \left.\frac{\partial q}{\partial \varepsilon}\right|_a = 0. \tag{0.10}$$

This proves the assertion.

Combining this result with Eq. (0.7) gives

$$\frac{\partial K}{\partial \varepsilon}\bigg|_{\varepsilon=0} = \int_a^b dx \left[\frac{\partial \mathscr{L}}{\partial q}\frac{\partial q}{\partial \varepsilon} - \frac{\partial q}{\partial \varepsilon}\frac{d}{dx}\left(\frac{\partial \mathscr{L}}{\partial q'}\right)\right]_{\varepsilon=0}. \tag{0.11}$$

Factoring out the common term $\partial q/\partial \varepsilon$, evaluating it at $\varepsilon = 0$ and using Eq. (0.9) gives

$$\frac{\partial K}{\partial \varepsilon}\bigg|_{\varepsilon=0} = \int_a^b dx \left[\frac{\partial \mathscr{L}}{\partial q} - \frac{d}{dx}\left(\frac{\partial \mathscr{L}}{\partial q'}\right)\right]\eta(x). \tag{0.12}$$

By our criterion (0.6) this is to be zero at the solution q. But the factor $\eta(x)$ is, by hypothesis, arbitrary. The only way the integral can be zero, then, is for the factor in square brackets to be zero at each x, that is,

$$\frac{d}{dx}\left(\frac{\partial \mathscr{L}}{\partial q'}\right) = \frac{\partial \mathscr{L}}{\partial q}. \tag{0.13}$$

This is the celebrated *Euler–Lagrange* solution to the problem. It is a differential equation whose solution clearly depends upon the function $\mathscr{L}$, called the 'Lagrangian', for the given problem. Some examples of its use follow.

Example 1 Return to the Lagrangian given below Eq. (0.2) where $x = t$ is the independent variable. We directly compute

$$\frac{\partial \mathscr{L}}{\partial q'} = mq' \text{ and } \frac{\partial \mathscr{L}}{\partial q} = -\frac{\partial V}{\partial q}. \tag{0.14}$$

Using this in Eq. (0.13) gives as the solution

$$mq'' = -\frac{\partial V}{\partial q}, \tag{0.15}$$

that is, Newton's law of motion for the particle.

It may be noted that Newton's law will not be derived in this manner in the text to follow. The EPI principle is covariant, i.e., treats time and space in the same way, whereas the above approach (0.14), (0.15) is not. Instead, the EPI approach will be used to derive the more general Einstein field equation, from which Newton's laws follow as a special case (the weak-field limit).

The reader may well question where this particular Lagrangian came from. The answer is that it was chosen merely because it 'works', i.e., leads to Newton's law of motion. It has no prior significance in its own right. This has been a well-known drawback to the use of Lagrangians. The next chapter addresses this problem in detail.

Example 2 What is the shortest path between two points in a plane? The integrated arc length between points $x = a$ and $x = b$ is

$$K = \int_a^b dx\, \mathscr{L}, \quad \mathscr{L} = \sqrt{1 + q'^2}. \tag{0.16}$$

Hence

$$\frac{\partial \mathscr{L}}{\partial q'} = \tfrac{1}{2}(1 + q'^2)^{-1/2}2q', \quad \frac{\partial \mathscr{L}}{\partial q} = 0 \tag{0.17}$$

here, so that the Euler–Lagrange Eq. (0.13) is

$$\frac{d}{dx}\left(\frac{q'}{\sqrt{1 + q'^2}}\right) = 0. \tag{0.18}$$

The immediate solution is

$$\frac{q'}{\sqrt{1 + q'^2}} = const., \tag{0.19}$$

implying that $q' = const.$, so that $q(x) = Ax + B$, A, $B = const.$, the equation of a straight line. Hence we have shown that the path of extreme (not necessarily shortest) distance between two fixed points in a plane is a straight line. We will show below that the extremum is a minimum, as intuition suggests.

Example 3 Maximum entropy problems (Jaynes, 1957a,b) have the form

$$\int dx\, \mathscr{L} = \text{max.}, \quad \mathscr{L} = -p(x)\ln p(x) + \lambda p(x) + \mu p(x)f(x) \tag{0.20}$$

with λ, μ constants and $f(x)$ a known 'kernel' function. The first term in the integral defines the 'entropy' of a probability density function (PDF) $p(x)$. (Notice we use the notation p in place of q here.) We will say a lot more about the concept of entropy in chapters to follow. Directly

$$\frac{\partial \mathscr{L}}{\partial p'} = 0, \quad \frac{\partial \mathscr{L}}{\partial p} = -1 - \ln p + \lambda + \mu f(x). \tag{0.21}$$

Hence the Euler–Lagrange Eq. (0.13) is

$$-1 - \ln p(x) + \lambda + \mu f(x) = 0, \text{ or } p(x) = A\exp[\mu f(x)]. \tag{0.22}$$

The answer $p(x)$ to maximum entropy problems is always of an exponential form. We will show below that the extremum obtained is actually a maximum, as required.

Example 4 Minimum Fisher information problems (Huber, 1981) are of the form

$$\int dx \, \mathscr{L} = min., \quad \mathscr{L} = 4q'^2 + \lambda q(x)f(x) + \mu q^2(x)h(x), \qquad (0.23)$$

$\lambda, \mu = const.$, where $f(x)$, $h(x)$ are known kernel functions. Also, the PDF $p(x) = q^2(x)$, i.e., $q(x)$ is a 'probability amplitude' function. The first term in the integral defines the Fisher information. Directly

$$\frac{\partial \mathscr{L}}{\partial q'} = 8q', \quad \frac{\partial \mathscr{L}}{\partial q} = \lambda f(x) + 2\mu q(x)h(x). \qquad (0.24)$$

The Euler–Lagrange Eq. (0.13) is then

$$q''(x) - (\mu/4)h(x)q(x) - (\lambda/8)f(x) = 0. \qquad (0.25)$$

That is, the answer $q(x)$ is the solution to a second-order differential equation. The particular solution will depend upon the form of the kernel functions and on any imposed boundary conditions. We will show below that the extremum obtained is a minimum, as required.

In comparing the maximum entropy solution (0.22) with the minimum Fisher information solution (0.25) it is to be noted that the former has the virtue of simplicity, always being an exponential. By contrast, the Fisher information solution always requires the solution to a differential equation: a bit more complicated a procedure. However, for purposes of deriving physical PDF laws the Fisher answer is actually preferred: the PDFs of physics generally obey differential equations (wave equations). We will further address this issue in later chapters.

We now proceed to generalize the variational problem (0.2) by degrees. For brevity, only the solutions (Korn and Korn, 1968) will be presented.

0.3.2 Multiple-curve problems

As a generalization of problem (0.2) with its single unknown function $q(x)$, consider the problem of finding N functions $q_n(x)$, $n = 1, \ldots, N$ that satisfy

$$K = \int dx \, \mathscr{L}[x, q_1, \ldots, q_N, q'_1, \ldots, q'_N] = extrem. \qquad (0.26)$$

The answer to this variational problem is that the $q_n(x)$, $n = 1, \ldots, N$ must obey N Euler–Lagrange equations

$$\frac{d}{dx}\left(\frac{\partial \mathscr{L}}{\partial q'_n}\right) = \frac{\partial \mathscr{L}}{\partial q_n}, \quad n = 1, \ldots, N. \qquad (0.27)$$

In the case $N = 1$ this becomes the one-function result (0.13).

0.3.3 Condition for a minimum solution

At this point we cannot know whether the solution (0.27) to the extremum problem (0.26) gives a maximum or a minimum value for the extremum. A simple test for this purpose is as follows.

Consider the matrix of numbers $[\partial^2 \mathscr{L}/\partial q'_i \partial q'_j]$. If this matrix is positive definite then the extreme value is a minimum; or, if negative definite the extreme value is a maximum. This is called *Legendre's Condition* for an extremum.

A particular case of interest is as follows.

0.3.4 Fisher information, multiple-component case

As will be shown in Chap. 2, the information Lagrangian is here

$$\mathscr{L} = 4 \sum_{n=1}^{N} q'^2_n. \tag{0.28}$$

Then

$$\frac{\partial \mathscr{L}}{\partial q'_i} = 8q'_i, \text{ so that } \frac{\partial^2 \mathscr{L}}{\partial q'_i \partial q'_j} = 8\delta_{ij} \tag{0.29}$$

where δ_{ij} is the Kronecker delta function. Thus the matrix $[\partial^2 \mathscr{L}/\partial q'_i \partial q'_j]$ is diag $[8, \ldots, 8]$ so that all its n-row minor determinants obey

$$\det \left[\frac{\partial^2 \mathscr{L}}{\partial q'_i \partial q'_j} \right] = 8^n > 0, \, n = 1, \ldots, N. \tag{0.30}$$

Then the matrix $[\partial^2 \mathscr{L}/\partial q'_i \partial q'_j]$ is positive definite (Korn and Korn, 1968, p. 420). By Legendre's condition the extremum is a minimum.

0.3.5 Exercise

Using the Lagrangian given below Eq. (0.2), show by Legendre's condition (Sec. 0.3.3) that the Newton's law solution (0.15) minimizes the integral K in Eq. (0.2).

0.3.6 Nature of extremum in other examples

Return to Example 2, the problem of the *minimum* path between two points. The solution $q(x) = Ax + B$ guarantees an extremum but not necessarily a minimum. Differentiating the first Eq. (0.17) gives

$$\frac{\partial^2 \mathscr{L}}{\partial q'^2} = \frac{1}{(1 + q'^2)^{3/2}} = \frac{1}{(1 + A^2)^{3/2}} > 0, \tag{0.31}$$

signifying a minimum by Legendre's condition.

Return to Example 3, maximum entropy solutions. The exponential solution (0.22) guarantees an extreme value to the integral (0.20) but not necessarily a maximum. We attempt to use the Legendre condition. But the Lagrangian (0.20) does not contain any dependence upon quantity $p'(x)$. Hence Legendre's rule gives the ambiguous result $\partial^2 \mathscr{L}/\partial p'^2 = 0$. This being neither positive nor negative, the nature of the extremum remains unknown.

We need to approach the problem in a different way. Temporarily replace the continuous integral (0.20) by a sum

$$K = \sum_{n=1}^{N} \Delta x(-p_n \ln p_n + \lambda p_n + \mu p_n f_n) = max., \tag{0.32}$$

$$p_n \equiv p(x_n), \ f_n \equiv f(x_n), \ x_n \equiv n\Delta x,$$

where $\Delta x > 0$ is small but finite. The sum approaches the integral as $\Delta x \to 0$. K is now an ordinary function (not a functional) of the N variables p_n and, hence, may be extremized in each p_n using the ordinary rules of differential calculus. The nature of such an extremum may be established by observing the positive- or negative-definiteness of the second derivative matrix $[\partial^2 K/\partial p_i \partial p_j]$. From Eq. (0.32) we have directly $[\partial^2 K/\partial p_i \partial p_j] = -\Delta x\, \mathrm{diag}\,[1/p_1, \dots, 1/p_N]$. Hence, all its n-row minor determinants obey

$$\det \left[\frac{\partial^2 K}{\partial p_i \partial p_j}\right] = -\prod_{i=1}^{n} \frac{\Delta x}{p_i} < 0 \tag{0.33}$$

since all $p_i \geq 0$. The matrix is negative definite, signifying a maximum as required. This result obviously holds in the (continuous) limit as $\Delta x \to 0$ through positive values.

0.3.7 Multiple-curve, multiple-variable, problems

The most general Lagrangian (0.1) has both a multiplicity of coordinates $x_1, \dots, x_M$ and of curves $q_1, \dots, q_N$. The solution to the extremization problem is the N Euler–Lagrange equations

$$\sum_{m=1}^{M} \frac{d}{dx_m} \left(\frac{\partial \mathscr{L}}{\partial q_{nm}}\right) = \frac{\partial \mathscr{L}}{\partial q_n}, \ n = 1, \dots, N, \text{ where } q_{nm} \equiv \frac{\partial q_n}{\partial x_m}. \tag{0.34}$$

We will make much use of this result in chapters to follow. In the case $M = 1$ of one coordinate it goes over into the solution (0.27).

0.3.8 Imposing constraints

In many problems of extremization there are some known constraints that must be obeyed by the unknowns **q**. How may they be imposed upon the extremum solution?

Suppose that the unknown functions **q** are to obey some constraints, such as

$$F_{nk} = \int d\mathbf{x} q_n^\alpha(\mathbf{x}) f_k(\mathbf{x}), \ \alpha = const., \ n = 1, \ldots, N; \ k = 1, \ldots, K_o. \quad (0.35)$$

The f_n are known functions. We seek the solution **q** that obeys these constraints and, simultaneously, extremizes a Lagrangian integral (0.1). This obeys the 'Lagrange method of undetermined multipliers', described as follows.

Given that $\mathscr{L}[\mathbf{q}, \mathbf{q}', \mathbf{x}]$ is the Lagrangian to be optimized, form the new problem

$$\overline{K} = \int d\mathbf{x} \mathscr{L}[\mathbf{q}, \mathbf{q}', \mathbf{x}] + \sum_{nk} \lambda_{nk} \left[\int d\mathbf{x} q_n^\alpha(\mathbf{x}) f_k(\mathbf{x}) - F_{nk} \right] = extrem. \quad (0.36)$$

Here the extremization is to be through variation of the **q** *and* the new unknowns $[\lambda_{nk}] \equiv \lambda$. The latter are the 'undetermined multipliers' spoken of above.

To show that the solution **q**, λ obeys the constraints, first impose the condition that $\overline{K}$ is extremized with respect to the λ, that is, $\partial \overline{K} / \partial \lambda_{nk} = 0$. From Eq. (0.36) this gives

$$\int d\mathbf{x} q_n^\alpha(\mathbf{x}) f_k(\mathbf{x}) - F_{nk} = 0, \quad (0.37)$$

which is the same as the constraint Eqs. (0.35).

Eq. (0.36) is also equivalent to the Euler–Lagrange problem of extremizing a net Lagrangian

$$\overline{\mathscr{L}} = \mathscr{L} + \sum_{nk} \lambda_{nk} q_n^\alpha(\mathbf{x}) f_k(\mathbf{x}). \quad (0.38)$$

That is, in Eq. (0.36), the terms in F_{nk} do not contribute to the Euler–Lagrange solution (0.34). Hence, the problem (0.36) may be re-posed more simply as

$$K + \sum_{nk} \lambda_{nk} \int d\mathbf{x} q_n^\alpha(\mathbf{x}) f_k(\mathbf{x}) = extrem. \quad (0.39)$$

That is, to incorporate constraints one merely weights and adds them to the 'objective' functional K. Some examples of interest are as follows.

With K as the entropy functional, and with moment constraints, the approach (0.39) was used by Jaynes (1957a,b) to estimate PDFs represented as $q^2(x)$.

Alternatively, with K as the Fisher information and with a constraint of mean kinetic energy, the approach (0.39) was used to derive the Schroedinger wave equation (Frieden, 1989) and other wave equations of physics (Frieden, 1990). Historically, the former was the author's first application of a principle of minimum Fisher information to a physical problem. The questions it raised, such as why *a priori* mean kinetic energy should be a constraint (ultimate answer: most generally it shouldn't), provoked an evolution of the theory which has culminated in this book.

0.3.9 Variational derivative, functional derivatives

The variation of a functional, the variational derivative, and the functional derivative are useful concepts that follow easily from the preceding. We shall have occasion to use the concept of the functional derivative later on. The concept of the variational derivative is also given, mainly so as to distinguish it from the functional variety.

We first define the concept of the variation of a functional. It was noted that K is a functional (Sec. 0.2). Multiply Eq. (0.12) through by a differential $d\varepsilon$. This gives

$$\frac{\partial K}{\partial \varepsilon}\bigg|_{\varepsilon=0} d\varepsilon = \int_a^b \left[\frac{\partial \mathscr{L}}{\partial q} - \frac{d}{dx}\left(\frac{\partial \mathscr{L}}{\partial q'}\right)\right]\left(\frac{\partial q}{\partial \varepsilon}\right)\bigg|_{\varepsilon=0} d\varepsilon\, dx. \tag{0.40}$$

We also used Eq. (0.9). Define the variation of K as

$$\delta K \equiv \left(\frac{\partial K}{\partial \varepsilon}\right)\bigg|_{\varepsilon=0} d\varepsilon. \tag{0.41}$$

This measures the change in functional K due to a small perturbation away from the stationary solution $q_0(x)$. Similarly

$$\delta q \equiv \left(\frac{\partial q}{\partial \varepsilon}\right)\bigg|_{\varepsilon=0} d\varepsilon \tag{0.42}$$

is the first variation in q.

The *first variational derivative* $\delta\mathscr{L}/\delta q$ is then defined such that Eq. (0.40) goes over into the simple form

$$\delta K = \int \left(\frac{\delta \mathscr{L}}{\delta q}\right)\delta q\, dx \tag{0.43}$$

after use of definitions (0.41) and (0.42). This is the definition

$$\frac{\delta \mathscr{L}}{\delta q} \equiv \frac{\partial \mathscr{L}}{\partial q} - \frac{d}{dx}\left(\frac{\partial \mathscr{L}}{\partial q'}\right). \tag{0.44}$$

The case $\mathscr{L} = \mathscr{L}[x, q(x), q'(x)]$ [see Eq. (0.5)] is presumed. Interesting background reading on this concept may be found in Goldstein (1950), p. 353, and Panofsky and Phillips (1955), p. 367.

Note: Although K is a functional (see below Eq. (0.1)) $\mathscr{L}$ is an ordinary function, since it has a definite value for *each* value of x. Hence, we do not call $\delta\mathscr{L}/\delta q$ in (0.44) a functional derivative, preferring instead to call it a 'variational' derivative. The derivative of a functional, $\delta K/\delta q$, will be defined below. (See also Feynman and Hibbs (1965), pp. 170–1.)

The concept of a functional derivative will be needed in our treatment of quantum gravity (Chap. 11). The functional there is $K \equiv \psi[\mathbf{q}(\mathbf{x})]$, the probability amplitude for the gravitational metric function $\mathbf{q}$ at *all* points $\mathbf{x}$ in four-space. Each possible functional form $\mathbf{q}(\mathbf{x})$ for the metric, defined over *all* $\mathbf{x}$, gives a single value for ψ.

We now derive the concepts of the first- and second-order *functional* derivative. The general approach is to proceed as in derivation of Eq. (0.44). That is, we develop an expression for the variation in (now) a functional and show that it takes a simple form like (0.43) only if the integrand is taken to be the suitable *functional* derivative. In contrast with the preceding, however, we expand out to *second*-order in the changes so as to bring in the concept of a second functional derivative. (By the way, the latter concept does not seem to appear in any other physical text.)

Consider a functional

$$K = K[x, q(x, \varepsilon)], \text{ all } x \qquad (0.45)$$

where $q(x, \varepsilon)$ obeys Eq. (0.3). By how much does the scalar value K change if function $q(x, \varepsilon)$ is perturbed by a small amount at *each* x?

In order to use the ordinary rules of calculus we first subdivide x-space as

$$x_{n+1} = x_n + \Delta x, \ n = 1, 2, \ldots \qquad (0.46)$$

in terms of which

$$K = K[x_1, x_2, \ldots, q(x_1, \varepsilon), q(x_2, \varepsilon), \ldots] \qquad (0.47)$$

(cf. Eq. (0.45)). Also, Eq. (0.3) is now discretized, to

$$q(x_n, \varepsilon) = q_0(x_n) + \varepsilon\eta(x_n). \qquad (0.48)$$

Note that the ordinary partial derivatives $\partial K/\partial q(x_n, \varepsilon)$, $n = 1, 2, \ldots$ are well-defined, by direct differentiation of Eq. (0.47).

In all of the preceding, the numbers $\eta(x_n)$ are presumed to be arbitrary but *fixed*. Then the perturbations in (0.48) are purely a function of ε. Consequently, K given by (0.47) is likewise purely a function $K(\varepsilon)$.

Next, consider the effect upon $K(\varepsilon)$ of a small change $d\varepsilon$ away from the stationary state $\varepsilon = 0$. This may simply be represented by a Taylor series in powers of $d\varepsilon$,

$$K(d\varepsilon) = dK(\varepsilon) \Bigg|_{\varepsilon=0} = \frac{\partial K}{\partial \varepsilon} \Bigg|_{\varepsilon=0} d\varepsilon + \frac{1}{2} \frac{\partial^2 K}{\partial \varepsilon^2} \Bigg|_{\varepsilon=0} d\varepsilon^2 + \ldots. \qquad (0.49)$$

The coefficients of $d\varepsilon$ and $d\varepsilon^2$ are evaluated as follows.

By the chain rule of differentiation

$$\frac{\partial K}{\partial \varepsilon} = \sum_n \frac{\partial K}{\partial q(x_n)} \frac{\partial q(x_n)}{\partial \varepsilon}. \qquad (0.50)$$

(For brevity, we used $q(x_n, \varepsilon) \equiv q(x_n)$.) Also,

$$\frac{\partial^2 K}{\partial \varepsilon^2} = \frac{\partial}{\partial \varepsilon} \left(\frac{\partial K}{\partial \varepsilon} \right) = \sum_m \frac{\partial}{\partial q(x_m)} \left[\sum_n \frac{\partial K}{\partial q(x_n)} \frac{\partial q(x_n)}{\partial \varepsilon} \right] \frac{\partial q(x_m)}{\partial \varepsilon}$$

after re-use of (0.50),

$$= \sum_{mn} \frac{\partial^2 K}{\partial q(x_m)\partial q(x_n)} \frac{\partial q(x_m)}{\partial \varepsilon} \frac{\partial q(x_n)}{\partial \varepsilon} \qquad (0.51)$$

after another derivative term drops out.

If we multiply Eq. (0.50) by $d\varepsilon$ and Eq. (0.51) by $d\varepsilon^2$, evaluate them at $\varepsilon = 0$ and use definition (0.41) of the first variation, we get

$$\frac{\partial K}{\partial \varepsilon} \Bigg|_{\varepsilon=0} d\varepsilon = \sum_n \frac{\partial K}{\partial q(x_n)} \delta q(x_n)$$

and $\qquad (0.52)$

$$\frac{\partial^2 K}{\partial \varepsilon^2} \Bigg|_{\varepsilon=0} d\varepsilon^2 = \sum_{mn} \frac{\partial^2 K}{\partial q(x_m)\partial q(x_n)} \delta q(x_m)\delta q(x_n),$$

respectively.

Following the plan, we substitute these coefficient values into Eq. (0.49). Next, multiply and divide the first sum by Δx and the second sum by Δx^2. Finally, take the continuous limit $\Delta x \to 0$. The sums approach integrals and we have

$$dK(\varepsilon) \Bigg|_{\varepsilon=0} = \int dx \left[\lim_{\Delta x \to 0} \frac{1}{\Delta x} \frac{\partial K}{\partial q(x_n)} \right] \delta q(x)$$

$$+ \frac{1}{2} \iint dx' \, dx \left[\lim_{\Delta x \to 0} \frac{1}{\Delta x^2} \frac{\partial^2 K}{\partial q(x_m)\partial q(x_n)} \right] \delta q(x')\delta q(x) + \ldots$$

$$(0.53)$$

where $x_n \to x$, $x_m \to x'$ in the limit. We demand this to take the simpler form (cf. Eq. (0.43))

$$dK(\varepsilon)\bigg|_{\varepsilon=0} = \int dx \, \frac{\delta K}{\delta q(x)} \delta q(x) + \frac{1}{2} \int\!\!\int dx' \, dx \, \frac{\delta^2 K}{\delta q(x')\delta q(x)} \delta q(x')\delta q(x) + \dots$$

(0.54)

By Eq. (0.53) this will be so if we define

$$\frac{\delta K}{\delta q(x)} \equiv \lim_{\Delta x \to 0} \frac{1}{\Delta x} \frac{\partial K}{\partial q(x_n)}$$

(0.55)

and

$$\frac{\delta^2 K}{\delta q(x')\delta q(x)} \equiv \lim_{\Delta x \to 0} \frac{1}{\Delta x^2} \frac{\partial^2 K}{\partial q(x_m)\partial q(x_n)}.$$

(0.56)

Eq. (0.55) is the first functional derivative of K in the case of a functional dependence (0.45). It answers the question 'By how much will the number K change if the function $q(x)$ changes by a small amount at all x?'

Eq. (0.56) defines the second mixed functional derivative of K. As noted before, we will have occasion to use this concept in Chap. 11 on quantum gravity. The dynamical equation of this phenomenon is not the usual second-order differential equation; but, rather, a second-order *functional* differential equation. The second functional derivative is with respect to (metric) functions $\mathbf{q}(\mathbf{x})$, where $\mathbf{x}$ is, now, a *four*-position.

Although the preceding derivation was for the case of a scalar coordinate x, it is easily generalized to the case of a four-vector $\mathbf{x}$ as well. (One merely replaces all scalars x by a four-vector $\mathbf{x}$, with all subscripts as before. Of course, ε is still a scalar.) This gives a definition

$$\frac{\delta^2 K}{\delta q(\mathbf{x}')\delta q(\mathbf{x})} \equiv \lim_{\Delta \mathbf{x} \to 0} \frac{1}{\Delta \mathbf{x}^2} \frac{\partial^2 K}{\partial q(\mathbf{x}_m)\partial q(\mathbf{x}_n)},$$

(0.57)

$$\mathbf{x}_n \to \mathbf{x}, \; \mathbf{x}_m \to \mathbf{x}', \; \Delta \mathbf{x} \equiv \Delta x \Delta y \Delta z c \Delta t$$

where c is the speed of light and $\Delta \mathbf{x}$ is the increment volume used in four-space.

Finally, we consider problems where *many* amplitude functions $q_k(\mathbf{x}, \varepsilon)$, $k = 1, 2, \dots$ exist. (We subscripted these with a single subscript, but results will hold for any number of subscripts as well.) In Eq. (0.45), the functional K now depends upon all of these functions. We now get the logical generalization of Eq. (0.57),

$$\frac{\delta^2 K}{\delta q_j(\mathbf{x}')\delta q_k(\mathbf{x})} = \lim_{\Delta \mathbf{x} \to 0} \frac{1}{\Delta \mathbf{x}^2} \frac{\partial^2 K}{\partial q_j(\mathbf{x}_m)\partial q_k(\mathbf{x}_n)}, \quad \mathbf{x}_n \to \mathbf{x}, \; \mathbf{x}_m \to \mathbf{x}'.$$

(0.58)

0.3.10 Exercise

Show this result, using the analogous steps to Eqs. (0.45)–(0.57). *Hint*: The right-hand functions in Eq. (0.48), and ε, must now be subscripted by k. Then Eq. (0.49) becomes a power series in changes $d\varepsilon_k$ including (now) second-order mixed-product terms $d\varepsilon_k d\varepsilon_l$ that are *summed* over k and l. Proceeding, on this basis, through to Eq. (0.57) now logically gives the definition (0.58).

The procedure (0.49)–(0.56) may be easily extended to allow third-, and all higher-orders of functional derivatives to be defined. The dots at the ends of Eqs. (0.53) and (0.54) indicate the higher-order terms, which may be easily added in.

0.3.11 Alternate form for functional derivative

Definition (0.55) is useful when the functional K has the form of a sum over discrete samples $q(x_n)$, $n = 1, 2, \ldots$. Instead, in many problems K is expressed as an action integral Eq. (0.2), where $q(x)$ is *continuously* sampled. Obviously, definition (0.55) is not directly useable for such a problem. For such continuous cases one may use, instead, the equivalent definition

$$\frac{\delta K}{\delta q(y)} \equiv \lim_{\varepsilon \to 0} \frac{K[q(x) + \varepsilon\delta(x - y)] - K[q(x)]}{\varepsilon} \tag{0.59}$$

where $\delta(x)$ is the Dirac delta function (Ryder, 1987, p. 177).

0.3.12 An observation

Suppose that two non-identical functionals $I[q(x)]$ and $J[q(x)]$ obey a relation

$$I - \kappa J = 0 \tag{0.60}$$

for some scalar function $q_1(x)$. Is there necessarily a solution $q_2(x)$ to the variational problem

$$\delta(I - \kappa J) = 0? \tag{0.61}$$

The latter is a problem requiring quantity $I - \kappa J$ to be *stationary* for some $q_2(x)$. As we saw, its solution would have to obey the Euler–Lagrange Eq. (0.13). This is, of course, a very different requirement on $q_2(x)$ than the zero-condition (0.60). Moreover, even if $q_1(x)$ exists there is no guarantee that any solution $q_2(x)$ exists. Thus, the fact that a functional is zero for some solution does not guarantee that that, or any other, solution extremizes the functional. Stated more simply, an algebraic zero-condition does not necessarily imply a calculus variational zero-condition.

0.4 Dirac delta function

A handy concept to use when evaluating integrals is that of the Dirac delta function $\delta(x)$. It is the continuous counterpart of the Kronecker delta function

$$\delta_{ij} = 0 \text{ for } i \neq j, \sum_{j=1}^{N} \delta_{ij} = 1, \tag{0.62a}$$

$$\sum_{j=1}^{N} f_j \delta_{ij} = f_i \tag{0.62b}$$

for any i on the interval $(1, N)$. Notice that δ_{ij}, for a fixed value of i, is a function of j that is a pure 'spike', i.e., zero everywhere except at the single point $i = j$ where it has a 'yield' of 1.

Similarly, $\delta(x)$ obeys

$$\delta(x - a) = 0 \text{ for } x \neq a, \int dx\, \delta(x - a) = 1, \tag{0.63a}$$

$$\int dx\, f(x)\delta(x - a) = f(a) \tag{0.63b}$$

for any real a and any function $f(x)$. (In these integrals and throughout the book, the limits are from $-\infty$ to ∞ unless otherwise stated.) It is useful to compare Eqs. (0.62a) with Eqs. (0.63a), and Eq. (0.62b) with Eq. (0.63b). What should function $\delta(x)$ look like?

From Eqs. (0.63a) it must be flat zero everywhere except at the point $x = a$, at which point it is so large a value that its area is still finite (at value 1). This is the continuous version of the Kronecker delta spike mentioned above.

Eq. (0.63b) shows what is called the 'sifting' property of the delta function. That is, the single value $f(a)$ is sifted out by the delta function during the integration over all x. Notice that Eq. (0.63b) *follows* from Eqs. (0.63a): because of the first property (0.63a) no value of $f(x)$ contributes to the integral (0.63b) except the value $f(a)$. Therefore it may be taken outside the integral. The remaining integral has unit area by the second property (0.63a), giving a net value of $f(a)$. The situation is sketched in Fig. 0.3.

Now a pure, discontinuous spike cannot represent a function in the ordinary sense of the word. However, it can represent the *limiting form* of a well-defined function. Functions that exist in this limit are called 'generalized functions' (Bracewell, 1965). For example, the ordinary Gaussian curve from probability theory when taken in the limit of vanishing variance obeys

$$\lim_{\sigma \to 0} p(x) = \lim_{\sigma \to \infty} \frac{1}{\sqrt{2\pi\sigma^2}} \exp\left(-\frac{x^2}{2\sigma^2}\right) = \delta(x) \tag{0.64}$$

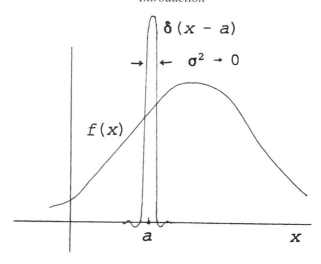

Fig. 0.3. The function $f(x)$ is sifted at value $f(a)$ by the Dirac delta function $\delta(x - a)$.

since it obeys the requirements (0.63a), the second of which becomes simple normalization of the probability law. By the same token, any well-behaved probability density (outside equality in (0.64)) approaches a Dirac delta function in this limit. There are, thus, an infinite number of different representations for the delta function.

An often-used representation grows out of the Fourier transform relation

$$F(k) = \frac{1}{\sqrt{2\pi}} \int dx\, f(x) \exp(-ikx), \quad i = \sqrt{-1} \qquad (0.65)$$

and its inverse relation

$$f(x) = \frac{1}{\sqrt{2\pi}} \int dk'\, F(k') \exp(ik'x). \qquad (0.66)$$

Substituting Eq. (0.66) into (0.65) gives

$$F(k) = \int dx \exp(-ikx) \frac{1}{2\pi} \int dk'\, F(k') \exp(ik'x)$$

$$= \int dk'\, F(k') \frac{1}{2\pi} \int dx \exp[-ix(k - k')] \qquad (0.67)$$

after switching orders of integration. Then by the sifting property Eq. (0.63b) it must be that

$$\frac{1}{2\pi} \int dx \exp[-ix(k - k')] = \delta(k - k'). \qquad (0.68)$$

Analogous properties to Eqs. (0.63b) and (0.68) exist for *multidimensional* functions $f(\mathbf{x})$ where $\mathbf{x}$ is a vector of dimension M. Thus there is a sifting property

$$\int d\mathbf{x}\, f(\mathbf{x})\delta(\mathbf{x} - \mathbf{a}) = f(\mathbf{a}) \tag{0.69}$$

and a Fourier representation

$$\frac{1}{(2\pi)^{M/2}}\int d\mathbf{x}\exp\left[-i\mathbf{x}\cdot(\mathbf{k} - \mathbf{k}')\right] = \delta(\mathbf{k} - \mathbf{k}'). \tag{0.70}$$

The latter is a multidimensional delta function. These relations may be derived as easily as were the corresponding scalar relations (0.63b) and (0.68).

A final relation that we will have occasion to use is

$$\delta(ax) = \frac{\delta(x)}{|a|}, \quad a = const. \tag{0.71}$$

Other properties of the delta function may be found in Bracewell (1965).

1

What is Fisher information?

Knowledge of Fisher information is not part of the educational background of most physicists. Why should a physicist bother to learn about this concept? Surely the (related) concept of entropy is sufficient to describe the degree of disorder of a given phenomenon. These important questions may be answered as follows.

(a) The point made about entropy is true, but does not go far enough. Why not seek a measure of disorder whose variation *derives* the phenomenon? The concept of entropy cannot do this, for reasons discussed in Sec. 1.3. Fisher information will turn out to be the appropriate measure of disorder for this purpose.

(b) Why should a physicist bother to learn this concept? Aside from the partial answer in (a): (i) Fisher information is a *simple* and intuitive concept. As theories go, it is quite elementary. To understand it does not require mathematics beyond differential equations. Even no prior knowledge of statistics is needed: this is easy enough to learn 'on the fly'. The derivation of the defining property of Fisher information, in Sec. 1.2.3, is readily understood. (ii) The subject has very little specialized jargon or notation. The beginner does not need a glossary of terms and symbols to aid in its understanding. (iii) Most importantly, once understood, the concept gives strong payoff – one might call it 'phenomen-all' – in scope of application. It's simply worth learning.

Fisher information has two basic roles to play in theory. First, it is a measure of the ability to estimate a parameter; this makes it a cornerstone of the statistical field of study called parameter estimation. Second, it is a measure of the state of disorder of a system or phenomenon. As will be seen, this makes it a cornerstone of physical theory.

Before starting the study of Fisher information, we take a temporary detour into a subject that will provide some immediate physical motivation for it.

Ronald A. Fisher, 1929, from a photograph taken in honor of his election to Fellow of the Royal Society. Sketch by the author.

1.1 On Lagrangians

The Lagrangian approach (Lagrange, 1788) to physics has been utilized now for over 200 years. It is one of the most potent and convenient tools of theory ever invented. One well-known proponent of its use (Feynman and Hibbs, 1965) calls it 'most elegant'. However, an enigma of physics is the question of where its Lagrangians come from. It would be nice to justify and derive them from a prior principle, but none seems to exist. Indeed, when a Lagrangian is presented in the literature, it is often with a disclaimer, such as (Morse and Feshbach, 1953) 'It usually happens that the differential equations for a given phenomenon are known first, and only later is the Lagrange function found, from which the differential equations can be obtained.' Even in a case where the differential equations are *not* known, often candidate Lagrangians are first constructed, to see if 'reasonable' differential equations result.

Hence, the Lagrange function has been principally a contrivance for getting the correct answer. It is the means to an end – a differential equation – but with no significance in its own right. One of the aims of this book is to show, in fact, that Lagrangians do have prior significance. A second aim is to present *a systematic approach to deriving* Lagrangians. A third is to clarify the role of the observer in a measurement. These aims will be achieved through use of the concept of Fisher information.

R. A. Fisher (1890–1962) was a researcher whose work is not well-known to physicists. He is renowned in the fields of genetics, statistics and eugenics. Among his pivotal contributions to these fields (Fisher, 1959) are the maximum likelihood estimate, the analysis of variance, and a measure of indeterminacy now called 'Fisher information.' (He also found it likely that the famous geneticist Gregor Mendel contrived the 'data' in his famous pea plant experiments. They were too regular to be true, statistically.) It will become apparent that his form of information has great utility in physics as well.

Table 1.1 shows a list of Lagrangians (most from Morse and Feshbach, 1953), emphasizing the common presence of a squared-gradient term. In quantum mechanics, this term represents mean kinetic energy, but why mean kinetic energy should be present remains a mystery: Schroedinger called it 'incomprehensible' (Schroedinger, 1926).

Historical note: As will become evident below, *Schroedinger's mysterious Lagrangian term was simply Fisher's data information.* May we presume from this that Schroedinger and Fisher, despite developing their famous theories nearly simultaneously, and with basically just the English channel between them, never communicated? If they had, it would seem that the mystery should have been quickly dispelled. This is an enigma.

What we will show is that, in general, the squared gradient represents a phenomenon that is natural to all fields, i.e., *information*. In particular, it is the amount of Fisher information residing in a variety of data called *intrinsic data*.

Table 1.1. *Lagrangians for various physical phenomena. Where do these come from and, in particular, why do they all contain a squared gradient term? (Reprinted from Frieden and Soffer, 1995.)*

Phenomenon	Lagrangian
Classical Mech.	$\frac{1}{2}m\left(\frac{\partial q}{\partial t}\right)^2 - V$
Flexible String or Compressible Fluid	$\frac{1}{2}\rho\left[\left(\frac{\partial q}{\partial t}\right)^2 - c^2 \nabla q \cdot \nabla q\right]$
Diffusion Eq.	$-\nabla\psi\cdot\nabla\psi^* - \ldots$
Schrödinger W. E.	$-\frac{\hbar^2}{2m}\nabla\psi\cdot\nabla\psi^* - \ldots$
Klein–Gordon Eq.	$-\frac{\hbar^2}{2m}\nabla\psi\cdot\nabla\psi^* - \ldots$
Elastic W. E.	$\frac{1}{2}\rho\dot{q}^2 - \ldots$
Electromagnetic Eqs.	$4\sum_{n=1}^{4}\Box q_n\cdot\Box q_n - \ldots$
Dirac Eqs.	$-\frac{\hbar^2}{2m}\nabla\psi\cdot\nabla\psi^* - \ldots = 0$
General Relativity (Eqs. of motion)	$\sum_{m,n=1}^{4}g_{mn}(q(\tau))\frac{\partial q_m}{\partial\tau}\frac{\partial q_n}{\partial\tau}$ $\uparrow$ metric tensor
Boltzmann Law	$4\left(\frac{\partial q(\mathrm{E})}{\partial\mathrm{E}}\right)^2 - \ldots, \; p(\mathrm{E}) \equiv q^2(\mathrm{E})$
Maxwell–Boltzmann Law	$4\left(\frac{\partial q(v)}{\partial v}\right)^2 - \ldots, \; p(v) \equiv q^2(v)$
Lorentz Transformation (special relativity)	$\partial_i q_n \partial_i q_n$ (invariance of integral)
Helmholtz W. E.	$-\nabla\psi\cdot\nabla\psi^* - \ldots$

The remaining terms of the Lagrangian will be seen to arise out of the information residing in the *phenomenon* that is under measurement. Thus, all Lagrangians consist entirely of two forms of Fisher information – data information and phenomenological information.

The concept of Fisher information is a natural outgrowth of classical measurement theory, as follows.

1.2 Classical measurement theory

1.2.1 The 'smart' measurement

Consider the basic problem of estimating a single parameter of a system (or phenomenon) from knowledge of some measurements. See Fig. 1.1. Let the parameter have value θ, and let there be N data values $y_1, \ldots, y_N \equiv \mathbf{y}$ in vector notation, at hand. The system is specified by a conditional probability law $p(\mathbf{y}|\theta)$ called the 'likelihood law'.

The data obey $\mathbf{y} = \theta + \mathbf{x}$, where the $x_1, \ldots, x_N \equiv \mathbf{x}$ are added noise values. The data are used in an estimation principle to form an estimate of θ which is an *optimal* function $\hat{\theta}(\mathbf{y})$ of all the data; e.g., the function might be the sample mean $N^{-1}\sum_n y_n$. The overall measurement procedure is 'smart' in that $\hat{\theta}(\mathbf{y})$ is on average a better estimate of θ than is any one of the data observables.

The noise $\mathbf{x}$ is assumed to be *intrinsic* to the parameter θ under measurement. For example, θ and $\mathbf{x}$ might be, respectively, the ideal position and quantum fluctuations of a particle. Data $\mathbf{y}$ are, correspondingly, called *intrinsic data*. No additional noise effects, such as noise of detection, are assumed present here. (We later allow for such additional noise in Sec. 3.8 and Chap. 10.) The system consisting of quantities $\mathbf{y}$, θ, $\mathbf{x}$ is a *closed*, or physically isolated, one.

1.2.2 Fisher information

This information arises as a measure of the expected error in a smart measurement. Consider the class of 'unbiased' estimates, obeying $\langle \hat{\theta}(\mathbf{y}) \rangle = \theta$; these are correct 'on average'. The mean-square error e^2 in such an estimate $\hat{\theta}$ obeys a relation (Van Trees, 1968; Cover and Thomas, 1991)

$$e^2 I \geqslant 1, \tag{1.1}$$

where I is called the Fisher 'information'. In a particular case of interest $N = 1$ (see below), this becomes

$$I = \int dx\, p'^2(x)/p(x), \; p' \equiv dp/dx. \tag{1.2}$$

(Throughout the book, integration limits are infinite unless otherwise specified.) Quantity $p(x)$ denotes the probability density function for the noise value

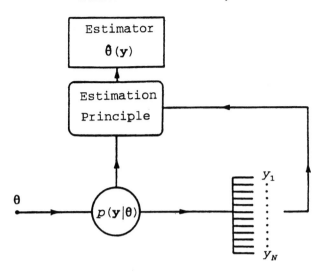

Fig. 1.1. The parameter estimation problem of classical statistics. An unknown but fixed parameter value θ causes intrinsic data **y** through random sampling of a likelihood law $p(\mathbf{y}|\theta)$. Then, the random likelihood law and the data are used to form the estimator $\hat{\theta}(\mathbf{y})$ via an estimation principle. (Reprinted from Frieden, 1991, by permission of Springer-Verlag Publishing Co.)

x. If $p(x)$ is Gaussian, then $I = 1/\sigma^2$ with σ^2 the variance (see derivation in Sec. 8.3.1).

Eq. (1.1) is called the Cramer–Rao inequality. It expresses *reciprocity* between the mean-square error e^2 and the Fisher information I in the intrinsic data. Hence, it is an expression of *intrinsic* uncertainties, i.e., in the absence of outside sources of noise. It will be shown at Eq. (4.53) that the reciprocity relation goes over into the Heisenberg uncertainty principle, in the case of a single measurement of a particle position value θ. Again, this ignores the possibility of noise of detection, which would add in additional uncertainties to the relation (Arthurs and Goodman, 1988; Martens and de Muynck, 1991).

The Cramer–Rao inequality (1.1) shows that estimation quality increases (e decreases) as I increases. Therefore, I is a quality metric of the estimation procedure. This is the essential reason why I is called an 'information'. Eqs. (1.1) and (1.2) derive quite easily, shown next.

1.2.3 Derivation

We follow Van Trees (1968). Consider the class of estimators $\hat{\theta}(\mathbf{y})$ that are unbiased, obeying

$$\langle \hat{\theta}(\mathbf{y}) - \theta \rangle \equiv \int d\mathbf{y}[\hat{\theta}(\mathbf{y}) - \theta] p(\mathbf{y}|\theta) = 0. \tag{1.3}$$

Probability density function (PDF) $p(\mathbf{y}|\theta)$ describes the fluctuations in data values $\mathbf{y}$ in the presence of the parameter value θ. PDF $p(\mathbf{y}|\theta)$ is called the 'likelihood law'. Differentiate Eq. (1.3) $\partial/\partial\theta$, giving

$$\int d\mathbf{y}(\hat{\theta} - \theta)\frac{\partial p}{\partial \theta} - \int d\mathbf{y}\, p = 0. \tag{1.4}$$

Use the identity

$$\frac{\partial p}{\partial \theta} = p\frac{\partial \ln p}{\partial \theta} \tag{1.5}$$

and the fact that p obeys normalization. Then Eq. (1.4) becomes

$$\int d\mathbf{y}(\hat{\theta} - \theta)\frac{\partial \ln p}{\partial \theta} p = 1. \tag{1.6}$$

Factor the integrand as

$$\int d\mathbf{y}\left[\frac{\partial \ln p}{\partial \theta}\sqrt{p}\right][(\hat{\theta} - \theta)\sqrt{p}] = 1. \tag{1.7}$$

Square the equation. Then the Schwarz inequality gives

$$\left[\int d\mathbf{y}\left(\frac{\partial \ln p}{\partial \theta}\right)^2 p\right]\left[\int d\mathbf{y}(\hat{\theta} - \theta)^2 p\right] \geq 1. \tag{1.8}$$

The left-most factor is defined to be the Fisher information I,

$$I \equiv I(\theta) \equiv \int d\mathbf{y}\left(\frac{\partial \ln p}{\partial \theta}\right)^2 p, \quad p \equiv p(\mathbf{y}|\theta), \tag{1.9}$$

while the second factor exactly defines the mean-squared error e^2,

$$e^2 \equiv \int d\mathbf{y}[\hat{\theta}(\mathbf{y}) - \theta]^2 p. \tag{1.10}$$

This proves Eq. (1.1).

It is noted that $I = I(\theta)$ in Eq. (1.9), i.e., in general I depends upon the (fixed) value of parameter θ. But note the following important exception to this rule.

1.2.4 Important case of shift invariance

Suppose that there is only $N = 1$ data value taken so that $p(\mathbf{y}|\theta) = p(y|\theta)$. Also, suppose that the PDF obeys a property

$$p(y|\theta) = p(y - \theta). \tag{1.11}$$

This means that the fluctuations in y from θ are invariant to the size of θ, a kind of shift invariance. (This becomes an expression of *Galilean invariance* when random variables y and θ are 3-vectors instead.) Using condition (1.11) and identity (1.5) in Eq. (1.9) gives

$$I = \int dy \left[\frac{\partial p(y - \theta)}{\partial (y - \theta)} \right]^2 \bigg/ p(y - \theta), \qquad (1.12)$$

since $\partial / \partial \theta = -\partial / \partial (y - \theta)$. Parameter θ is regarded as fixed (see above), so that a change of variable $x = y - \theta$ gives $dx = dy$. Equation (1.12) then becomes Eq. (1.2), as required. Note that I no longer depends upon θ. This is convenient since θ was unknown.

1.3 Comparisons of Fisher information with Shannon's form of entropy

A related quantity to I is the Shannon entropy (Shannon, 1948) H (called Shannon 'information' in this book). This has the form

$$H \equiv - \int dx p(x) \ln p(x). \qquad (1.13)$$

Like I, H is a functional of an underlying probability density function (PDF) $p(x)$. Historically, I predates the Shannon form by about 25 years (1922 vs. 1948). There are some known relations connecting the two information concepts (Stam, 1959; Blachman, 1965; Frieden, 1991) but these are not germane to our purposes. H can be, but is not always, the thermodynamic, Boltzmann entropy.

The analytic properties of the two information measures are quite different. Thus, whereas H is a *global* measure of smoothness in $p(x)$, I is a *local* measure. Hence, when extremized through variation of $p(x)$, Fisher's form gives a differential equation while Shannon's always gives directly the same form of solution, an exponential function. These are shown next.

1.3.1 Global vs. local nature

For our purposes, it is useful to work with a discrete form of Eq. (1.13),

$$H = -\Delta x \sum_n p(x_n) \ln p(x_n) \equiv \delta H, \ \Delta x \to 0. \qquad (1.14)$$

(Notation δH emphasizes that Eq. (1.14) represents an *increment* in information.) Of course, the sum in Eq. (1.14) may be taken in any order. Graphically, this means that if the curve $p(x_n)$ undergoes a rearrangement of its points $(x_n, p(x_n))$, although the shape of the curve will drastically change the value of

H remains constant. H is then said to be a *global* measure of the behavior of $p(x_n)$.

By comparison, the discrete form of Fisher information I is, from Eq. (1.2),

$$I = \Delta x^{-1} \sum_n \frac{[p(x_{n+1}) - p(x_n)]^2}{p(x_n)}. \tag{1.15}$$

If the curve $p(x_n)$ undergoes a rearrangement of points x_n as above, dis-continuities in $p(x_n)$ will now occur. Hence the local slope values $[p(x_{n+1}) - p(x_n)]/\Delta x$ will change drastically, and so the sum (1.15) will also change strongly. Since I is thereby sensitive to local rearrangement of points, it is said to have a property of *locality*.

Thus, H is a global measure, while I is a local measure, of the behavior of the curve $p(x_n)$. These properties hold in the limit $\Delta x \to 0$, and so apply to the continuous probability density $p(x)$ as well.

This global vs. local property has an interesting ramification. Because the integrand of I contains a squared derivative p'^2 (see Eq. (1.2)), when the integrand is used as part of a Lagrangian the resulting Euler–Lagrange equation will contain second-order derivative terms p''. Hence, a second-order differential equation results (see Eq. (0.25)). This dovetails with nature, in that the major fundamental differential equations that define probability densities or amplitudes in physics are second-order differential equations. Indeed, the thesis of this book is that the correct differential equations result when the informa-tion I-based EPI principle of Chap. 3 is followed.

By contrast, the integrand of H in (1.13) does not contain a derivative. Therefore, when this integrand is used as part of a Lagrangian the resulting Euler–Lagrange equation will not contain any derivatives (see Eq. (0.22)); it will be an algebraic equation, with the immediate solution that $p(x)$ has the exponential form Eq. (0.22) (Jaynes, 1957a,b). This is not, then, a differential equation, and hence cannot represent a general physical scenario. The excep-tions are those distributions which happen *to be* of an exponential form, as in statistical mechanics. (In these cases, I gives the correct solutions anyhow; see Chap. 7.)

It follows that, if one or the other of global measure H or local measure I is to be used in a variational principle in order to derive the physical law $p(x)$ describing a *general* scenario, the preference is to the local measure I.

As all of the preceding discussion implies, H and I are two distinct functionals of $p(x)$. However, quite the contrary is true in comparing I with an entropy that is closely related to H, namely, the Kullback–Leibler entropy. This is discussed in Sec. 1.4.

1.3.2 Additivity properties

It is of further interest to compare I and H in the special case of mutually isolated systems, which give rise to independent data. As is well-known, the entropy H obeys additivity in this case. Indeed, many people have been led to believe that, because H has this property, it is *the only* functional of a probability law that obeys additivity. In fact, information I obeys additivity as well. This will be shown in Sec. 1.8.11.

1.4 Relation of *I* to Kullback–Leibler entropy

The Kullback–Leibler entropy G (Kullback, 1959) is a functional of (now) two PDFs $p(x)$ and $r(x)$,

$$G \equiv -\int dx p(x) \ln [p(x)/r(x)]. \tag{1.16}$$

This is also called the 'cross entropy' or 'relative entropy' between $p(x)$ and a reference PDF $r(x)$. Note that if $r(x)$ is a constant, G becomes essentially the entropy H. Also, $G = 0$ if $p(x) = r(x)$. Thus, G is often used as a measure of the 'distance' between two PDFs $p(x)$ and $r(x)$. Also, in a multidimensional case $x \rightarrow (x, y)$ the information G can be used to define the mutual information of Shannon (1948).

Now we show that the Fisher information I relates to G. Using Eq. (1.15), with $x_{n+1} = x_n + \Delta x$, I may be expressed as

$$I = \Delta x^{-1} \sum_n p(x_n) \left[\frac{p(x_n + \Delta x)}{p(x_n)} - 1 \right]^2 \tag{1.17}$$

in the limit $\Delta x \rightarrow 0$. Now quantity $p(x_n + \Delta x)/p(x_n)$ is close to *unity* since Δx is small. Therefore, the $[\cdot]$ quantity in (1.17),

$$p(x_n + \Delta x)/p(x_n) - 1 \equiv \nu, \tag{1.18}$$

is small. Now for small ν the expansion

$$\ln (1 + \nu) = \nu - \nu^2/2 \tag{1.19}$$

holds, or equivalently,

$$\nu^2 = 2[\nu - \ln (1 + \nu)]. \tag{1.20}$$

Then by Eqs. (1.18) and (1.20), Eq. (1.17) becomes

$$I = -2\Delta x^{-1} \sum_n p(x_n) \ln \frac{p(x_n + \Delta x)}{p(x_n)}$$

$$+ 2\Delta x^{-1} \sum_n p(x_n + \Delta x) - 2\Delta x^{-1} \sum_n p(x_n). \tag{1.21}$$

But each of the two far-right sums is Δx^{-1}, by normalization, so that their difference cancels, leaving

$$I = - (2/\Delta x) \sum_n p(x_n) \ln \frac{p(x_n + \Delta x)}{p(x_n)} \tag{1.22a}$$

$$\rightarrow - (2/\Delta x^2) \int dx p(x) \ln \frac{p(x + \Delta x)}{p(x)} \tag{1.22b}$$

$$= - (2/\Delta x^2) G[p(x), \, p(x + \Delta x)] \tag{1.22c}$$

by definition (1.16). Thus, I is proportional to the cross-entropy between the PDF $p(x)$ and its shifted version $p(x + \Delta x)$.

1.4.1 Historical note

Vstovsky (1995) first proved the converse of the preceding, that I is an approximation to G. However, the expansion contains lower-order terms as well, in distinction to the effect in (1.21) where our lower-order terms cancel out exactly.

1.4.2 Exercise

One notes that the form (1.22b) is indeterminate $0/0$ in the limit $\Delta x \rightarrow 0$. Show that one use of l'Hôpital's rule does not resolve the limit, but two does, and the limit is precisely the form (1.2) of I.

1.4.3 Fisher information as a 'mother' information

Eq. (1.22b) shows that I is the cross-entropy between a PDF $p(x)$ and its infinitesimally shifted version $p(x + \Delta x)$. It has been noted (Caianiello, 1992) that I more generally results as a 'cross-information' between $p(x)$ and $p(x + \Delta x)$ for a host of *different* types of information measures. Some examples are as follows:

$$R_\alpha \equiv \ln \int dx \, p(x)^\alpha p(x + \Delta x)^{1-\alpha} \rightarrow -\Delta x^2 2^{-1} \alpha(1 - \alpha)I, \tag{1.22d}$$

for $\alpha \neq 1$, where R_α is called the 'Renyi information' measure (Amari, 1985); and

$$W \equiv \cos^{-1} \left[\int dx \, p^{1/2}(x) p^{1/2}(x + \Delta x) \right], \quad W^2 \rightarrow \Delta x^2 4^{-1} I, \tag{1.22e}$$

called 'Wootters information' measure (Wootters, 1981). To derive these results, one only has to expand the indicated function of $p(x + \Delta x)$ in the integrand out to *second order* in Δx, and perform the indicated integrations, using the identities $\int dx p'(x) = 0$ and $\int dx p''(x) = 0$.

Hence, Fisher information is the limiting form of many different measures of information; it is a kind of 'mother' information.

1.5 Amplitude form of *I*

In definition (1.2), the division by $p(x)$ is bothersome. (For example, is *I* undefined since necessarily $p(x) \to 0$ at certain x?) A way out is to work with a real 'amplitude' function $q(x)$,

$$p(x) = q^2(x). \tag{1.23}$$

(Interestingly, probability amplitudes were used by Fisher (1943) independent of their use in quantum mechanics. The purpose was to discriminate among population classes.) Using form (1.23) in (1.2) directly gives

$$I = 4 \int dx q'^2(x). \tag{1.24}$$

This is of a simpler form than (1.2) (no more divisions), and shows that *I* *simply measures the gradient content in q(x)* (and hence in $p(x)$). The integrand $q'^2(x)$ in (1.24) is the origin of the squared gradients in Table 1.1 of Lagrangians, as will be seen.

Representation (1.24) for *I* may be computed independent of the preceding. One measure of the 'distance' between an amplitude function $q(x)$ and its displaced version $q(x + \Delta x)$ is the quadratic measure (Braunstein and Caves, 1994)

$$L^2 \equiv \int dx [q(x + \Delta x) - q(x)]^2 \to \Delta x^2 \int dx\, q'^2(x) = \Delta x^2 4^{-1} I \tag{1.25}$$

after expanding out $q(x + \Delta x)$ in first-order Taylor series about point x (cf. Eqs. (1.22c–e) preceding).

1.6 Efficient estimators

Classically, the main use of information *I* has been as a measure of the ability to estimate a parameter. This is through the Cramer–Rao inequality (1.1), as follows.

If the equality can be realized in Eq. (1.1), then the mean-square error will go inversely with *I*, indicating that *I* determines how small (or large) the error can be in any particular scenario. The question is, then, when is the equality realized?

The left-hand side of Eq. (1.7) is actually an inner product between two 'vectors' $A(\mathbf{y})$ and $B(\mathbf{y})$,

$$A(\mathbf{y}) = \frac{\partial \ln p}{\partial \theta} \sqrt{p}, \quad B(\mathbf{y}) \equiv (\hat{\theta} - \theta)\sqrt{p}. \tag{1.26a}$$

Here the continuous index $\mathbf{y}$ defines the $\mathbf{y}$th component of each such vector (in contrast to the elementary case where vector components are discrete). The inner product of two vectors A, B is always less than or equal to its value when the two vectors are *parallel*, i.e., when all their $\mathbf{y}$-components are proportional,

$$A(\mathbf{y}) = k(\theta)B(\mathbf{y}), \quad k(\theta) = const. \tag{1.26b}$$

(Note that function $k(\theta)$ remains constant since the parameter θ is, of course, constant.) Combining Eqs. (1.26a) and (1.26b) then provides a necessary condition (i) for attaining the equality in Eq. (1.1),

$$\frac{\partial \ln p(\mathbf{y}|\theta)}{\partial \theta} = k(\theta)[\hat{\theta}(\mathbf{y}) - \theta]. \tag{1.27}$$

A condition (ii) is the previously used unbiasedness assumption (1.3).

A PDF scenario where (1.27) is satisfied causes a minimized error e_{min}^2 that obeys

$$e_{min}^2 = 1/I. \tag{1.28}$$

The estimator $\hat{\theta}(y)$ is then called 'efficient'. Notice that in this case the error varies inversely with information I, so that the latter becomes a well-defined quality metric of the measurement process.

1.6.1 Exercise

It is noted that only certain PDFs $p(\mathbf{y}|\theta)$ obey condition (1.27), among them (a) the independent normal law $p(\mathbf{y}|\theta) = A\prod_n \exp\left[-(y_n - \theta)^2/2\sigma^2\right]$, $A = const.$, and (b) the exponential law $p(\mathbf{y}|\theta) = \prod_n e^{-y_n/\theta}/\theta$, $y_n \geqslant 0$. On the other hand, with $N = 1$, (c) a PDF of the form

$$p(y|\theta) = A\sin^2(y - \theta), \quad A = const., \quad |y - \theta| \leqslant \pi$$

does not satisfy (1.27). Note that this PDF arises when the position θ of a one-dimensional quantum mechanical particle within a box is to be estimated. Hence, this fundamental measurement problem does not admit of an efficient estimate. Show these effects (a)–(c).

Also show that the estimators in (a) and (b) are unbiased, as required.

1.6.2 Exercise

If the condition (1.27) *is* obeyed, and if the estimator is unbiased, then the estimator function $\hat{\theta}(\mathbf{y})$ that *attains* efficiency is the one that maximizes the likelihood function $p(\mathbf{y}|\theta)$ through choice of θ (Van Trees, 1968). This is called the *maximum likelihood* (ML) estimator. As an example, the ML estimators for the problems (a) and (b) preceding are both the simple average of the data. Show this.

Note the simplification that occurs if one maximizes, instead of the likelihood, the *logarithm* of the likelihood. This *log-likelihood* law is also of fundamental importance to quantum measurement theory; see Chap. 10.

1.7 Fisher *I* as a measure of system disorder

We showed that information I is a quality metric of an efficient measurement procedure. Now we will find that I is also a measure of the degree of disorder of a system. *High disorder* means a lack of predictability of values x over its range, i.e., a uniform or 'unbiased' probability density function $p(x)$. Such a curve is shown in Fig. 1.2b. The curve has small gradient content (if it is physically meaningful, i.e., is piecewise continuous). Simply stated, *it is broad and smooth*. Then by (1.24) the Fisher information I *is small*.

Conversely, if a curve $p(x)$ shows bias to particular x values then it exhibits *low disorder*. See Fig. 1.2a. Analytically, the curve will be *steeply sloped* about these x values, and so the value of I *becomes large*. The net effect is that I measures the degree of disorder of the system.

On the other hand, the ability to measure disorder is usually associated with the word 'entropy'. For example, the Shannon entropy H is known to measure the degree of disorder of a system. (Example: By direct use of Eq. (1.13), if $p(x)$ is normal with variance σ^2 then $H = \ln \sigma + \ln \sqrt{2\pi e}$. This shows that H monotonically increases with the 'width' σ of the PDF, i.e., with the degree of disorder in the system.)

Since we found that I likewise measures the disorder of a system this suggests that I ought to likewise be regarded as an 'entropy'. However, the entropy H has another important property: When H is, as well, the *Boltzmann* entropy, it obeys the Second Law of thermodynamics, increasing monotonically *with time*,

$$\frac{dH(t)}{dt} \geq 0. \tag{1.29}$$

Does I, then, also change monotonically with time? A particular scenario suggests that this is so.

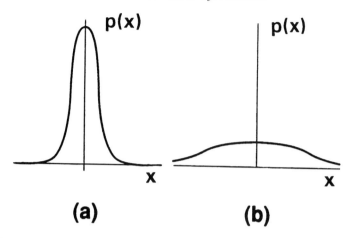

Fig. 1.2. Degree of disorder measured by I values. In (a), random variable x shows relatively low disorder and large I (gradient content). In (b), x shows high disorder and small I. (Reprinted from Frieden and Soffer, 1995.)

1.8 Fisher I as an entropy

We next show that, for a particular isolated system, I monotonically decreases with time (Frieden, 1990). All measurements and PDF laws are now taken at a specified time t, so that the system PDF now has the form $p(x|t)$, i.e., the probability of reading $(x, x + dx)$ conditional upon a time $(t, t + dt)$.

1.8.1 Paradigm of the broken urn

Consider a scenario where many particles fill a small urn. Imagine these to be ideal, point masses that collide elastically and that are not in an exterior force field. We want a smart measurement of their mean horizontal position θ. Accordingly, a particle at horizontal position y is observed, $y = \theta + x$, where x is a random fluctuation from θ. Define the mean-square error $e^2(t) = \langle[\theta - \hat{\theta}(y)]^2\rangle$ due to repeatedly forming estimates $\hat{\theta}(y)$ of θ within a small time interval $(t, t + dt)$. How should error e vary with t?

Initially, at $t = 0$, the particles are within the small urn. Hence, any observed value y should be near to θ; then, any good estimate $\hat{\theta}(y)$ will likewise be close to θ, and resultingly $e^2(0)$ will be small. Next, the walls of the container are broken, so that the particles are free to randomly move away. They will follow, of course, the random walk process which is called Brownian motion (Papoulis, 1965).

Consider a later time interval $(t, t + dt)$. For Brownian motion, the PDF $p(x|t)$ is Gaussian with a variance $\sigma^2 = Dt$, $D = const.$, $D \geqslant 0$. For a Gaussian PDF, $I = 1/\sigma^2$ (see derivation in Sec. 8.3.1). Then $I = I(t) = 1/Dt$, or I decreases with t.

Can this result be generalized?

1.8.2 The 'I-theorem'

Eq. (1.29) states that H increases monotonically with time. This result is usually called the 'Boltzmann H-theorem.' In fact there is a corresponding 'I-theorem'

$$\frac{dI(t)}{dt} \leq 0. \tag{1.30}$$

1.8.3 Proof

Start with the cross-entropy representation (1.22b) of $I(t)$,

$$I(t) = -2 \lim_{\Delta x \to 0} \Delta x^{-2} \int dx\, p \ln (p_{\Delta x}/p) \tag{1.31}$$

$$p \equiv p(x|t),\ p_{\Delta x} \equiv p(x + \Delta x|t).$$

Under certain physical conditions, e.g., 'detailed balance', short-term correlation, shift-invariant statistics (Gardiner, 1985; Reif, 1965; Risken, 1984) p obeys a *Fokker–Planck* differential equation

$$\frac{\partial p}{\partial t} = -\frac{d}{dx}[D_1(x,\,t)p] + \frac{d^2}{dx^2}[D_2(x,\,t)p] \tag{1.32}$$

where $D_1(x,\,t)$ is a drift function and $D_2(x,\,t)$ is a diffusion function. Suppose that $p_{\Delta x}$ also obeys the equation (Plastino and Plastino, 1996). Risken (1984) shows that two PDFs, such as p and $p_{\Delta x}$, that obey the Fokker–Planck equation have a cross-entropy

$$G(t) \equiv -\int dx\, p \ln (p/p_{\Delta x}) \tag{1.33}$$

that obeys an H-theorem (1.29),

$$\frac{dG(t)}{dt} \geq 0. \tag{1.34}$$

It follows from Eq. (1.31) that I, likewise, obeys an I-theorem (1.30). Thus, the I-theorem and the H-theorem both hold under certain physical conditions.

There also is a possibility that physical conditions exist for which one theorem holds to the exclusion of the other. From the empirical viewpoint that the I-theorem leads to the derivation of a much wider range of physical laws (as in Chaps. 4–11) than does the *H*-theorem, such conditions must exist; however, they are yet to be found.

It should be remarked that the *I*-theorem was first proven (Plastino and Plastino, 1996) from the direct defining form (1.2) for *I* (i.e., *without* recourse to the cross-entropy form (1.22b)).

1.8.4 Ramification to definition of time

The *I*-theorem (1.30) is an extremely important result. It states that the Fisher information of a physical system can only decrease (or remain constant) in time. Combining this with Eq. (1.28) indicates that $e^2_{\min}$ must increase, so that even in the presence of efficient estimation *the quality of estimates must decrease with time*. This seems to be a reasonable alternative statement of the Second Law of thermodynamics. If, by the Second Law, the disorder of a system (as measured by the Boltzmann entropy) must increase, then the disorder of any measuring system must increase as well. This must degrade its use as a measuring instrument, causing the error $e^2_{\min}$ to increase. On this basis, one could estimate the age of an instrument by simply observing how well it measures.

Thus, *I* is a measure of physical disorder that has its mathematical roots in estimation theory. By the same token, one may regard the Boltzmann entropy *H* to be a measure of physical disorder that has its mathematical roots in communication theory (Shannon, 1948). Communication theory plays a complementary role to estimation theory: the former describes how well messages can be *transmitted*, in the presence of given errors in the channel (system noise properties); the latter describes how accurately messages may be *estimated*, also in the presence of given errors in the channel.

If *I* really is a physical measure of system disorder, it ought to somehow relate to temperature, pressure, and all other extrinsic parameters of thermodynamics. This is, in fact, the subject of Secs. 1.8.5–1.8.7.

Next, consider the concept of the flow of thermodynamic time (Zeh, 1992; Halliwell *et al.*, 1994). This concept is intimately tied to that of the Boltzmann entropy: an increase in the latter *defines* the positive passage of Boltzmann time. The *I*-theorem suggests an alternative definition to Boltzmann time: a decrease in *I* defines an increase in 'Fisher time'. However, whether the two times always agree is an open question. In numerical experiments on randomly perturbed PDFs (Frieden, 1990), usually the resulting perturbations δI went down when δH went up, i.e., both measures agreed that disorder (and time) increased. They also usually agreed on decreases of disorder. However, there were disagreements about 1% of the time.

1.8.5 Ramification to temperature

The Boltzmann temperature (Reif, 1965) T is defined as $1/T \equiv \partial H_B/\partial E$, where H_B is the Boltzmann entropy of an isolated system and E is its energy. Consider two systems A and A' that are in thermal contact, but are otherwise isolated, and are approaching thermal equilibrium. The Boltzmann temperature has the important property that after thermal equilibrium is attained, a situation

$$T = T', \frac{1}{T} = \frac{\partial H_B}{\partial E}, \frac{1}{T'} = \frac{\partial H'_B}{\partial E'} \tag{1.35}$$

of equal temperature results. Let us now look at the phenomenon from the standpoint of information I, i.e. *without* recourse to the Boltzmann entropy.

Denote the total information in system A by I, and that of system A' by I'. The parameters θ, θ' to be measured are the total energies E and E' of the two systems. The corresponding measurements are Y_E, $Y_{E'}$. Because of the I-theorem (1.30), *both I and I' should approach minimum values as time increases*. We will show later that, since the two systems are physically separated and hence independent in their energy data Y_E, $Y_{E'}$, the Fisher information state of the two is the sum of the two I values. Hence, the I-theorem states that, after an infinite amount of time, the information of the combined system is

$$I(E) + I'(E') = Min. \tag{1.36}$$

On the other hand, energy is conserved, so that

$$E + E' \equiv C, \tag{1.37}$$

C = constant. (Notice that this is a deterministic relation between the two ideal parameter values, and not between the data; if it held for the data, then the prior assumption of independent data would have been invalid.)

The effect of (1.37) on (1.36) is

$$I(E) + I'(C - E) = Min. \tag{1.38}$$

We now define a generalized 'Fisher temperature' T_θ as

$$\frac{1}{T_\theta} \equiv -k_\theta \frac{\partial I}{\partial \theta}. \tag{1.39}$$

Notice that θ is any parameter under measurement. Hence, there is a Fisher 'temperature' associated with any parameter to be measured. From (1.39), T_θ simply measures the sensitivity of information level I to a change in system parameter θ. The constant k_θ gives each T_θ value the same units. A relation between the two temperatures T and T_θ is found below for a perfect gas.

Consider the case in point, $\theta = E$, $\theta' = C - \theta$. The temperature T_θ is now an energy temperature T_E. Differentiating Eq. (1.38) $\partial/\partial E$ gives

$$\frac{\partial I}{\partial E} + \frac{\partial I'}{\partial E'}(-1) = 0 \text{ or } T_E = T_{E'} \tag{1.40}$$

by (1.39). At equilibrium both systems attain a common Fisher energy temperature. This is analogous to the Boltzmann (conventional) result (1.35).

1.8.6 Exercise

The right-hand side of Eq. (1.39) is impractical to evaluate (although still of theoretical importance) if I is close to independent of θ. This occurs in close to a shift invariant case (1.11), (1.12) where the resulting I is close to the form (1.2). The key question is, then, whether the shift invariance condition Eq. (1.11) holds when $\theta = E$ and a measurement y_E is made. The total number N of particles comprising the system is critical here. If $N \approx 10$ or more, then (a) the PDF $p(y_E|E)$ will tend to obey the central limit theorem (Frieden, 1991) and, hence, be close to Gaussian in the shifted random variable $y_E - E$. An I results that is close to the form (1.2). At the other extreme, (b) for small N the PDF can typically be χ^2 (assuming that the $N = 1$ law is Boltzmann, i.e., exponential). Here, shift invariance would not hold. Show (a) and (b).

1.8.7 Perfect gas law

So far we have defined concepts of time and temperature on the basis of Fisher information. We now show that the perfect gas law may likewise be derived on this basis. This will also permit the (so far) unknown parameter k_E to be evaluated from known parameters of the system.

Consider an ideal gas consisting of M identical molecules confined to a volume V and kept at Fisher temperature T_E. We want to know how the pressure in the gas depends upon the extrinsic parameters V and T_E. The plan is to first compute the temporal mean pressure $\overline{p}$ within a small volume $dV = A dx$ of the gas and then integrate through to get the macroscopic answer.

Suppose that the pressure results from a force F that is exerted normal to area A and through the distance dx, as in the case of a moving piston. Then (Reif, 1965)

$$\overline{p} = \frac{F \, dx}{A \, dx} = -\frac{\partial E}{\partial V} \tag{1.41}$$

where the minus sign signifies that energy E is stored in reaction to work done by the force. Using the chain rule, Eq. (1.41) becomes

$$\overline{p} = -\frac{\partial E}{\partial I}\frac{\partial I}{\partial V} = k_E T_E \frac{\partial I}{\partial V}, \tag{1.42}$$

the latter by definition (1.39) with $\theta = E$. Here dI is the information in a data reading dy_E of the ideal energy value dE. In general, quantities $\overline{p}$, dI and T_E can be functions of the position r of volume dV within a gas. Multiplying (1.42) by dV gives

$$\overline{p}(r)dV = k_E T_E(r)dI(r) \qquad (1.43)$$

with the r-dependence now noted. Near equilibrium the gas should be well mixed and homogeneous, such that $\overline{p}$ and T are independent of position r. Then Eq. (1.43) may be directly integrated to give

$$\overline{p}V = k_E T_E I. \qquad (1.44)$$

Note that $I = \int dI(r)$ is simply the total information due to many independent data readings dy_E. This again states that the information adds under independent data conditions.

The dependence (1.44) of $\overline{p}$ upon V and T_E is of the same form as the known equation of state of the gas

$$\overline{p}V = MkT, \qquad (1.45)$$

where k is the Boltzmann constant and T is the *ordinary* (Boltzmann) temperature. Comparing Eqs. (1.44) and (1.45), exact compliance is achieved if $k_E T_E$ is related to kT as

$$\left(\frac{kT}{k_E T_E} \right) = I/M, \qquad (1.46)$$

the information per molecule. The latter should be a constant for a well-mixed gas.

These considerations seem to imply that thermodynamic theory may be developed completely from the standpoint of Fisher entropy, without recourse to the well-known properties of the Boltzmann entropy. At this point in time, the question remains an open one.

1.8.8 Ramification to derivations of physical laws

The uni-directional nature of the *I*-theorem (1.30) implies that, as $t \to \infty$,

$$I(t) = 4 \int dx q'^2(x|t) \to Min. \qquad (1.47)$$

Here we used the shift-invariant form (1.24) of *I*. The minimum would be achieved through variation of the amplitude function $q(x|t)$. It is convenient, and usual, to accomplish this through use of an Euler–Lagrange equation (see Eq. (0.13)). The result would define the form of $q(x|t)$ at temporal equilibrium.

In order for this approach to be tenable it would need to be modified by appropriate input constraint properties of $q(x|t)$ such as normalization of

$p(x|t)$. Other constraints, describing the particular physical scenario, must also be tacked on. Examples are fixed values of the means of certain physical quantities (case $\alpha = 2$ below). Such constraints may be appended to principle (1.47) by using the method of Lagrange undetermined multipliers, Eq. (0.39):

$$I + \sum_{k=1}^{K_\circ} \lambda_k \int dx q^\alpha(x|t) f_k(x) = Extrem., \qquad (1.48a)$$

$$\int dx q^\alpha(x|t) f_k(x) = F_k, \ k = 1, \ldots, K_\circ, \ \alpha = Const. \qquad (1.48b)$$

The kernel functions $f_k(x)$, constraint exponent α and data values F_k are assumed known. The multipliers λ_k are found such that the constraint equations (1.48b) are obeyed. See also Huber (1981).

The most difficult step in this approach is deciding what constraints to utilize (called the 'input' constraints). The solution depends critically upon the choice of input constraints, and yet they cannot simply be all that are known to the user. They must be the particular subset of constraints that are *actually imposed* by nature. In general, this is difficult to know *a priori*. Our own approach – the EPI principle described in Chap. 3 and applied in Chaps. 4–9 and 11 – is, in fact, of the Lagrange form (1.48a). However, it attempts to free the problem of the arbitrariness of the constraint terms. For this purpose, a physical rationale for the terms is utilized.

It is important to verify that a minimum (1.47) will indeed be attained in solution of the constrained variational problem. A maximum or point of inflection could conceivably result instead, defeating our aims. For this purpose, we may use *Legendre's condition* for a minimum (Sec. 0.3.3): Let $\mathscr{L}$ denote the integrand (or Lagrangian) of the total integral to be extremized. In our scalar case, if

$$\frac{\partial^2 \mathscr{L}}{\partial q'^2} > 0 \qquad (1.49)$$

the solution will be a minimum. From Eqs. (1.47) and (1.48a), our Lagrangian is

$$\mathscr{L} = 4q'^2 + \sum_k \lambda_k q^\alpha f_k. \qquad (1.50)$$

Using this in Eq. (1.49) gives

$$\frac{\partial^2 \mathscr{L}}{\partial q'^2} = +8, \qquad (1.51)$$

showing that a minimum is indeed attained.

The foregoing assumed that coordinate x is real and a scalar. However, most scenarios will involve *multiple* coordinates due to Lorentz covariance requirements. One or more of these are purely imaginary. For example, in Chap. 4 we use a space coordinate x that is purely imaginary. The same analysis as the preceding shows that in this case the second derivative (1.51) is negative, so that a maximum is instead attained in this coordinate. However, others of the coordinates are real and, hence, tend to give a mimimum. Obviously Legendre's condition cannot give a unique answer in this scenario, and other criteria for the determination must be used. The question of maximum or minimum under such general conditions remains an open question.

Note that these results apply to all physical applications of our variational principle (as formed in Chap. 3). These applications only differ in their effective constraint terms, none of which contains terms in q'^2.

When I is minimized by the variational technique, $q(x|t)$ tends to be maximally smooth (Sect. 1.7). We saw that this describes a situation of maximum disorder. The Second Law of thermodynamics causes, as well, increased disorder. In this behavior, then, the I-theorem (1.30) acts like the Second Law.

1.8.9 Is the I-theorem equivalent to the H-theorem?

We showed at Eq. (1.34) that if a PDF $p(x)$ and its *infinitesimally* shifted version $p(x + \Delta x)$ both obey the Fokker–Planck equation, then the I-theorem follows. On the other hand, Eq. (1.34) with Δx *finite* is an expression of the H-theorem. Hence, the two theorems have a common pedigree, so to speak. Are, then, the two theorems equivalent? In fact they are not equivalent because the I-theorem is a *limiting form* (as $\Delta x \to 0$) of Risken's H-theorem. Taking the limit introduces the derivative p' of the PDF into the integrand of what was H, transforming it into the form Eq. (1.2) of I. The presence of this derivative in I, and its absence in H, has strong mathematical implications. One example is as follows.

The equilibrium solutions for $p(x)$ that are obtained by extremizing Fisher I (called the EPI principle below) are, in general, different from those obtained by the corresponding use $H = max.$ of entropy. See Sec. 1.3.1. In fact, EPI solutions and $H = max.$ solutions agree only in statistical mechanics; this is shown in Appendix A.

It is interesting that correct solutions via EPI occur even for PDFs that do not obey the Fokker–Planck equation. By the form of Eq. (1.32), the time rate of change of p only depends upon the present value of p. Hence, the process has short-term memory (see also Gardiner, 1991, p. 144). However, EPI may be used to derive the $1/f$ power spectral noise effect (Chap. 8), a law famous for exhibiting long-term memory. Also, the relativistic electron obeys an equation of continuity of flow $\partial p/\partial t = c\nabla\cdot(\psi^*[\alpha]\psi)$, $p \equiv \psi^*\psi$ (Schiff, 1955), where all quantities are defined in Chap. 4. This does not quite have the form of a Fokker–Planck Eq. (1.32) (compare right-hand sides). However, EPI may indeed be used to derive the Dirac equation of the electron (Chap. 4).

These considerations imply that the Fokker–Planck equation is a sufficient, but not necessary, condition for validity of the EPI procedure. An alternative condition of wider scope must exist. Such a one is the unitary condition to be discussed in Secs. 3.8.5 and 3.8.7.

1.8.10 Flow property

Since information I obeys an I-theorem Eq. (1.30), temperature effects Eqs. (1.39) and (1.40), and a gas law Eq. (1.44), indications are that I is every bit as 'physical' an entity as is the Boltzmann entropy. This includes, in particular, a property of temporal *flow* from an information source to a sink. This property is used in our physical information model of Sec. 3.3.2.

1.8.11 Additivity property

A vital property of the information I is that of additivity: the information from mutually isolated systems adds. This is shown as follows.

Suppose that we have N copies of the urn mentioned in Sec. 1.8.1. See Fig. 1.3. As before, each urn contains particles that are undergoing Brownian motion. (This time the urns are not broken.) Each sits rigidly in place upon a table that moves with an unknown velocity θ in the X-direction, relative to the laboratory. A particle is randomly selected in each urn, and its total X-component laboratory velocity value y_n is measured. Let x_n denote the particle's *intrinsic* speed, i.e., relative to its urn, with $(x_n, n = 1, \ldots, N) \equiv \mathbf{x}$. The $\mathbf{x}$ are random because of the Brownian motion. Assuming nonrelativistic speeds, the intrinsic data $(y_n, n = 1, \ldots, N) \equiv \mathbf{y}$ obey simply

$$\mathbf{y} = \theta + \mathbf{x}. \tag{1.52}$$

Assume that the urns are physically isolated from one another by the use of barriers B (see Fig. 1.3), so that there is no interaction between particles from different urns. Then the data $\mathbf{y}$ are independent. This causes the likelihood law to break up into a product of factors (Frieden, 1991)

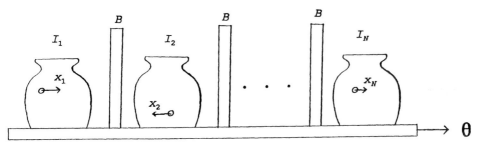

Fig. 1.3. N urns are moving at a common speed θ. Each contains particles in Brownian motion. Barriers B physically isolate the urns. Measurements $y_n = \theta + x_n$ of particle velocities are made, one to an urn. Each y_n gives rise to an information amount I_n. The total information I over all the data $\mathbf{y}$ is the sum of the individual informations I_n, $n = 1, \ldots, N$ from the urns.

$$p(\mathbf{y}|\theta) = \prod_{n=1}^{N} p_n(y_n|\theta), \tag{1.53}$$

where p_n is the likelihood law for the nth observation.

We may now compute the information I in the independent data $\mathbf{y}$. By Eq. (1.53)

$$\ln p = \sum_n \ln p_n, \ p \equiv p(\mathbf{y}|\theta), \ p_n \equiv p_n(y_n|\theta),$$

so that
$$\frac{\partial \ln p}{\partial \theta} = \sum_n \frac{1}{p_n} \frac{\partial p_n}{\partial \theta}. \tag{1.54}$$

Squaring the latter gives

$$\left(\frac{\partial \ln p}{\partial \theta}\right)^2 = \sum_{\substack{mn \\ m \neq n}} \frac{1}{p_m} \frac{1}{p_n} \frac{\partial p_m}{\partial \theta} \frac{\partial p_n}{\partial \theta} + \sum_n \frac{1}{p_n^2} \left(\frac{\partial p_n}{\partial \theta}\right)^2 \tag{1.55}$$

where the last sum is for indices $m = n$. Then the defining Eq. (1.9) for I gives, with the substitutions Eqs. (1.53), (1.55),

$$I = \int d\mathbf{y} \prod_k p_k \left[\sum_{\substack{mn \\ m \neq n}} \frac{1}{p_m} \frac{1}{p_n} \frac{\partial p_m}{\partial \theta} \frac{\partial p_n}{\partial \theta} + \sum_n \frac{1}{p_n^2} \left(\frac{\partial p_n}{\partial \theta}\right)^2\right]. \tag{1.56}$$

Now use the fact that, in this equation, the probabilities p_k for $k \neq m$ or n integrate through as simply factors 1, by normalization. The remaining factors in $\prod_k p_k$ are then $p_m p_n$ for the first sum, and just p_n for the second sum. The result is, after some cancellation,

$$I = \sum_{\substack{mn \\ m\neq n}} \iint dy_m dy_n \frac{\partial p_m}{\partial \theta} \frac{\partial p_n}{\partial \theta} + \sum_n \int dy_n \frac{1}{p_n} \left(\frac{\partial p_n}{\partial \theta} \right)^2. \qquad (1.57)$$

This simplifies, drastically, as follows. The first sum separates into a product of a sum

$$\sum_n \int dy_n \frac{\partial p_n}{\partial \theta} \qquad (1.58)$$

with a corresponding one in index m. But

$$\int dy_n \frac{\partial p_n}{\partial \theta} = \frac{\partial}{\partial \theta} \int dy_n p_n = \frac{\partial}{\partial \theta} 1 = 0 \qquad (1.59)$$

by normalization. Hence the first sum in Eq. (1.57) is zero.

The second sum in Eq. (1.57) is, by Eq. (1.5),

$$\sum_n \int dy_n p_n \left(\frac{\partial}{\partial \theta} \ln p_n \right)^2 \equiv \sum_n I_n \qquad (1.60)$$

by the definition Eq. (1.9) of I. Hence, we have shown that

$$I = \sum_{n=1}^{N} I_n \qquad (1.61)$$

in this scenario of independent data. This is what we set out to prove.

It is well-known that the Shannon entropy H obeys additivity, as well, under these conditions. That is, with

$$H = -\int dy\, p(\mathbf{y}|\theta) \ln p(\mathbf{y}|\theta), \qquad (1.62)$$

under the independence condition Eq. (1.53) it gives

$$H = \sum_{n=1}^{N} H_n, \quad H_n = -\int dy_n p_n(y_n|\theta) \ln p_n(y_n|\theta). \qquad (1.63)$$

1.8.12 Exercise

Show this. *Hint*: The proof is much simpler than the preceding. One merely uses the argument below Eq. (1.56) to collapse the multidimensional integrals into the one in y_n as needed.

One notes from all this that a requirement of *additivity* does not in itself uniquely identify the appropriate measure of disorder. It could be entropy or, as shown above, Fisher information. This is despite the identity $\ln (fg) = \ln (f) + \ln (g)$, which seems to uniquely imply entropy as the measure. Undoubtedly many other measures satisfy additivity as well.

1.8.13 *I = Min. from statistical mechanics viewpoint*

According to a basic premise of statistical mechanics (Reif, 1965), the PDF for a system that *will occur* is the one that is maximum probable to occur.

A general image-forming system is shown in Fig. 1.4. It consists of a source S of particles – any type will do, whether electrons, photons, etc. – a focussing device L of some sort and an image plane M for receiving the particles. Plane M is subdivided into coordinate positions $(x_n, n = 1, \ldots, N)$ with a constant, small spacing ε. An 'image event' x_n is the receipt of a particle within the interval $(x_n, x_n + \varepsilon)$. The number m_n of image events x_n is noted, for each $n = 1, \ldots, N$. There are M particles in all, with M very large. What is the joint probability $P(m_1, \ldots, m_n)$?

Each image event is a possible position x_n, of which there are N. Therefore the image events comprise an *N*ary events sequence. This obeys a multinomial probability law (Frieden, 1991) of order N,

$$P(m_1, \ldots, m_n) = M! \prod_{n=1}^{N} \frac{r(x_n)^{m_n}}{m_n!}. \tag{1.64}$$

The quantities $r(x_n)$ are the 'prior probabilities' of events x_n. These are considered next.

The ideal source S for the experiment is a very small aperture that is located on-axis. This situation would give rise to ideal (prior) probabilities $r(x_n)$, $n = 1, \ldots, N$. However, in performing the experiment, we really cannot know exactly where the source is. For example, for quantum particles, there is an ultimate uncertainty in position of at least the Compton length (Sec. 4.1.17). Hence, in general, the source S will be located at a small position Δx off-axis. The result is that the particles will, in reality, obey a different set of probabilities $P(x_n) \neq r(x_n)$. These can be evaluated. Assuming shift invariance (Eq. (1.11)) and 1:1 magnification in the system,

$$p(x_n) = r(x_n - \Delta x), \text{ or, } r(x_n) = p(x_n + \Delta x). \tag{1.65}$$

By the law of large numbers (Frieden, 1991), since M is large the probabilities $p(x_n)$ agree with the occurrences m_n, by the simple rule

$$m_n = Mp(x_n). \tag{1.66}$$

(This takes the conventional, von Mises viewpoint that probabilities measure the frequency of occurrence of actual – not ideal – events (Von Mises, 1936)). Using Eqs. (1.65) and (1.66) in Eq. (1.64) and taking the logarithm gives

$$\ln P = C + \sum_n Mp(x_n) \ln p(x_n + \Delta x) - \sum_n \ln \left[Mp(x_n)! \right] \tag{1.67}$$

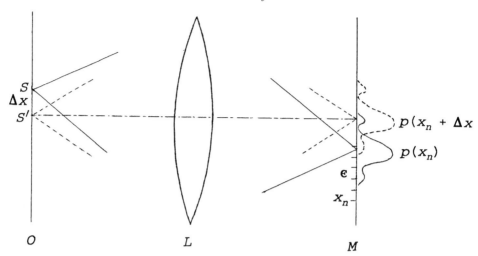

Fig. 1.4. A statistical mechanics view of Fisher information. The ideal point source position S' gives rise to the ideal PDF $p(x_n + \Delta x)$, while the actual point source position S gives rise to the empirical PDF $p(x_n)$. Maximizing the logarithm of the probability of the latter PDF curve implies a condition of minimum Fisher information, $I[p(x_n)] = I = Min.$

where C is an irrelevant constant. Since M is large we may use the Stirling approximation $\ln u! \approx u \ln u$, so that

$$\ln P \approx B + M \sum_n p(x_n) \ln \frac{p(x_n + \Delta x)}{p(x_n)}, \qquad (1.68)$$

where B is an irrelevant constant. The normalization of $p(x_n)$ was also used. Multiplying and dividing Eq. (1.68) by the fine spacing ε allows us to replace the sum by an integral. Also, since P is to be a maximum, so will be $\ln P$. The result is that Eq. (1.68) becomes

$$\ln P \approx \int dx p(x) \ln \frac{p(x + \Delta x)}{p(x)} = Max. \qquad (1.69)$$

after ignoring all multiplicative and additive constants. Noticing the minus sign in Eq. (1.22b), we see that Eq. (1.69) states that

$$I[p(x)] \equiv I = Min., \qquad (1.70)$$

agreeing with Eq. (1.47).

This approach can be generalized. Regardless of the physical nature of coordinate x, there will always be uncertainty Δx in the actual value of the origin of a PDF $p(x)$. As we saw, this uncertainty is naturally expressed as a 'distance measure' I between $p(x)$ and its displaced version $p(x + \Delta x)$ (Eq. (1.69)).

It is interesting to compare this approach with the derivation of the *I*-theorem in Sec. 1.8.3. That was based purely on the assumption that the Fokker–Planck equation is obeyed. By comparison, here the assumptions are that (i) maximum probable PDFs actually occur (basic premise of statistical mechanics) and (ii) the system admits of an ultimate resolution 'length' Δx of finite extent.

The two derivations may be further compared on the basis of effective 'resolution lengths'. In Sec. 1.8.3 the limit $\Delta x \to 0$ is rigorously taken, since Δx is, there, just a *mathematical* artifact (which enables I to be expressed as the cross-entropy via Eq. (1.22b)). Also, the approach by Plastino *et al.* to the *I*-theorem that is mentioned in that section does not even use the concept of Δx. By contrast, in the current derivation Δx is not merely of mathematical origin. It originates physically, as an ultimate resolution length and, hence, is small but intrinsically *finite*. This means that the transition from the cross-entropy on the right-hand side of Eq. (1.69) to information I via Eq. (1.22b) is, here, only an approximation.

If one takes this derivation seriously, then an important effect follows. Since I is only an approximation on the scale of Δx, the use of $I[q(x)]$ in any variational principle (such as EPI) must give solutions $q(x)$ that *lose their validity at scales finer than* Δx. For example, Δx results as the Compton length in the EPI derivation of quantum mechanics (Chap. 4). A ramification is that quantum mechanics is not represented by its famous wave equations at such scales.

This is a somewhat moot point, since then observations at that scale could not be made anyhow. Nevertheless, it suggests that a different kind of mechanics ought to hold at scales finer than Δx. Such considerations of course lead one to thoughts of quantum gravity (Misner *et al.*, 1973, p. 1193); see also Chap. 11. This is a satisfying transition from a physical point of view. Also, from the statistical viewpoint, it says that EPI is a complete theory, defining the limits of its range of validity.

Likewise, the electromagnetic wave equation (Chap. 5) would break down at scales Δx finer than the vacuum fluctuation length given by Eq. (5.4); suggesting a transition to a quantum electrodynamics. And the classical gravitational field equation (Chap. 6) would break down at scales finer than the Planck length, Eq. (6.22); again suggesting a transition to quantum gravity (see preceding paragraph). The magnitudes of these ultimate resolution lengths are, in fact, predicted by the EPI approach; see Chaps. 5, 6 and Sec. 3.4.15.

However, this kind of reasoning can be repeated to endlessly finer and finer scales. Thus, the theory of quantum gravity that is derived by EPI in Chap. 11 must break down beyond a finest scale Δx defined by the (presumed) approximate nature of I at that scale. The length Δx would be much smaller than the Planck length. This, in turn, suggests the need for a *new* 'mechanics' that would hold at finer scales than the Planck length, etc., to ever-finer scales. The same reasoning applies to electromagnetic and gravitational theories. Perhaps all three theories would converge to a common theory at a finest resolution length to be determined, at which 'point' the endless subdivision process would terminate.

On the other hand, if one does not take this derivation seriously then, based upon the derivation of Sec. (1.8.2), the wave equations of quantum mechanics are valid *down to all scales*, and a transition to quantum gravity is apparently not needed. Which of the two views to take is, at this time, unknown.

The statistical mechanics approach of this section is based partly on work by Shirai (1998).

1.8.14 Multiple PDF cases

In all of the preceding, there was one, scalar parameter θ to be estimated. This implies an information Eq. (1.2) that may be used to predict a single-component PDF $p(x)$ on scalar fluctuations x, as sketched in Sec. 1.8.8. Many phenomena are indeed describable by such a PDF. For example, in statistical mechanics the Boltzmann law $p(E)$ defines the single-component PDF on scalar energy fluctuations E (cf. Chap. 7).

Of course, however, nature is not that simple. There are physical phenomena that require *multiple* PDFs $p_n(\mathbf{x})$, $n = 1, \ldots, N$ or amplitude functions $q_n(\mathbf{x})$, $n = 1, \ldots, N$ for their description. Also, the fluctuations $\mathbf{x}$ might be vector quantities (as indicated by the boldface). For example, in relativistic quantum mechanics there are four wave functions and, correspondingly, four PDFs to be determined (actually, we will find eight real wave functions $q_n(\mathbf{x})$, $n = 1, \ldots, 8$, corresponding to the real and imaginary parts of the four complex wave functions). To derive a multiple-component, vector phenomenon, it turns out, requires use of the Fisher information defining the estimation quality of *multiple* vector parameters. This is the subject of the next chapter.

2

Fisher information in a vector world

2.1 Classical measurement of four-vectors

In the preceding chapter, we found that the accuracy in an estimate of a single parameter $\boldsymbol{\theta}$ is determined by an information I that has some useful *physical* properties. It provides new definitions of disorder, time and temperature, and a variational approach to finding a single-component PDF law $p(x)$ of a single variable x. However, many physical phenomena are describable only by multiple-component PDFs, as in quantum mechanics, and for vector variables $\mathbf{x}$ since worldviews are usually four-dimensional (as required by covariance). *Our aim in this chapter, then, is to form a new, scalar information I that is appropriate to this multi-component, vector scenario.* The information should be intrinsic to the phenomenon under measurement and not depend, e.g., upon exterior effects such as the noise of the measuring device.

2.1.1 The 'intrinsic' measurement scenario

In Bayesian statistics, a *prior* scenario is often used to define an otherwise unknown prior probability law; see, e.g., Good (1976), Jaynes (1985) or Frieden (1991). This is a model scenario that permits the prior probability law to be computed on the basis of some ideal conditions, such as independence of data, and/or 'maximum ignorance' (see below), etc. We will use the concept to define our unknown information expression.

For this purpose we proceed to analyze a particular prior scenario. This is an ideal, $4N$-dimensional measurement scenario – the vector counterpart of the scalar parameter problem of Sec. 1.2.1. Here, N *four*-vectors $\boldsymbol{\theta}_n$, $n = 1, \ldots, N$ of any physical nature (four-positions or potentials, etc.) are observed as intrinsic data

$$\mathbf{y}_n = \boldsymbol{\theta}_n + \mathbf{x}_n, \, n = 1, \ldots, N. \tag{2.1}$$

51

As before, the data are 'intrinsic' in that their fluctuations $\mathbf{x}_n$ are presumed to characterize solely the phenomenon under measurement. There is, e.g., no additional fluctuation due to noise in the instrument providing the measurements. An information measure I of fluctuations $\mathbf{x}_n$ will likewise be intrinsic to the phenomenon, as we required at the outset.

How realistic is this model? In Chap. 10 we will find that immediately before *real* measurements are made the physics of the intrinsic fluctuations $\mathbf{x}_n$ are independent of instrument noise. Indeed, how could flucuations before measurement depend upon the measuring system? Hence, ignoring the instrumental errors in this *model, prior* measurement scenario agrees with the *real, prior* measurement scenario. We call this scenario the 'intrinsic' data scenario.

For simplicity of notation, it is convenient to define 'grand' vectors $\boldsymbol{\theta}$, $\mathbf{y}$, $d\mathbf{y}$ over all n as

$$\boldsymbol{\theta} = (\boldsymbol{\theta}_1, \ldots, \boldsymbol{\theta}_N)$$

$$\mathbf{y} = (\mathbf{y}_1, \ldots, \mathbf{y}_N) \tag{2.2}$$

$$d\mathbf{y} = d\mathbf{y}_1 \cdots d\mathbf{y}_N.$$

2.1.2 Aim of the data collection

In chapters to follow, the intrinsic data $\mathbf{y}_n$ will be analyzed presuming either of two general aims: (i) to estimate $\boldsymbol{\theta}_n$ *per se*, as when $\boldsymbol{\theta}_n$ is the nth four-position of a material particle; or (ii) to estimate a *function* of each $\boldsymbol{\theta}_n$, as when $\boldsymbol{\theta}_n$ is an ideal four-position and the electromagnetic four-potential $\mathbf{A}(\boldsymbol{\theta}_n)$ is required.

For either scenario (i), (ii), we want to form a scalar information measure that defines the quality of the $\mathbf{y}_n$. The answer should, intuitively, resemble the form of Eq. (1.2).

2.1.3 Assumption of independence

Suppose that the $\mathbf{y}_n$ are collected efficiently, i.e., independently. This is the usual goal in data collection. It can be accomplished by two different experimental procedures: (a) N independent repetitions of the experiment under the same initial conditions, measuring $\boldsymbol{\theta}_n$ at each; or (b) in the case of measurements upon particles, one experiment upon N particles, measuring the N different parameters $\boldsymbol{\theta}_n$ that ensue from one set of initial conditions. In scenario (a), independence is automatically satisfied. Scenario (b) tries to induce ergodicity in the data $\mathbf{y}_n$, e.g., by measuring particles that are sufficiently separated in one or more coordinates.

2.1.4 Real or imaginary coordinates

Each coordinate x_n of a four-vector $\mathbf{x}$ is, in general, either purely real or purely imaginary (Frieden and Soffer, 1995); also see Sec. 2.2.2. An example of 'mixed' real and imaginary coordinates is given in Sec. 4.1.2.

2.2 Optimum, unbiased estimates

An observer's aim is generally to learn as much as possible about the parameters $\boldsymbol{\theta}$. For this purpose, an optimum estimate

$$\hat{\boldsymbol{\theta}}_n = \hat{\boldsymbol{\theta}}_n(\mathbf{y}) \tag{2.3}$$

of each four-parameter $\boldsymbol{\theta}_n$ may be fashioned. Each estimate is, thus, a general function of all the data. An example of such an estimator is simply $\mathbf{y}_n$, i.e. the corresponding data vector, but this will usually not be optimum. One well-known class of optimum estimators is the 'maximum likelihood' estimator class discussed in Chap. 1.

2.2.1 Unbiased estimators

As with the case of 'good' experimental apparatus, the estimators are assumed to be unbiased, i.e., to obey

$$\langle \hat{\boldsymbol{\theta}}_n(\mathbf{y}) \rangle \equiv \int d\mathbf{y} \hat{\boldsymbol{\theta}}_n(\mathbf{y}) p(\mathbf{y}|\boldsymbol{\theta}) = \boldsymbol{\theta}_n, \tag{2.4}$$

where $p(\mathbf{y}|\boldsymbol{\theta})$ is the conditional probability of all data $\mathbf{y}$ in the presence of all parameters $\boldsymbol{\theta}$. Eq. (2.4) says that, although a given estimate will generally be in error, on the average it will be correct. How *small* the error may be, is next established. This introduces the vital concept of information.

2.2.2 Cramer–Rao inequalities

We temporarily suppress index n and focus attention on the four components of any one (n fixed) foursome of *scalar* values θ_ν, y_ν, x_ν, $\nu = 0, 1, 2, 3$. The mean-square errors from the true values θ_ν are

$$e_\nu^2 \equiv \int d\mathbf{y}[\hat{\boldsymbol{\theta}}_\nu(\mathbf{y}) - \boldsymbol{\theta}_\nu]^2 \, p(\mathbf{y}|\boldsymbol{\theta}). \tag{2.5}$$

Since the data are independent, each mean-square error obeys complementarity with an 'information' quantity I_ν,

$$e_\nu^2 I_\nu \geq 1, \tag{2.6}$$

where

$$I_\nu \equiv \int d\mathbf{y} \left[\frac{\partial \ln p(\mathbf{y}|\boldsymbol{\theta})}{\partial \boldsymbol{\theta}_\nu} \right]^2 p(\mathbf{y}|\boldsymbol{\theta}). \tag{2.7}$$

(See Appendix B for derivation, where $I_\nu \equiv F_{\nu\nu}$). Eqs. (2.6) and (2.7) comprise Cramer–Rao inequalities for our vector quantities. They hold for either real or *imaginary* components $\boldsymbol{\theta}_\nu$; see Appendix C. When equality is attained in (2.6), the minimum possible error e_ν^2 is attained. Then the estimator is called 'efficient'. Quantity I_ν is the νth element along the diagonal of the Fisher information matrix. The I_ν thus comprise a *vector* of informations.

2.3 Stam's information

We are now in a position to decide how to construct, from the vector of informations I_ν, the single scalar information quantity I that we seek. Regaining subscripts n and summing on Eq. (2.6) gives

$$\sum_n \sum_\nu 1/e_{n\nu}^2 \leq \sum_n \sum_\nu I_{n\nu}. \tag{2.8}$$

Each term in the left-hand sum was called an 'intrinsic accuracy' by Fisher. Stam (1959) proposed using the sum as a scalar information measure I_s for a vector scenario (as here),

$$I_s \equiv \sum_n \sum_\nu 1/e_{n\nu}^2 \leq \sum_n \sum_\nu I_{n\nu}, \tag{2.9}$$

where the inequality is due to Eq. (2.8). Stam's information is promising, since it is a scalar quantity. We adapt it to our purposes.

2.3.1 Exercise

Stam's information I_s, in depending explicity upon the error variances, ignores all possible error cross-correlations. But, for our additive error case (2.1), where the data y_n are independent and the estimators are unbiased (2.4), all error cross-correlations are zero. Show this.

2.3.2 Trace form, channel capacity, efficiency

The right-hand side of Eq. (2.9) is seen to be an upper-bound to I_s. Assume that efficient estimators are used. Then, the equality is attained in Eq. (2.6), so that each left-hand term $1/e_{nv}^2$ of Eq. (2.8) equals its corresponding information value I_{nv}. This means that the upper bound in Eq. (2.9) is realized. An analogous situation arises in the theory of Shannon information. There, the channel capacity, denoted as C, denotes the maximum possible amount of information that may be passed by a channel (Reza, 1961). Hence, we likewise define a capacity C for the estimation procedure to convey Fisher information I_s about the intrinsic system,

$$I \equiv C = \sum_n \int d\mathbf{y}\, p(\mathbf{y}|\boldsymbol{\theta}) \sum_v \left(\frac{\partial \ln p(\mathbf{y}|\boldsymbol{\theta})}{\partial \boldsymbol{\theta}_{nv}} \right)^2, \qquad (2.10)$$

the latter due to (2.7). It is interesting that this is the trace of the Fisher information matrix (see Appendix B, Eq. (B7)). This information also satisfies our goal of measuring the intrinsic *disorder* of the phenomenon under measurement (see Sec. 2.8).

2.3.3 Exercise

Taking the trace of the Fisher information matrix ignores, of course, all off-diagonal elements. But, because we have assumed independent data, the off-diagonal elements are, in fact, zero. Show this.

The trace operation (sum over n) in Eq. (2.10) has many physical connotations, in particular *relativistic invariance*, as shown in Sec. 3.5.1.

2.4 Physical modifications

The channel capacity Eq. (2.10) simplifies, in steps, due to various physical aspects of the intrinsic measurement scenario.

2.4.1 Additivity of the information

This was previously shown (Sec. 1.8.11) for the case of a single parameter. The generalization to a vector of parameters is now taken up. As will be seen, because of the lack of 'cross talk' between different data and parameters, the proof below is a little easier.

Since the intrinsic data are collected independently (Sec. 2.1.3), the joint probability of all the data separates into a product of marginal laws

$$p(\mathbf{y}|\boldsymbol{\theta}) = \prod_{n=1} p_n(\mathbf{y}_n|\boldsymbol{\theta}) = \prod_{n=1} p_n(\mathbf{y}_n|\boldsymbol{\theta}_n). \qquad (2.11)$$

The latter equality follows since by Eq. (2.1), $\boldsymbol{\theta}_m$ has no influence on $\mathbf{y}_n$, $m \neq n$. Taking the logarithm of Eq. (2.11) and differentiating then gives

$$\frac{\partial \ln p(\mathbf{y}|\boldsymbol{\theta})}{\partial \boldsymbol{\theta}_{n\mu}} = \frac{1}{p_n}\frac{\partial p_n}{\partial \boldsymbol{\theta}_{n\mu}}, \quad p_n \equiv p_n(\mathbf{y}_n|\boldsymbol{\theta}_n). \qquad (2.12)$$

Substitution of Eqs. (2.11) and (2.12) into Eq. (2.10) gives

$$I = \sum_n \int d\mathbf{y} \prod_m p_m(\mathbf{y}_m|\boldsymbol{\theta}_m) \sum_\nu \frac{1}{p_n^2}\left(\frac{\partial p_n}{\partial \boldsymbol{\theta}_{n\nu}}\right)^2, \qquad (2.13)$$

$$= \sum_n \int d\mathbf{y}_n p_n(\mathbf{y}_n|\boldsymbol{\theta}_n) \sum_\nu \frac{1}{p_n^2}\left(\frac{\partial p_n}{\partial \boldsymbol{\theta}_{n\nu}}\right)^2 \qquad (2.14)$$

after integrating out $d\mathbf{y}_m$ for terms in $m \neq n$, using normalization of each probability p_m. After an obvious cancellation, we get

$$I = \sum_n \int d\mathbf{y}_n \frac{1}{p_n} \sum_\nu \left(\frac{\partial p_n}{\partial \boldsymbol{\theta}_{n\nu}}\right)^2. \qquad (2.15)$$

This is the first simplification.

2.4.2 Shift invariance property

Most physical PDFs are independent of an absolute origin. This is generally called 'shift invariance'. For example, the Schroedinger wave equation, which describes the PDF on position and time, is invariant to constant shifts in these coordinates. This particular application of shift invariance is called Galilean invariance (Sec. 3.5.3). Applying shift invariance to our general PDF, any fluctuation $\mathbf{y}_n - \boldsymbol{\theta}_n \equiv \mathbf{x}_n$ should occur with a probability that is independent of the absolute size of $\boldsymbol{\theta}_n$,

$$p_n(\mathbf{y}_n|\boldsymbol{\theta}_n) = p_{x_n}(\mathbf{y}_n - \boldsymbol{\theta}_n|\boldsymbol{\theta}_n) = p_{x_n}(\mathbf{y}_n - \boldsymbol{\theta}_n) = p_{x_n}(\mathbf{x}_n), \; \mathbf{x}_n \equiv \mathbf{y}_n - \boldsymbol{\theta}_n. \qquad (2.16)$$

We see from the notation that the $p_{x_n}(\mathbf{x}_n)$ are independent of absolute origins $\boldsymbol{\theta}_n$. Substituting the $p_{x_n}(\mathbf{x}_n)$ into Eq. (2.15) and changing integration variables to $\mathbf{x}_n$, gives

$$I = \sum_n \int d\mathbf{x}_n \frac{1}{p_{x_n}(\mathbf{x}_n)} \sum_\nu \left(\frac{\partial p_{x_n}(\mathbf{x}_n)}{\partial x_{n\nu}}\right)^2. \qquad (2.17)$$

Observing the disappearance of absolute origins $\boldsymbol{\theta}_n$ from the expression, the information likewise obeys shift invariance.

2.4.3 Use of probabability amplitudes

Equation (2.17) further simplifies if we introduce real probability 'amplitudes' $q_n(\mathbf{x}_n)$,

$$I = 4 \sum_n \int d\mathbf{x}_n \sum_\nu \left(\frac{\partial q_n}{\partial x_{n\nu}}\right)^2, \quad p_{x_n}(\mathbf{x}_n) \equiv q_n^2(\mathbf{x}_n). \qquad (2.18)$$

The subscript n of $\mathbf{x}$ can now be suppressed, since each x_n ranges over the same values. Then Eq. (2.18) becomes

$$I = 4 \int d\mathbf{x} \sum_n \boldsymbol{\nabla} q_n \cdot \boldsymbol{\nabla} q_n,$$

$$\mathbf{x} = (x_0, \ldots, x_3), \quad d\mathbf{x} \equiv |dx_0| dx_1\, dx_2\, dx_3, \qquad (2.19)$$

$$\boldsymbol{\nabla} \equiv \partial/\partial x_\nu, \quad \nu = 0, 1, 2, 3.$$

(Note the boldface 'del' notation, indicating a four-dimensional gradient.) Derivation of this equation was the main aim of the chapter; see also Frieden and Cocke (1996). This is the channel capacity expression that will be used in all information-based derivations of physical laws to follow.

2.4.4 Information I as a sum of covariant informations

It is interesting to go one step further, using the fact that $\mathbf{x}$ is a four-vector, so that its first component is linear in the imaginary unit i. Then the first term in the sum (2.19) is negative, and using covariant notation (Sec. 6.2.8), (2.19) becomes

$$I = 4 \int d^4x\, q_{n,\lambda} q_n{}^\lambda. \qquad (2.20)$$

This has implied sums over position index n and derivative index λ. The derivative indices λ in this equation form a covariant pair (one raised, one lowered), according to Eq. (6.6). Thus, *for fixed n*, the integrand – call it $i_n(\mathbf{x})$ – is invariant to Lorentz transformation (obeying Eq. (3.34a)).

However, the indices n in the equation do not themselves form a covariant pair (both are lowered). Thus, the vector q_n does not enter covariantly into the implied sum over n. Hence, the integrand of Eq. (2.20) – call it the information density $i(\mathbf{x})$ – is not invariant to Lorentz transformation.

Of course, the simple expedient of multiplying one component of q_n by the imaginary number $i = \sqrt{-1}$ (as in Eq. (3.34b)) would make the resulting vector invariant. However, this would correspond to a negative probability (cf.

second Eq. (2.18)), which would be an inconsistency in the theory. Hence, we do not do this.

For fixed n, the right-hand side of (2.20) represents the Fisher information at the data point $\mathbf{y}_n$. This is by the additivity of information property in Sec. 2.4.1. Thus, Eq. (2.20) may be re-expressed as

$$I = \int d^4x\, i(\mathbf{x}), \quad i(\mathbf{x}) = \sum_n i_n(\mathbf{x}) \tag{2.21a}$$

$$i_n(\mathbf{x}) \equiv 4q_{n,\lambda}q_n{}^{\lambda} \; (no \; sum \; on \; n). \tag{2.21b}$$

Quantity $i_n(\mathbf{x})$ is the Fisher information *density* for the data point $\mathbf{y}_n$. We found, above, that each component $i_n(\mathbf{x})$ is Lorentz invariant. Thus, these information densities are, in a sense, more fundamental than the information I itself (whose density is *not* Lorentz invariant, as we found). This will be borne out by many density effects to follow; see, e.g., Eqs. (3.20) and (3.42). Also, note the following.

(1) The wave equations of physics, and the basic distribution functions, describe *point processes*. These define wave/probability amplitudes at a single point n, as opposed to joint statistics over pairs (or higher-order combinations) of points. Hence, they likewise ought to derive from information considerations at a point n, namely information $i_n(\mathbf{x})$.

(2) Each Euler–Lagrange equation that follows from the use of Eq. (2.20) in EPI is a differential equation in a component q_n, n fixed. *Each such equation is, in fact, always covariant.* That is, the fact that the vector q_n is not Lorentz invariant (as discussed above) is irrelevent to these solutions.

These considerations suggest that we ought to be able, somehow, to derive physical laws from an information that avoids the sum over n in Eq. (2.20). This would avoid the need for independent data in the intrinsic scenario – an advantage because of the increased generality of the approach. However, this cannot be done. At least a two-point sum over $n = 1, 2$ is needed in Eq. (2.20) in order to accomodate the presence of *complex* probability amplitudes ψ_n (see Sec. 3.7.1). Hence, at least two independent data points have to be assumed.

The impediment could be overcome by simply *re-defining* the information I in terms of complex amplitudes in the first place, as $I = \sum_n I_n$, $I_n = 4 \int d^4x \psi_{n,\lambda}^* \psi_n{}^{\lambda}$ (no sum on n). Such informations I_n could be used as the basis for a one-point EPI approach, replacing cumulative informations I, J, K in the EPI principle of Chap. 3 with corresponding single-point informations I_n, J_n, K_n. The information densities corresponding to these would be covariant, as was found desirable, and the output Euler–Lagrange solutions to this EPI approach would be the same as those derived by the approach using I, J, K (as in subsequent chapters). However, we do not accept the re-definition.

The grounds are that it is an *ad hoc* way of bringing in complex amplitudes. The manner by which we bring in complex amplitudes is, instead, a natural consequence of the classical statistics of independent data points (Sec. 3.7.1). This seems preferable. In the end, then, we stick with representation (2.20) for *I*.

Tensor generalizations of Eq. (2.20) for *I* are given at Eqs. (11.9) and (11.17). These allow for multiply subscripted coordinates embedded in a generally curved space.

2.5 Net probability $p(x)$ for a particle

The intrinsic measurement scenario also makes a prediction on the overall PDF $p(\mathbf{x})$ for a single particle. The single particle scenario was indentified in Sec. 2.1.3 as (a), where one particle is repeatedly measured. Then all particle events x_n are now being 'binned' in a common event space $\mathbf{x}$. Equation (2.16) now gives

$$p_n(\mathbf{y}_n|\boldsymbol{\theta}_n) = p(\mathbf{y}|\boldsymbol{\theta}_n) = p_x(\mathbf{x}|\boldsymbol{\theta}_n) = q_n(\mathbf{x})^2. \tag{2.21c}$$

The latter is by the second Eq. (2.18). To proceed further, we have to further quantify the meaning of the $\boldsymbol{\theta}_n$. These are N unknown ideal parameter values that we name $\boldsymbol{\theta}_1, \boldsymbol{\theta}_2, \ldots$, one for each experiment (a). However, in reality we have no control over which one of the $\boldsymbol{\theta}_n$ occurs in a given experiment – the numbering choice is completely due to chance. Hence, the $\boldsymbol{\theta}_n$ obey a PDF wherein each is equally probable,

$$P(\boldsymbol{\theta}_n) = Const. = \frac{1}{N}, \tag{2.22}$$

the latter by normalization.

Knowledge of the quantities $p_x(\mathbf{x}|\boldsymbol{\theta}_n)$ and $P(\boldsymbol{\theta}_n)$ from the preceding equations suggests use of the ordinary law of total probability (Frieden, 1991).

$$p(\mathbf{x}) = \sum_{n=1}^{N} p_x(\mathbf{x}|\boldsymbol{\theta}_n)P(\boldsymbol{\theta}_n) = \frac{1}{N}\sum_{n=1}^{N} q_n^2(\mathbf{x}). \tag{2.23}$$

We specialize, next, to the case of the relativistic electron. Define complex amplitudes as (Frieden and Soffer, 1995)

$$\psi_n \equiv \frac{1}{\sqrt{N}}(q_{2n-1} + iq_{2n}), \, i = \sqrt{-1}, \, n = 1, \ldots, N/2. \tag{2.24}$$

Fluctuations $\mathbf{x}$ are now, in particular, those of the space-time coordinates of the electron. Using Eq. (2.24) and then Eq. (2.23) gives

$$\sum_{n=1}^{N/2} \psi_n^* \psi_n = \frac{1}{N} \sum_n q_n^2 = p(\mathbf{x}). \tag{2.25}$$

Hence, the familiar dependence (2.25) of $p(\mathbf{x})$ upon $\psi_n(\mathbf{x})$ is a straightforward expression of the law of total probability of statistics. By (2.24), the $\psi_n(\mathbf{x})$ also have the significance of being complex *probability* amplitudes. The famous Born assumption to this effect (Born, 1926; 1927) does not have to be made. A further property of the $\psi_n(\mathbf{x})$ defined in (2.24) is of course that they obey the Dirac equation, which will be established by the use of EPI in Chap. 4.

2.6 The two types of 'data' used in the theory

We have shown that Fisher channel capacity form Eq. (2.19) follows from a scenario where N 'data values' are measured and utilized under ideal conditions (independence, efficiency). These were the 'intrinsic' data of Sec. 2.1.1. As was noted, such data values are not actually taken in the EPI process (which is activated by *real* data). *Their sole purpose is to define the channel capacity Eq. (2.19) for the system.* The channel capacity will be seen to be the key information quantity in an approach for deriving the physical law that governs the phenomenon under measurement.

This leads one to ask how real data fit into the information approach. Real data are formed when an intrinsic data value is detected by a generally imperfect instrument. The instrument will contribute additional noise to the intrinsic data value. *Mathematically,* real data are *appended to* the approach as data constraints, since real data are known (since Heisenberg) to constrain, or affect, the physical process that is under measurement. It is shown in Chap. 10 that the taking of real data has a dual action: it both activates the information form (2.19) *and* constrains the information by data constraints. It is this dual action that, in fact, gives rise to *both* the Schroedinger squared-gradient terms and Mensky data terms in the Feynman–Mensky measurement model of Chap. 10.

2.7 Alternative scenarios for the channel capacity I

(This section may be skipped in a first reading.) There are alternative scenario models for arriving at the channel capacity form Eq. (2.19) of information. These will give alternative meanings to the form from a, perhaps, more familiar statistical mechanics viewpoint.

Suppose that the same system is repeatedly prepared in the same way, and measured but once at each repetition. This guarantees independence of the data, and is essentially the scenario called (a) in Sec. 2.1.3. What will be the *ensemble average* of the information in the one measurement? Eq. (2.19) may be put in the form

$$I = \sum_{n=1}^{N} I_n, \quad I_n \equiv 4 \int d\mathbf{x} \nabla q_n \cdot \nabla q_n. \tag{2.26}$$

As at Eq. (2.23), the mean information $\bar{I}$ in a measurement is the average of the I_n values weighted with respect to their probabilites $P(\boldsymbol{\theta}_n) = 1/N$ of occurring (see Eq. (2.22)). Thus

$$\bar{I} = \sum_n I_n P(\boldsymbol{\theta}_n) = I/N \tag{2.27}$$

by Eq. (2.26). We might call this the 'single measurement scenario'. Even though N measurements were obtained, the information $\bar{I}$ represents the information that a single 'average', ideal measurement would see.

An alternative scenario is where there is no measurement at all. Regard the system to be specified by a choice of real probability amplitudes $q_n \equiv q(\mathbf{x}|\boldsymbol{\theta}_n)$, $n = 1, \ldots, N$, where one is physically chosen according to an 'internal degree of freedom' number $\boldsymbol{\theta}_n$. Suppose, further, that it is impossible to individually distinguish the states $\boldsymbol{\theta}_n$ by any experiment (as is the case in quantum mechanics). Then the numbers $\boldsymbol{\theta}_n$ are random events, and occur with probabilities $P(\boldsymbol{\theta}_n) = 1/N$ once again. Each such event is disjoint with one another, and defines an information $I_n \equiv I(\boldsymbol{\theta}_n)$ by the second Eq. (2.26). Then the average information obeys, once again, Eq. (2.27). Note that the disjointness of the events here replaces the independence of the data in preceding models in producing the simple sum in Eq. (2.27). Since no data were taken, the information represents an amount that is intrinsic to the system, just as we wanted. The approach resembles that of statistical mechanics, and so might be called a 'statistical mechanics model'.

This model also has the virtue of representing a situation of *maximum prior ignorance*. The 'maximum ignorance' comes about through two effects: (1) the Fisher information Eq. (2.27) has resulted from allowing an intrinsic *indistinguishability* of the states; and (2) information (2.27) will be extremized in the derivations that follow; most often the extremum will be a minimum (Sec. 1.8.8), and minimized Fisher information defines a situation of maximum disorder (or ignorance) about the system (Sec. 1.7).

2.8 Multiparameter *I*-theorem

The *I*-theorem (1.30) has been found to hold for a single PDF of a single random variable x. We want to generalize this to the multiparameter scenario. For the present, let the joint random variable $\mathbf{x}$ define all component fluctuations of the phenomenon *except for* the time t. Suppose that each pair of probability densities

$$p_n(\mathbf{x}|t),\; p_n(\mathbf{x}+\Delta\mathbf{x}|t),\; n = 1,\, \ldots,\, N \qquad (2.28)$$

obeys the multidimensional Fokker–Planck differential equation

$$\frac{\partial p_n}{\partial t} = -\sum_i \frac{\partial}{\partial x_i}[D_i(\mathbf{x},\, t)p_n] + \sum_{ij} \frac{\partial^2}{\partial x_i \partial x_j}[D_{ij}(\mathbf{x},\, t)p_n]. \qquad (2.29)$$

Functions D_i and D_{ij} are arbitrary drift- and diffusion functions, respectively. Risken (1984) shows that, under these conditions, the cross entropy G_n between each pair (2.28) of PDFs obeys an *H*-theorem,

$$\frac{\partial G_n(t)}{\partial t} \geqslant 0. \qquad (2.30)$$

On the other hand, the derivation Eqs. (1.17)–(1.22b, c) is easily generalized to a multidimensional PDF $p_n(\mathbf{x}|t)$ and information I_n, $n = 1,\, \ldots,\, N$ as well, with

$$I_n = -\frac{2}{\Delta x^2} \int d\mathbf{x}\, p_n(\mathbf{x}|t) \ln \frac{p_n(\mathbf{x}+\Delta\mathbf{x}|t)}{p_n(\mathbf{x}|t)}$$

$$= -\frac{2}{\Delta x^2} G_n[p_n(\mathbf{x}|t),\, p_n(\mathbf{x}+\Delta\mathbf{x}|t)]. \qquad (2.31)$$

It follows from Eqs. (2.30) and (2.31) that *each component* I_n obeys an *I*-theorem

$$\frac{\partial I_n}{\partial t} \leqslant 0. \qquad (2.32)$$

Then summing Eq. (2.32) over n and using the notation (2.26) gives our *multiparameter I-theorem*,

$$\frac{\partial I}{\partial t} \leqslant 0. \qquad (2.33)$$

This implies that, as $t \to \infty$,

$$I(t) \to Min, \qquad (2.34)$$

as at Eq. (1.47) for the single-component case.

In summary, the multiparameter information I *measures the physical state of disorder of the system*. This fact is important to the theory that follows.

3

Extreme physical information

3.1 Covariance, and the 'bound' information J

3.1.1 Transition to variational principle

Consider a system, or phenomenon, that is specified by amplitudes $q_n(\mathbf{x}|t)$ as in Chap. 2. The corresponding probabilities $p_n(x|t)$ are (temporarily) assumed to obey the Fokker–Planck Eq. (2.29) and, hence, the I-theorem Eq. (2.33). This theorem implies that, as time $t \to \infty$ the amplitudes $q_n(\mathbf{x}|t)$ approach equilibrium values such that I is a minimum. This translates into a variational principle

$$\delta I[\mathbf{q}(\mathbf{x}|t)] = 0, \ \mathbf{q}(\mathbf{x}|t) \equiv q_1(\mathbf{x}|t), \ \ldots, \ q_N(\mathbf{x}|\mathbf{t}). \tag{3.1}$$

As indicated at Eq. (0.45), I is a *functional* of the amplitudes $q_n(\mathbf{x}|t)$, whose variations cause the zero in the principle. The variation could instead be with respect to the probabilities (see Eq. (2.17)), but for our purposes it is more useful to regard the amplitudes as fundamental.

It is important to ask why the information I might change in the first place. In other words, what drives the I-theorem? As in thermodynamic theory, any interaction with the given system would cause the $q_n(\mathbf{x}|t)$ to change. Then, by Eq. (2.19), I would change as well. *The interaction that is of central interest to us in this book is that due to a real measurement.* Hence, from this point on, we view the system or phenomenon under study in the context of a measurement of its parameters $\boldsymbol{\theta}$.

Looking ahead, it will be shown in Sec. 3.8 that for certain measurement apparatus the Fokker–Planck equation requirement mentioned above can be dropped. That is, the overall approach taken – the EPI principle – will hold independent of this requirement. This will give EPI a much wider scope of application.

63

3.1.2 *On covariance*

A basic property of phenomena is that of covariance (see Sec. 3.5). Covariance requires that the system functionally depends upon all its coordinates $\mathbf{x}$ (including the time, if that is one of the coordinates) in the same way. If a phenomenon is to obey covariance, then any variational principle that implies the phenomenon should likewise be covariant. Let us examine variational principle (3.1) in this context. The conditional notation $(\mathbf{x}|t)$ in the principle suggests that t is not treated covariantly with $\mathbf{x}$. This may be shown as well. By definition of a conditional probability $p(\mathbf{x}|t) = p(\mathbf{x}, t)/p(t)$ (Frieden, 1991). This implies that the corresponding amplitudes (cf. the second Eq. (2.18)) obey $q(\mathbf{x}|t) = \pm q(\mathbf{x}, t)/q(t)$. The numerator treats $\mathbf{x}$ and t covariantly, but the denominator, in only depending upon t, does not. Thus, principle (3.1) is not covariant.

From a statistical point of view, principle (3.1) is objectionable as well, because it treats time as a deterministic, or known, coordinate while treating space as random. Why should time be *a priori* known any more accurately than space?

These problems can be remedied if we simply *make* (3.1) covariant. This may readily be done, by replacing it with the more general principle

$$\delta I[\mathbf{q}(\mathbf{x})] = 0, \; \mathbf{q}(\mathbf{x}) \equiv q_1(\mathbf{x}), \, \ldots, \, q_N(\mathbf{x}). \tag{3.2}$$

Here I is given by Eq. (2.19) and the $q_n(\mathbf{x})$ are to be varied. Coordinates $\mathbf{x}$ are, now, any *four-vector* of coordinates. In the particular case of space-time coordinates, $\mathbf{x}$ now *includes* the time.

3.1.3 *Need for a second term J*

As a principle of estimation for the $\mathbf{q}(\mathbf{x})$, however, (3.2) is lacking; I is of the fixed form (2.19) so that there is no room in the principle for the injection of physical information about the measurement scenario. This dovetails with the need for Lagrange constraint terms in the one-dimensional principle (1.48a,b): constraints, after all, stand for physical prior knowledge. The question is how to do this from a physical (and not *ad hoc*) standpoint. In other words, what constraints upon the principle (3.2) are actually imposed by nature?

This suggests that we need a revised principle of the form

$$\delta(I - J) = 0, \text{ or } I - J = Extrem., \tag{3.3}$$

where the term J is to embody all constraints that are imposed by the physical phenomenon under measurement. Thus, J describes, or characterizes, the physical phenomenon. Both I and J are functionals

$$I[\mathbf{q}], \; J[\mathbf{q}] \tag{3.4}$$

of the unknown functions $\mathbf{q}(\mathbf{x})$ with $I[\mathbf{q}]$ given by (2.19) and $J[\mathbf{q}]$ to be found.

At this point, Eq. (3.3) is merely a hypothesis. However, it will be shown, below, to follow from properties for I and J that are analogous to those of the Boltzmann and Shannon entropies.

We note from the form of Eq. (3.3) that J and I have the same units. This implies that J is an information too. We call it the 'bound' information, in analogy to a concept of Brillouin (1956) that is discussed next.

3.2 The equivalence of entropy and Shannon information

Brillouin (1956), building on the work of Szilard (1929), discovered an intriguing connection between the Boltzmann entropy and Shannon information. Consider a measurement made upon an isolated system. After the measurement a certain amount of Shannon information δH is gained. But the act of measurement requires some physical interaction with the system under measurement, so that the system entropy H_B increases as well. In fact, by the Second Law of thermodynamics, the increase in entropy must exceed (or equal) the gain in information,

$$\delta H_B \geqslant \delta H, \text{ or } H_B \geqslant H \tag{3.5}$$

as a sum over many such measurements. We call this 'Brillouin's equivalence principle'. It is exemplified in the following *gedanken* experiment (Brillouin, 1956).

3.2.1 *Particle location experiment*

A closed cylinder of volume V is maintained at a constant temperature T. It is known to contain a single molecule. The cylindrical volume is divided into two volumes V_1 and V_2 by means of a partition. See Fig. 3.1. An observer wants to determine whether the molecule is in V_1 or in V_2. To do this he places a photocell P_1 within V_1 and shines a beam of light B_1 into V_1 such that, if the particle is in V_1, it might deflect a photon into the photocell. In order for the photon to be distinguished from background radiation, its energy $h\nu$ must, at least, equal the background mean energy due to the ambient temeperature T,

$$h\nu \geqslant kT, \tag{3.6}$$

where h and k are the Planck and Boltzmann constants, respectively. Alternatively, even if the particle is in V_1 it might scatter photons that all miss P_1.

Likewise a beam of light B_2 and a photocell P_2 may be used in the volume V_2 in order to identify the molecule as lying within *that* volume.

Leon Brillouin, from an undated photograph at the National Academy of Sciences. Sketch by the author.

The molecule is located as follows. First the beam B_1 is used. If P_1 registers a photon then the particle is known to be located in V_1. This ends the experiment. If not, then the beam B_2 is used in the same way. If P_2 does not register a photon then beam B_1 is used again, etc. Eventually a photon will be

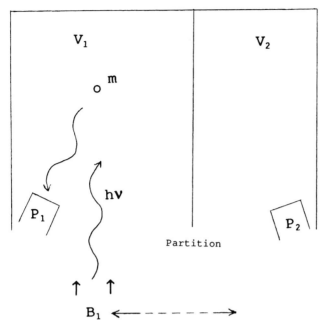

Fig. 3.1. Brillouin's *gedanken* experiment for finding the subvolume V_1 or V_2 within which a molecule m resides.

detected and the particle will be located (correctly) in one or the other of V_1 and V_2.

Locating the particle causes an increase δH in the Shannon information (1.14). But the location also involved the absorption of a photon into the cylinder system and hence an increase δH_B in its Boltzmann entropy. We calculate δH and δH_B below.

The absorption of energy $h\nu$ by one of the photocells causes an entropy increase

$$\delta H_B = h\nu/T \geqslant k, \tag{3.7}$$

the latter by (3.6).

We now compute the increase in Shannon information. As with Brillouin, we use a multiplier k of Eq. (1.14) and $\Delta x \equiv 1$ so as to give the information the same units as the Boltzmann entropy. Then, with only two different possible location events, Eq. (1.14) becomes

$$\delta H = -k(p_1 \ln p_1 + p_2 \ln p_2), \; p_i \equiv p(V_i), \; i = 1, 2. \tag{3.8}$$

The probabilities p_i are *a priori* probabilities, i.e., representing the state of knowledge about location before the experiment is performed. Before the experiment, the particle is equally probable to be anywhere in the cylinder, so that

$$p_i = V_i/V, \ i = 1, 2; \ V_1 + V_2 \equiv V. \tag{3.9}$$

This verifies that normalization $p_1 + p_2 = 1$ is obeyed. Now an important property of Eq. (3.8) is found.

3.2.2 Exercise

Using Eq. (3.8) and the normalization constraint, shows that the maximum value of δH over all possible (p_1, p_2) (or, by Eq. (3.9), geometries (V_1, V_2)) obeys

$$\delta H_{\max} = k \ln 2 \approx 0.7k. \tag{3.10}$$

Comparing Eqs. (3.7) and (3.10) shows that Eq. (3.5) is obeyed, as was required. Brillouin called H_B the 'bound information', in the sense of being intrinsic, or bound, to nature. Then (3.5) becomes a theorem involving purely informations: the change in bound information must always exceed (or equal) the acquired Shannon information.

The preceding location experiment is one example of the equivalence principle (3.5); numerous others are constructed by Brillouin (1956). Lindblad (1983) and Zeh (1992) critically discuss the principle and provide numerous other references on the subject.

Current thought seems to be that Eq. (3.5) describes, as well, the entropy changes that take place during the *erasure* of a stored message (Landauer, 1961) prior to its replacement with another message. This scenario is, thus, complementary to the Brillouin one, which deals instead with the *measurement* or acquisition of a message. Landauer's principle is that if the stored message contains an amount H of Shannon information, then it requires for its erasure an entropy expenditure of at least the amount H_B. As Caves notes (Halliwell *et al.*, 1994), this is effectively just an application of the Second Law to the erasure process. Overall, then, there is entropy expenditure at *both ends* of an overall information retrieval/storage process – when acquiring the message, and, when erasing a previous message record in order to store the acquired one.

During either the Brillouin or the Landauer phase of the information retrieval process, H in Eq. (3.5) is the Shannon information contained within a message. In the Brillouin phase, the message is the one arising out of measurement. In the Landauer, the message is the one that is erased or, equivalently, its replacement (which, for a fixed storage system, must retain the same number of bits). We will be primarily interested in the Brillouin, or measurement, phase of the process.

The equivalence principle (3.5) shows a complementarity between the information in a measured message and an equivalent expenditure of Boltzmann entropy. Such complementarity is intriguing, and serves to clarify what role the quantity J must play in our theory. (I thank B. H. Soffer (personal communication, 1994) for first noticing the tie-in between the Szilard–Brillouin work and the EPI formalism.)

3.3 System information model

At this point it is useful to review the status of our theory. We have found that I is an information or channel capacity, and that J is another information. The distinction between I and J is the distinction between the general and the specific: functional I always has the same form, Eq. (2.19), regardless of the physical parameter that is being measured, whereas functional J characterizes the specific physical parameter under measurement (see discussion below Eqs (3.2) and (3.3)).

In this section, we will gradually 'zero in' on the exact meaning, and form(s), to be taken by J, simultaneous with developing the overall EPI system model. For readers who don't mind looking ahead, Secs. 3.4.5 and 3.4.6 specify how J is formed for a specific problem.

We found that I relates to the cross-entropy G via Eq. (1.22c). We also saw that H_B is a 'bound' information quantity that defines the effect of the measurement upon the physical scenario. This prompts us to make the identifications

$$I \to G \text{ and } J \to H_B. \tag{3.11}$$

3.3.1 Bound Fisher information J

The first identification (3.11) was derived in Sec. 1.4. The second, however, is new: it states that the role of J is to be analogous to that of H_B, i.e., to define the effect of the measurement upon the physical scenario (or 'parameter' or 'system'; we use the terms interchangeably). This will serve to bring the particulars of the physical scenario into the variational principle, as was required above.

But, how can J be actually computed? Brillouin's equivalence principle Eq. (3.5) suggests a *natural* way of doing so, i.e., without the need for *ad hoc* Lagrange constraints as at Eqs (1.48a,b). It suggests that the intrinsic information I must have a *physical equivalence* in terms of the measured phenomenon (as information H and physical entropy H_B were related in the preceding example). Represent by J the physical effect, or manifestation, of I. Thus, J is the information that is intrinsic (or, 'bound') to the phenomenon under measurement. This is why we call J the 'bound' Fisher information. Computationally, it is to be evaluated – not directly by (2.19) – but equivalently in terms of physical parameters of the scenario (more on this later). The precise tie-in between I and J is quantified as follows.

3.3.2 Information transferral effects

Suppose that a real measurement of a four-parameter $\boldsymbol{\theta}$ is made. The parameter is any physical characteristic of a system. The system is necessarily *perturbed* (Sec. 3.8, Chap. 10) by the measurement. More specifically, its amplitude functions $\mathbf{q(x)}$ are perturbed. An analysis follows.

As at Eq. (0.3), denote the perturbations in $\mathbf{q(x)}$ as the vector $\varepsilon\boldsymbol{\eta}(\mathbf{x})$, $\varepsilon \to 0$, with the $\boldsymbol{\eta}(\mathbf{x})$ arbitrary perturbation functions. By Eq. (2.19), the perturbations in $\mathbf{q(x)}$ cause a perturbation δI in the intrinsic information (see Eq. (0.41)). Since J will likewise be a functional of the $\mathbf{q(x)}$, a perturbation δJ also results. Is there a relation between δI and δJ?

We next find such a relation, and others, for informations I and J. This is by the use of an *axiomatic model* for the measurement process. There are three axioms as given below.

Prior to measurement of parameter $\boldsymbol{\theta}$, the system has a bound information level J. Then a real measurement $\bar{\mathbf{y}}$ is initiated – say, by shining light upon the system. This causes an information transformation or transition $J \to I$ to take place, where I is the intrinsic information of the system. The measurement perturbs the system (Sec. 3.8), so that informations I and J are perturbed by amounts δI and δJ. How are the four informations related?

First, by the correspondence (3.11) and the Brillouin effect (3.5), I and J should obey

$$J \geq I. \tag{3.12}$$

This suggests the possibility of some loss of information during the information transition. We also postulate the perturbed amounts of information to obey

$$axiom\ 1:\ \delta I = \delta J. \tag{3.13}$$

This is a *conservation law*, one of conservation of information change. It is a predicted new effect. Its validity is demonstrated many times over in this book; as a vital part of the information approach that is used to derive the fundamental physical laws in Chaps. 4–9 and 11. It also will be seen to occur whenever the measurement space has a conjugate space that connects with it by unitary transformation (Sec. 3.8).

The basic premises (3.12), (3.13) are further discussed in Secs. 3.4.5 and 3.4.7, and are exemplified in Sec. 3.8. We next build on these premises.

3.4 Principle of extreme physical information (EPI)

3.4.1 Physical information K and variational principle

Eq. (3.13) is a statement that

$$\delta I - \delta J \equiv \delta(I - J) = 0. \tag{3.14}$$

Define a new quantity.

$$K \equiv I - J. \tag{3.15}$$

Quantity K is called the 'physical information' of the system. As with I and J, it is a functional of the amplitudes $\mathbf{q}(\mathbf{x})$. By its definition and Eq. (3.12), K is a loss, or defect, of information. The zero in Eq. (3.14) implies that this defect of information is the same over a *range* of perturbations $\varepsilon\boldsymbol{\eta}(\mathbf{x})$. Eqs. (3.14) and (3.15) are the basis for the EPI principle, Eqs. (3.16) and (3.18), as shown next.

Combining Eqs. (3.14) and (3.15) gives $\delta K = 0$ or

$$K = I - J = Extrem. \tag{3.16}$$

This is a variational principle for finding the $\mathbf{q}$ (see Sec. 0.3). It states that at the solution $\mathbf{q}$ the information K is an extremum. As was postulated at Eq. (3.13), the variational principle is in reaction to a measurement. Variational principle (3.16) is one-half of our overall information approach.

3.4.2 Zero-conditions, on micro- and macrolevel

Eq. (3.15) shows that generally

$$I \neq J. \tag{3.17}$$

In fact, from inequality (3.12), a 'lossy' situation exists: the information I in the data never exceeds the amount J in the phenomenon,

$$I = \kappa J \text{ or } I - \kappa J = 0, \ \kappa \leqslant 1. \tag{3.18}$$

This may be shown to follow from the *I*-theorem; see Sec. 3.4.7. Eq. (3.18) comprises the second half of our overall information approach.

Interestingly, only the values $\kappa = 1/2$ or 1 have occurred in those physical-law derivations that fix κ at specific values (see subsequent chapters). These verify that $I \leqslant J$ or, equivalently, $\kappa \leqslant 1$ in (3.18). The nature of κ is taken up further in Sec. 3.4.5.

Returning to the question of information J, Eq. (3.18) provides a partial answer. It shows that J is proportional to I, in numerical value. However, as mentioned before, in contrast to I its *functional form* will depend upon physical parameters of the scenario (e.g., c or $\hbar$) and not solely upon the amplitude functions $\mathbf{q}$. This is further taken up in Sec. 3.4.5.

Equation (3.18) is a zero-condition on the *macroscopic* level, i.e., in terms of the information in physical observables (the data). On the other hand, (3.18) may be derived from viewing the information transfer procedure on the *microscopic* level. Postulate the existence of *information densities* $i_n(\mathbf{x})$ and $j_n(\mathbf{x})$ such that

$$\text{axiom 2: } I \equiv \int d\mathbf{x} \sum_n i_n(\mathbf{x}) \text{ and } J \equiv \int d\mathbf{x} \sum_n j_n(\mathbf{x}),$$

$$(3.19)$$

$$\text{where } i_n(\mathbf{x}) = 4\nabla\mathbf{q}_n\cdot\nabla\mathbf{q}_n$$

by Eq. (2.19). Also, $j_n(\mathbf{x})$ is a function that depends upon the particular scenario (see subsequent chapters). In contrast with I and J, densities $i_n(\mathbf{x})$ and $j_n(\mathbf{x})$ exist on the microscopic level, i.e., within local intervals $(\mathbf{x}, \mathbf{x} + d\mathbf{x})$.

Let us demand a zero-condition on the microscopic level,

$$\text{axiom 3: } i_n(\mathbf{x}) - \kappa j_n(\mathbf{x}) = 0, \text{ all } \mathbf{x}, n. \qquad (3.20)$$

Summing and integrating this $d\mathbf{x}$, and using Eqs. (3.19), verifies Eq. (3.18), which was our aim.

Condition (3.20) will often be used to implement physical-law derivations in later chapters. It is to be used in the sense that quantities $[i_n(\mathbf{x}) - \kappa j_n(\mathbf{x})]$ comprise the *net* integrands in Eq. (3.19) after possible partial integrations are performed. By $[i_n(\mathbf{x}) - \kappa j_n(\mathbf{x})]$ we mean the terms that remain.

Information densities $i_n(\mathbf{x})$ and $j_n(\mathbf{x})$ have the added physical significance of defining the field dynamics of the scenario; see Sec. 3.7.3.

3.4.3 EPI principle

In summary, the information model consists of the axioms (3.13), (3.19) and (3.20). As we saw, these imply the variational principle (3.16) and the zero-condition (3.18). In fact, *the variational principle and the zero-condition comprise the overall principle we will use to derive major physical laws.* Since, by Eq. (3.16), physical information K is extremized the principle is called 'extreme physical information' or EPI. In some derivations, the microlevel condition Eq. (3.20) (which, we saw, implied condition (3.18)) will be used as well.

Self portrait of the author, 1998.

3.4.4 Role of the coordinates

The physics of the measurement scenario lies in the fluctuations **x**. Hence, the choice of coordinates **x** (or equivalently, parameter **θ**) in EPI is crucial to its use. For purposes of using EPI the following considerations apply: **x** must be a four-vector, on relativistic grounds (Sec. 3.5).

EPI solutions **q** for scenarios in which **x** is purely real are fundamentally different from scenarios (a 'mixed' case) where some components of **x** are real and some are imaginary (Sec. 1.8.8). Solutions **q** in the purely real **x** case obey maximum disorder (minimum *I*). But in a mixed real–imaginary case a state of maximum disorder may not result. Usually, the mixed case leads to a wave equation in the coordinate space (Chaps. 4–6 and 11). Basically, the two cases differ in the way that a diffusion equation differs from a wave equation. All-real coordinates lead to diffusion-type equations, where disorder must increase, whereas mixed real–imaginary cases lead to wave equations, where the state of disorder cycles with time.

3.4.5 Defining J from a statement of invariance; nature of κ

The use of EPI in a given measurement scenario requires definition of the information functional $J[\mathbf{q}]$. In fact, since $I[\mathbf{q}]$ is of a fixed form (Eq. (2.19)), the functional $J[\mathbf{q}]$ *uniquely determines* the solution $\mathbf{q}$. How may it be found?

In general, the functional form $J[\mathbf{q}]$ follows from a statement of *invariance* about the system. Our overall stance is that the measuring instrument affects the physical law. Hence, ideally, the statement of invariance should be suggested by the internal workings of the measuring instrument (see Sec. 3.8 for an example). If this cannot be accomplished, then at least the invariance principle should be as 'weak' as is necessary to define the system or phenomenon. For example, it might be the expression of continuity of flow (Chaps. 5, 6). Overly strong invariance principles lead, in fact, to departures from correct EPI answers; see Sec. 7.4.11.

Examples of invariance principles, as used in later chapters, are: (i) a unitary transformation, such as that between direct- and momentum-space in quantum mechanics; (ii) the gauge invariance of classical electromagnetic theory or gravitational theory; (iii) an equation of continuity (invariance) of flow, usually involving the sources.

Information $J[\mathbf{q}]$ and coefficient κ are always *solved for* by the combination of EPI Eqs. (3.16) and (3.18), and the invariance principle. It is interesting that the resulting forms for $J[\mathbf{q}]$ turn out to be *the simplest* possible functionals of the sources (e.g., linear) that are consistent with the invariance principle. This simplicity is not forced by, e.g., recourse to Ockham's razor. Rather, it is a natural consequence of the EPI analytical approach.

The invariance principle plays a central role in the implementation of EPI. EPI is a principle for defining the physics of the system through its amplitudes $\mathbf{q}$. As mentioned before, the answer $\mathbf{q}$ that EPI gives for a problem is completely dependent upon the particular $J[\mathbf{q}]$ for that problem. This, in turn, depends completely upon the invariance principle that is used. Hence, the answer $\mathbf{q}$ that EPI gives can only be as valid as is the presumed invariance principle. If the principle is not sufficiently 'strong' in defining the system, then we can expect the EPI output $\mathbf{q}$ to be only approximately correct. Information J

reflects this situation, in that J is the level that information I acquires when the invariance principle is optimally strong. Hence, $I \leq J$ generally but $I = J$ at an optimally strong invariance principle.

We can discuss the meaning of the coefficient κ in this context as well. By its definition (3.18) $\kappa = I/J$, so that κ measures the efficiency of the EPI process in *transferring* Fisher information from the phenomenon (specified by J) to the output (specified by I). Thus, κ is an efficiency coefficient. From the preceding paragraph, a value of $\kappa < 1$ should indicate that the answer $\mathbf{q}$ is only approximate. We will verify, in subsequent chapters, that such κ values only occur in EPI uses for which the invariance principle is *explicitly* incomplete: use of a non-quantum theory (Chaps. 5, 6). In all such cases, the lost information is therefore associated with ignored quantum effects.

When the invariance principle is the statement of a unitary transformation between measurement $\mathbf{x}$ space and a conjugate coordinate space, then the solution to requirement Eq. (3.18) is that *functional J be simply the re-expression of I in the conjugate space.* We will later prove (Sec. 3.8) that, due to the unitary transformation *identically $I = J$.* This satisfies Eq. (3.18) with the particular value of $\kappa = 1$. In this case, the invariance principle is optimally strong as well (see preceding paragraphs). Then, as was discussed, the output $\mathbf{q}$ of the invariance principle will be 'correct' (i.e., not explicitly incorrect due to ignored quantum effects). Unitary transformations are used in Chap. 4 to define relativistic quantum mechanics and in Chap. 11 to define quantum gravity.

Conversely, it will turn out that when the invariance principle *is not* that of a unitary transformation then, depending upon the phenomenon, often the efficiency value is $\kappa < 1$. Hence, a unitary transformation seems to be the hallmark of an accurate EPI output.

Interestingly, there are also *non-quantum* (and non-unitary) theories for which κ is unity (Chap. 7) or, even, any value in the continuum (Chaps. 8 or 9). It seems, then, that a unitary theory is sufficient, but not necessary, for producing an optimally accurate output $\mathbf{q}$. The nature of κ is still not fully understood. It is further discussed in each of the applications chapters. See, e.g., Secs. 4.4, 5.1.18 and 6.3.18.

3.4.6 *Unique or multiple solutions?*

The EPI conditions (3.16), (3.18) are to be satisfied through variation of the $\mathbf{q}$. Depending upon the nature of the invariance principle just discussed, the two conditions will either give (a) a unique solution $\mathbf{q}$ or (b) two generally different solutions.

Case (a) occurs when the invariance principle is not of the *direct* equality form Eq. (3.18), e.g., *is not* the statement of a unitary transformation between coordinate spaces I and J. Instead, it might be a statement of continuity of flow for the **q** and/or the sources. Since, in any event, Eq. (3.18) must be satisfied, *here we must solve for the parameters κ and the functional J such that (3.18) (or more fundamentally, microlevel Eq. (3.20)) is true.* This is done by seeking *a common solution* **q** to conditions (3.16) and (3.20). Examples of the common solution case are found in Chaps. 5–8.

Case (b) occurs when the invariance principle is precisely in the form of Eq. (3.18). *An example is when the invariance principle is that of a unitary transformation (see above).* The parameter κ and information functional $J[\mathbf{q}]$ are known *a priori*, and do not have to be solved for as in (a) preceding. Then Eq. (3.18) *by itself* gives a solution **q**, and this solution is generally different from that of the variational principle (3.16). However, it is as valid a physical solution as the solution to (3.16). An example is in Chap. 4, where the solution to (3.16) is the Klein–Gordon equation while the solution to (3.18) is the Dirac equation. Each is, of course, equally valid within its domain of application.

3.4.7 Independent routes to EPI

It is important to note that the EPI Eq. (3.18) was arrived at by analogy with the Brillouin effect. It would be better yet to confirm (3.18) from other points of view.

Suppose that an object (say, a particle) is to be measured. The particle is in the input space to a measuring instrument. Before the measurement is initiated, the particle has 'bound' information level J. Then the measurement is initiated, by the use of a probe particle that interacts with the particle (see Sec. 3.8). The interaction perturbs the particle and causes it to have the new information level I. The result is an information transition $J \rightarrow I$, and this occurs in measurement input space. Next, the I-theorem states that $\Delta I \equiv I - J \leqslant 0$ over the duration Δt of the interaction. *This implies the EPI condition (3.12) or (3.18).*

The other equation of EPI, Eq. (3.16), was seen to arise out of axiom 1, Eq. (3.13). Is axiom 1 purely an assertion, or, can it be shown to be physically obeyed? In fact, this axiom is obeyed by the class of measuring instruments that are described in Sec. 3.8. The unitary transformations that occur during internal operation of the instruments directly lead to axiom 1 and, hence, EPI Eq. (3.16). It also will be shown in Sec. 3.8 that the other half of EPI – Eq. (3.18) – is obeyed as well in this measurement scenario. This means that the EPI principle will follow independent of the validity of the I-theorem, greatly widening its scope of application.

Thus, the overall EPI principle may be confirmed by alternative lines of reasoning.

Of course, the ultimate test of a theory is the validity of its consequences. On this basis, the EPI Eqs. (3.16) and (3.18) constitute one of the most well-verified principles of physics. It has as its outputs the Schroedinger wave equation (Chap. 4 and Appendix D), Dirac and Klein–Gordon equations (Chap. 4), Boltzmann- and Maxwell–Boltzmann distribution functions (Chap. 7), and many other phenomena as found in corresponding chapters.

3.4.8 Physical laws as a reaction by nature to measurement

The solution to the extremum principle (3.16) is an Euler–Lagrange equation (0.34) in the amplitudes **q**. This differential equation expresses the law of physics (e.g., the Klein–Gordon equation) governing the real measurement of **θ**. The law is in the form that holds in the input space of the measuring instrument (see Sec. 3.8 and Chap. 10). One may conclude, then, that *measurement elicits physical law*. This is a most interesting effect, and, in fact, is one of the ideas expressed by Wheeler in the Introduction, that of 'observer participancy'. Physics has been called the 'science of measurement' (Kelvin, 1889) in the sense that a phenomenon is not understood until it is measured and quantified. Now we find that this statement can be strengthened to the extent that the phenomenon is formed *in reaction* to the measurement.

3.4.9 The intrinsic scenario defines independent degrees of freedom

Eq. (2.19) shows that information I is defined by N amplitude functions **q**. By Eqs. (3.16) and (3.18) so are informations J and K. The **q** comprise 'degrees of freedom' of the observed phenomenon (Sec. 2.7). Alternatively, the theory of Chap. 2 models the same number N of artificial measurements to have been made in an ideal, 'intrinsic' scenario. The result is that *each such intrinsic measurement gives rise (via EPI) to a new degree of freedom q_n of the physical theory.* (For example, in non-relativistic quantum mechanics where the wave amplitude is described by an amplitude pair (q_1, q_2), exactly two intrinsic measurements define the physical theory.) This is an interesting significance for these model measurements.

3.4.10 The question of the size of N

The number N of degrees of freedom for a given problem appears to be arbitrary: the intrinsic scenario model (Chap. 2) does not fix N to a specific value. On this basis, the number of amplitude components $\mathbf{q}$ needed to describe a given physical phenomenon could conceivably be of unlimited value. However, the value of N will also describe the number of mathematical solutions *that are sufficient* for satisfying EPI conditions (3.16) and (3.18). Typically, this is the number of independent solutions to a set of Euler–Lagrange equations. In any physical circumstance this will be a well-defined, finite number. For example, $N = 2$ for quantum mechanics as described by the Klein–Gordon equation, or, $N = 8$ for the Dirac equation (see Chap. 4). In this way, each physical phenomenon will be specified by a definite, finite number N of amplitude functions.

3.4.11 A mathematical game

In certain scenarios the EPI principle may be regarded as the workings of a 'mathematical game' (Morgenstern and von Neumann, 1947). In fact the simplest such game arises. This is a game with *discrete*, deterministic moves i and j defined below. The transition to continuous values of i and j may be made without changing the essential results.

Those two arch-rivals A and B are vying for a scarce commodity (information). See Table 3.1. Player A can make either of the row value moves $i = 1$ or 2; and player B can make either of the column value moves $j = 1$ or 2. Each move pair (i, j) constitutes a play of the game. Each such move pair defines a location in the table, and a resulting payout of the commodity to A, *at the expense of* B. ('Thus A is happy, and B is not.') For example, the move (1, 2) denotes a gain by A of 2.0 and a loss by B of 2.0. Such a game is called 'zero-sum', since at any move the total gain of both players is zero. Assume that both players know the payoff table. What are their respective optimum moves?

Consider player A. If he chooses a row $i = 1$ then he can gain either 3.0 or 2.0. Since these are at the expense of B, B will choose a column $j = 2$ so as to minimize his loss to 2.0. So, A knows that if he chooses $i = 1$ he will gain 2.0. Or, if A chooses the row $i = 2$ then he will gain, in the same way, 4.0. Since this exhausts his choices, to maximize his gain he chooses $i = 2$, with the certainty that he will gain 4.0.

Next, consider player B. If he chooses a column $j = 1$ then he will pay out either 3.0 or 5.0. But then A will choose $i = 2$ so as to maximize his gain to 5.0. So, if B chooses $j = 1$ he will lose 5.0. Or, if B chooses column $j = 2$ then he will lose, in the same way, 4.0. Since this exhausts his choices, to minimize his loss he chooses $j = 2$, with the certainty that he will lose 4.0.

Notice that the resulting optimum play of the game, from *either* the viewpoint of A or B, is the move (2, 2). This is the extreme lower-right item in the

Table 3.1. *A 2 × 2 payoff matrix.*

i \ j	1	2
1	3.0	2.0
2	5.0	4.0

table. This item results because the point (2, 2) is locally a *saddle point* in the payouts, i.e., a local minimum in j and a maximum in i. Also, note that there is nothing random about this game; every time it is played the result will be the same. Such a game is called 'fixed-point'.

3.4.12 *EPI as a game of knowledge acquisition*

We next construct a mathematical model of the EPI process that has the form of such a mathematical game (Frieden and Soffer, 1995). Note that this game model is an explanatory device, and not a distinct, physical derivation of EPI. Also, delegating human-like attributes to the players is merely part of the anthropomorphic model taken. The 'game that is played' is merely a descriptive device. EPI is a physical process (Chap. 10).

The game model can also be regarded as an epistemological model of EPI. It will show that EPI arises, in certain circumstances, *as if it were* the result of a quest for information and knowledge.

In many problems the Fisher coordinates are mixed real and imaginary quantities. Denote by **x** the subset of real coordinates of the problem. We want to deal, in this section, with problems having only real coordinates. Thus, let the amplitudes **q(x)** of the problem represent either the fully dimensioned amplitudes for an all-real case, or the *marginal* probability amplitudes in the real **x** for a mixed case. Eq. (2.19) shows that, for real coordinates **x**, I monotonically *decreases* as the amplitudes **q** are monotonically broadened or blurred. This effect should hold for any fixed state of correlation in the intrinsic data.

On the other hand, we found (Sec. 2.3.2) that the form Eq. (2.19) for I represents a model scenario of *maximized* information due to efficient estimation and independent data. The latter effect is illustrated by the following example.

Example Suppose that there are two Fisher variables (x, y) and these obey a Gaussian bivariant PDF, with a common variance σ^2, a *general correlation coefficient* ρ, and a common mean value θ. Regard the latter as the unknown parameter to be estimated. Then the Fisher information Eq. (1.9) gives $I = 2\sigma^{-2}(1 + \rho)^{-1}$. This shows how the information in the variables (x, y) depends upon their degree of correlation. Suppose that the variables are initially correlated, with $\rho > 0$. Then as $\rho \to 0$, $I \to 2/\sigma^2 = \text{max. in } \rho$. As a check, this is twice the information in a single variable (see Sec. 1.2.2), as is required by the additivity of independent data (Sec. 2.4.1) in the case of independent data. Hence, independent data have maximal I values.

This effect holds in the presence of any fixed state of blur, or half-width, of the joint amplitude function $\mathbf{q}(\mathbf{x})$ of all the data fluctuations.

We have, then, determined the qualitative dependence of I upon the data correlation and the state of blur. This is summarized in Fig. 3.2. As in Table 3.1, the vertical and horizontal coordinates are designated by i and j, respectively, although they are continuous variables here. Coordinate i increases with the degree of independence (by any measure) of the intrinsic data, and each coordinate j represents a possible trial solution $\mathbf{q}(\mathbf{x})$ to EPI in the particular sequence defined below. Since, as we saw, the I values increase with increasing independence, they must increase with increasing i.

Now, the solution $\mathbf{q} = \mathbf{q}_0$ to EPI is given by the simultaneous solution to Eqs. (3.16) and (3.18). Also, Eq. (2.19) for I presumes maximal independence (vertical coordinate) of the intrinsic data (Sec. 2.1.3). Then the EPI solution $\mathbf{q}_0$ is represented by a particular coordinate value j located along the *bottom row* of Fig. 3.2.

The trial solutions $\mathbf{q}$ are sequenced according to the coordinate j as follows. Let each $q_n(\mathbf{x})$ function of a given vector $\mathbf{q}$ monotonically decrease in blur from the solution 'point' $j \to \mathbf{q}_0(\mathbf{x})$ on the far right to an initial sharp state at the far left, corresponding to $j = 1$. Then, by the form Eq. (2.19) of I, the values I decrease along that row as j *increases* to the right. On the other hand, we found before that, for a given j, values I increase with increasing i (independence) values. Then the solution point $\mathbf{q}_0$ designates a local maximum in i but a minimum in j, i.e., a *saddle point*.

We found, by analysis of the game in Table 3.1, that a saddle point represents the outcome of a zero-sum mathematical game. Hence, the EPI solution point $\mathbf{q}_0$ represents the outcome of such a game. Here, the commodity that the two players are vying for is the amount of Fisher information I in a trial solution $\mathbf{q}$. Then the game is one of Fisher 'information hoarding' between a player A, who controls the vertical axis choice of *correlation*, and a player B, who controls the horizontal axis choice of *the degree of blur*. Who are the opposing players?

The choice of correlation is made by the *observer* (Sec. 2.1.3) in a prior scenario, so this identifies player A. Player B is associated with the degree of

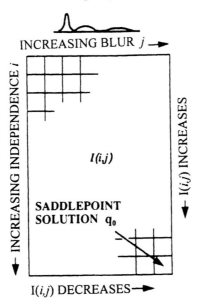

Fig. 3.2. The information game. The observer bets on a state of independence i (a row value). The information demon bets on a state of blur j (a column value). The payoff from the demon to the observer is the information at the saddlepoint, defining the EPI solution q_0. (Reprinted from Frieden and Soffer, 1995.)

blur. This is obviously out of the hands of the observer. Therefore it is adjusted by 'nature'. This may be further verified by the zero-sum nature of the game. By Eqs. (3.13) or (3.18) we notice that any information I that is acquired by the observer is at the expense of the phenomenon under measurement. Hence, player B is identified to be the phenomenon, or, nature in general. Then, according to the rules of such a game, nature has the 'aim' of increasing the degree of blur.

Hence, for real coordinates $\mathbf{x}$ the EPI principle represents a game of information hoarding between the observer and nature. The observer wants to maximize I while nature wants to minimize it. Each different physical scenario, defined by its dependence $J[\mathbf{q}]$, defines a particular play of the game, leading to a different saddle point solution $\mathbf{q}_0$ along the bottom row of Fig. 3.2. This is the game-theory aspect of EPI that we sought.

Such real coordinates $\mathbf{x}$ occur in the Maxwell–Boltzmann law derivation in Chap. 7, in the time-independent Schroedinger wave equation derivation in Appendix D, and in the $1/f$ power-law derivation in Chap. 8. These EPI derivations are, then, equivalent to a play of the knowledge acquisition game.

It is interesting that, because the information transfer efficiency factor $\kappa \leqslant 1$ in Eq. (3.18), $I \leqslant J$ or $K \leqslant 0$. This indicates that, of the two protagonists, nature always wins or at least breaks even in the play of the game. This effect agrees in spirit with Brillouin's equivalence principle Eq. (3.5). That is, in acquiring data information there is an inevitable *larger* loss of phenomeno-logical ('bound' in his terminology) information. You can't win.

Exercise

The scope of application of the game can be widened. Suppose that a particle has a complex amplitude function $\psi(r, t) \equiv q_1(r, t) + iq_2(r, t)$, by the use of Eq. (2.24). Suppose that the particle is subjected to a conservative force field, due to a time-independent potential function $V(r)$. Then energy is conserved, and we may regard the particle as being in a definite energy state E. It results that the amplitude function separates, as $\psi(r, t) = u(r) \exp(iEt/\hbar)$. Suppose that all time fluctuations t are constrained to lie within a *finite*, albeit large, interval $(-\Delta T, \Delta T)$. Show that, under these circumstances, the time-coordinate contribution to information I is merely an additive constant, and that the space coordinates give rise to positive information terms, so that the game may be played with the space coordinates. Hence, the knowledge game applies, more broadly, to the scenario of a quantum mechanical particle in a conservative force field.

3.4.13 The information 'demon'

The solution $\mathbf{q}_0$ to the game defines a physical law (e.g., the Maxwell–Boltzmann law). The preceding shows, then, that a physical law is the result of a lossy information transfer game between an observer and the phenomenon under observation. Since we can sympathize with the observer's aims of increased information and knowledge, and since he always loses the game to the phenomenon (or at best breaks even), it seems fitting to regard the phenomenon as an all-powerful, but malevolent, force: an information 'demon'. (Note: this demon is not the Maxwell demon.)

In summary, for real Fisher coordinates $\mathbf{x}$ the EPI process amounts to carrying through a game. The game is a zero-sum contest between the observer and the information demon for a limited resource I of intrinsic information. The demon generally wins.

We note with interest that the existence of such a demon implies the *I*-theorem, Eq. (1.30). (This is the converse of the proof in Sec. 3.4.7.) Information $I - J = K$ represents the change of Fisher information ΔI due to the EPI process $J \to I$ that takes place over an interaction time interval Δt. Since the game is played and the demon wins, i.e., $K \leq 0$, and since $\Delta I = K$, necessarily $\Delta I \leq 0$. Then since $\Delta t \geq 0$, the *I*-theorem follows.

3.4.14 Absolute nature of extremum; why physical constants are constant

In the presence of fields and sources, one would expect the extremum that is attained in principle (3.16) to depend upon their functional forms. Such is the case. Conversely, in their absence (i.e., free-field situations) the extremum will be an absolute number that can only be a function of whatever free parameters the functional K contains (e.g., h, k, c). Everything else is integrated out in Eq. (3.16). These parameters occur, in particular, within the functional J part of K. The absolute nature of the scenario suggests, then, that we regard the extreme value of J as a *universal physical constant* (Hughes, 1994). This has an unusual 'bootstrap' effect.

Suppose, e.g., that in a given scenario the extremum is a known function of parameter c, the speed of light in vacuum. Then since the extremum is a universal constant this implies that c is a universal constant. Next, proceed to another scenario whereby the extremum is a known function of (say) c and another parameter, e.g., h. Since c was already fixed this, in turn, fixes h. In this manner, EPI and the smart measurer (Sec. 1.2.1) proceed through physical phenomena, fixing one constant after another.

A corollary is that the physical constants do not change in time. The contrary had been suggested by Dirac (Dirac, 1937; 1938).

Although EPI demands that these quantities be constants, it (somewhat paradoxically) regards their values as having been chosen *a priori* from a probability law. The form of this law is found in Chap. 9.

3.4.15 Ultimate resolution lengths

The maximized nature (Sec. 2.3.2) of the information I suggests the possibility that ultimate resolution lengths could be determined from EPI solutions. This is the case. The solution $\mathbf{q}$ to EPI for a given measurement scenario allows $I[\mathbf{q}]$ to be calculated in terms of physical constants; see preceding section. By the inverse relation (1.28) between I and the error e_{min}^2, this permits the latter to be calculated. If the parameter $\boldsymbol{\theta}$ under measurement is a length, the result is an expression for the ultimate resolution length e_{min} intrinsic to the measured phenomenon. In this way, EPI computes the Compton resolution length of quantum mechanics (Sec. 4.1.17) and the Planck length of quantum gravity (Sec. 6.3.22).

3.4.16 Unified probabilistic view of phenomena

The basic elements of EPI theory are the *probability amplitudes* $\mathbf{q}$. All phenomena that are derived by EPI in this book are expressed in terms of these amplitudes. This includes intrinsically statistical phenomena such as quantum mechanics but, also, ostensibly non-statistical phenomena such as classical electromagnetic theory and classical gravitational theory. EPI regards all phenomena as being fundamentally statistical in origin.

The Euler–Lagrange solution (0.34) for each phenomenon is a differential equation in function **q**. Since **q** is a probability amplitude, *EPI regards each such equation as describing a kind of 'quantum mechanics' for the particular phenomenon*. Among other things, this viewpoint predicts that the classical electromagnetic four-potential **A** can be regarded as a probability amplitude for photons (Chap. 5); and that the classical gravitational metric tensor $g_{\mu\nu}$ can be regarded as a probability amplitude for gravitons (Chap. 6). As discussed in those chapters, these views are consistent with other known properties of the phenomena. These views also permit the Planck length L and the cosmological constant Λ of gravitation to be computed (Sec. 6.3.23). See also Secs. 1.8.13, 3.4.15 and the last paragraphs of Sec. 5.3.

3.5 Derivation of Lorentz group of transformations

It may now be shown that the particular use of the trace operation in definition (2.19) of I leads, via a 'relativistic postulate' of EPI theory (Eq. (3.21) below), to the Lorentz transformation of special relativity; and to the requirement that the individual components of the Lagrangians (i.e., the integrands) of all information quantities I, J and K be covariant.

Consider the intrinsic measurement scenario mapped out in Sec. 2.1.1. Suppose these measurements to be carried through, and viewed, in a flat space, laboratory coordinate system O. The same measurements also are viewed in a reference system O′ that is moving with constant velocity u along the x-direction. As usual, we denote with primes quantities that are observed from the moving system. Thus, the extremized Fisher information as viewed from O is value I, but as viewed from O′ is value I'.

3.5.1 Is there a preferred reference frame for estimation?

The channel capacity I defines the ultimate ability to estimate from given data (see Secs. 2.1, 2.6). Consider, next, the following basic question. Should there be a preferred speed for optimally estimating parameters? In particular, should the accuracy with which parameter $\boldsymbol{\theta}$ can be estimated depend upon the speed u? It seems plausible that the answer is no. Let's find where this leads. Assume that the estimation errors $e_{n\nu}^2$ obey invariance to reference frame, as expressed by invariance of the total accuracy (2.9),

$$\sum_{nv} 1/e'^2_{nv} = \sum_{nv} 1/e^2_{nv}. \tag{3.21}$$

This is a relativistic postulate of EPI theory. Then, by Eqs. (2.9) and (2.10), the invariance principle (3.21) becomes a statement that information I obeys invariance to reference frame,

$$I'[\mathbf{q}'] = I[\mathbf{q}]. \tag{3.22}$$

This translates into a requirement on the amplitude functions $\mathbf{q}$ as follows.

3.5.2 Requirement on amplitude functions

From representation (2.19) for I, requirement (3.22) becomes

$$\int d\mathbf{x}' \sum_{nv} \left(\frac{\partial q'_n}{\partial x'_v}\right)^2 = \int d\mathbf{x} \sum_{nv} \left(\frac{\partial q_n}{\partial x_v}\right)^2. \tag{3.23a}$$

Since the data are independent, this will hold only if it holds for each measurement number n. Suppressing n, the new requirement is

$$\int d\mathbf{x}' \sum_{v} \left(\frac{\partial q'}{\partial x'_v}\right)^2 = \int d\mathbf{x} \sum_{v} \left(\frac{\partial q}{\partial x_v}\right)^2. \tag{3.23b}$$

At this point it becomes convenient to use the Einstein implied summation notation whereby repeated indices connote a summation. Also, derivatives follow the notation

$$\partial_v \equiv \partial/\partial x_v, \; v = 0, \ldots, 3. \tag{3.24}$$

Requirement (3.23a) becomes

$$\int d\mathbf{x}' \partial'_v q' \partial'_v q' = \int d\mathbf{x} \partial_v q \partial_v q. \tag{3.25}$$

How can this be satisfied?

To be definite, we now regard coordinates $\mathbf{x}$ of the problem to be the position and time of a particle. The nature of $\mathbf{x}$ is later generalized to be that of any physical 'four-vector' (as defined).

3.5.3 The transformation sought is linear

As in conventional relativity theory, assume that relation (3.25) will hold because of the special nature of the relation between coordinates $\mathbf{x}$ and coordinates $\mathbf{x}'$. In ordinary Galilean (non-relativistic) mechanics, that relation is, of course,

$$t' = t, \; x' = x - ut, \; y' = y, \; z' = z. \tag{3.26}$$

This may be placed in a convenient matrix form by the use of new coordinates

$$x_0 = ct, \ x_1 = x, \ x_2 = y, \ x_3 = z \tag{3.27}$$

and corresponding primed coordinates. The matrix form of Eqs. (3.26) is then

$$\mathbf{x}' = [B_0]\mathbf{x}, \quad [B_0] = \begin{bmatrix} 1 & 0 & 0 & 0 \\ -u/c & 1 & 0 & 0 \\ 0 & 0 & 1 & 0 \\ 0 & 0 & 0 & 1 \end{bmatrix}. \tag{3.28}$$

This emphasizes that the relation between $\mathbf{x}'$ and $\mathbf{x}$ is a linear one.

3.5.4 Reduced problem

As in Eqs. (3.28), we seek a linear relation

$$\mathbf{x}' = [B]\mathbf{x}, \tag{3.29}$$

where the new matrix [B] is to be found. By direct substitution, Eq. (3.25) will be true if [B] obeys

$$d\mathbf{x}' = d\mathbf{x} \text{ and } \partial'_\nu q' \partial'_\nu q' = \partial_\nu q \partial_\nu q. \tag{3.30}$$

The later, in particular, states that the (squared) length of a vector $\mathbf{v}$ with components $v_\nu \equiv \partial_\nu q$ should be invariant in the two spaces, i.e.,

$$v'^2 = v^2. \tag{3.31}$$

By Eq. (3.29) and the chain rule of differentiation, the vector $\mathbf{v}$ of derivatives transforms as

$$\mathbf{v}' = B^T \mathbf{v}, \tag{3.32}$$

where the T denotes the transpose. The two requirements (3.30), (3.31) constitute our reduced problem.

3.5.5 Solution

The problem asks for a linear transformation of coordinates that maintains length. Of course a vector that is rotated to a new position maintains its length. Or equivalently, the length remains constant when instead the vector remains fixed while the coordinate system rotates about it. Hence, a rotation matrix [B] will suffice as a solution. This defines a subgroup of solutions called 'proper' transformations. A second subgroup, called 'improper', defines inversion of coordinates as a transformation (Jackson, 1975). We will ignore this type of solution.

Recall that our I is the trace (2.10) of the Fisher information matrix. It is well-known that the trace of a matrix is invariant to rotation of coordinates. This is the essential reason for the rotation matrix solution mentioned previously.

The most well-known rotation matrix solution is the Lorentz transformation (Jackson, 1975). This obeys

$$[B] = \begin{bmatrix} \gamma & -(\gamma u/c) & 0 & 0 \\ -(\gamma u/c) & \gamma & 0 & 0 \\ 0 & 0 & 1 & 0 \\ 0 & 0 & 0 & 1 \end{bmatrix}, \tag{3.33}$$

where $\gamma \equiv (1 - u^2/c^2)^{-1/2}$ and c is a universal physical constant.

Matrix [B] obeying Eq. (3.33) defines the 'Lorentz transformation'. This may be shown to satisfy requirement (3.31), as follows. Generate quantities $\mathbf{v}'$ in terms of the $\mathbf{v}$ according to Eqs. (3.32) and (3.33). Squaring each component of $\mathbf{v}'$ then shows that they obey

$$\sum_{\nu=1}^{3} v_\nu'^2 - v_0'^2 = \sum_{\nu=1}^{3} v_\nu^2 - v_0^2. \tag{3.34a}$$

This satisfies the requirement (3.31) of invariant length (and, hence, the second requirement of Eqs. (3.30)) if we define new quantities

$$\mathbf{v} \equiv (iv_0, v_1, v_2, v_3) \text{ or } (v_0, iv_1, iv_2, iv_3), \ i = \sqrt{-1}, \tag{3.34b}$$

and analogously for primed quantities v_ν'. By the definitions $v_\nu \equiv \partial q/\partial x_\nu$ with x_ν given by Eqs. (3.27), condition (3.34b) corresponds to the use of new coordinates

$$\mathbf{x} \equiv (ict, x, y, z) \text{ or } (ct, ix, iy, iz). \tag{3.34c}$$

Each vector $\mathbf{v}$ and $\mathbf{x}$ given by the last two equations defines a Minkowski space and is called a 'four-vector'. In summary, *the second requirement (3.30) is satisfied by recourse to Fisher coordinates that are four-vectors.*

3.5.6 Volume invariance requirement

We turn next to satisfying the first requirement of Eqs. (3.30). All proper matrices [B] have a unit determinant,

$$\det[B] = 1. \tag{3.35a}$$

Also, the four-space volumes transform as

$$d\mathbf{x}' = \det[B]d\mathbf{x}. \tag{3.35b}$$

By Eqs. (3.35a,b) the first requirement (3.30) is satisfied as well.

With both requirements (3.30) now satisfied, we have satisfied the original information requirement (3.22).

3.5.7 *Invariance of the physical information K*

We have found that the Lorentz transformation (3.33) satisfies the requirement (3.22) that the Fisher information I in the measurements be invariant to reference frame. It follows from Eq. (3.18), then, that the bound information J likewise obeys such invariance. In turn, by definition (3.16), K likewise obeys the invariance. The upshot is that all information components of the EPI approach obey invariance to reference frame,

$$I', J', K' = I, J, K. \tag{3.36}$$

As we found, such invariance is consistent with the assumption that there is no preferred frame for accuracy.

3.5.8 *Ramifications to EPI*

The requirement of invariance of errors to reference frame has hinged upon the use of Fisher coordinates $\mathbf{x}$ that constitute a *four*-vector. Then, by Eq. (2.1) all coordinates $\mathbf{x}$, $\mathbf{y}$ and $\boldsymbol{\theta}$ of EPI theory must be four-vectors. These, of course, place the time coordinate on an equal basis with the space coordinates. As we saw, a benefit of this is that both time and space transform by a common rule, Eq. (3.29).

EPI theory extends the common treatment of space and time to, now, regard both as *random variables*. Thus, in the EPI use of the Fisher theory of Chaps. 1 and 2, a space measurement is always accompanied jointly by a time measurement; and the time component is allowed to be as uncertain as are the space components. (Why should one regard time measurements as any more accurate *a priori* than space measurements?) A ramification is that *conditional* PDFs such as $p(r|t)$, which regard time t as fundamentally different from space coordinates r (deterministic vs random), are not directly derivable by EPI.

We showed the need for Fisher variables that are four-vectors of space and time, in particular. But, in the same way, it can be shown that Fisher variables must be four-vectors generally: four-momenta, four-potentials, etc. In each instance, all four coordinates are placed on an equal footing, both regarding transformation properties *and insofar as being fundamentally statistical quantities* (cf. Sec. 3.1.2).

The invariance property (3.36) that is obeyed by all information quantities I, J, K of the theory implies that their corresponding information *densities* (Lagrangians) should be *covariant*, i.e., keep the same functional form under Lorentz transformation (3.29), (3.33). It is interesting that this requirement followed, ultimately, from a postulate (3.21) that errors of estimation should be equal, regardless of reference frame. By comparison, the covariance requirement conventionally follows from the 'first postulate' of *relativity theory* (Einstein, 1956), that the laws of nature should keep the same form regardless of reference frame. Correspondingly, then, Eq. (3.21) is a relativistic postulate of *EPI theory*.

This relativistic postulate could be obeyed, in the first place, because the Fisher information density in Eq. (2.20) is mathematically covariant in form (at fixed index n). An information density that was not covariant could not have implied the Lorentz transformation as in Sec. 3.5.5.

The sense by which information I (and consequently informations J and K) should be covariant is of interest. The requirements (3.30) were to hold for each distinct value of index n (suppressed); see below Eq. (3.23a). Hence, the Lorentz invariance/covariance requirements that were implied must, likewise, hold for each value of n. Therefore, the requirement of covariance that was previously discussed is to hold separately for each component number n of the information densities of I, J and K. That is, it is to hold for components i_n, j_n and $k_n = i_n - j_n$ (see Eq. (3.19)).

This confirms a previous observation. It was noted below Eq. (2.20) that (2.20) *is* a superposition over index n of covariant forms $i_n(\mathbf{x})$.

It is emphasized that the need (Eq. (3.34c)) for either imaginary time or imaginary space coordinates arises out of these purely relativistic considerations. Such mixed real–imaginary coordinates will be seen to be decisive in forming the d'Alembertian operator $\Box$ (Eq. (5.12b)) that is at the root of all wave equations derived by EPI (Chaps. 4–6 and 10).

In conclusion, the postulate (3.21) of invariance to error requires the entire EPI approach to be covariant. As a result, the output physical laws of EPI, which have the general form of wave equations, will likewise obey covariance. This satisfies the first postulate of special relativity (Einstein, 1905), that the laws of physics remain invariant under a change of reference frame. Covariance also requires mixed real and imaginary coordinates, which will give rise to the all-important d'Alembertian operator in the derived wave equations.

3.6 Gauge covariance property

3.6.1 Is there a preferred gauge for estimation?

We found in the preceding that a condition of Lorentz invariance follows from the demand (3.21) that accuracy be invariant to reference frame. As with reference frame, the choice of gauge (Sec. 5.1.21) is arbitrary for a given physical scenario. Gauges are constructed so that their choice has no effect upon observable field quantities (cf. Jackson, 1975). Now, mean-square errors are observable quantities. It is therefore reasonable to postulate, as well, that mean-square errors be insensitive to such choice. Then, as in transition from Eq. (3.21) to (3.22), the Fisher information should likewise be invariant to choice of gauge.

3.6.2 Ramification to I

Now, with $\boldsymbol{\theta}$, $\mathbf{y}$ and $\mathbf{x}$ representing four-positions, in the presence of arbitrary electromagnetic potentials A, ϕ the form (2.19) for I is not gauge invariant (Lawrie, 1990). However, the replacement of all derivatives as

$$\nabla \rightarrow \nabla - ieA/c\hbar, \ \partial/\partial t \rightarrow \partial/\partial t + ie\phi/\hbar \qquad (3.37)$$

renders it gauge covariant. Here, e is the particle charge and the non-boldface gradient ∇ indicates a *three*-dimensional gradient.

For convenience, these replacements in EPI can be delayed until after $J[\mathbf{q}]$ is formed; this is because functional $J[\mathbf{q}]$ never depends upon derivatives $\partial \mathbf{q}/\partial x_\nu$. Then the Euler–Lagrange solution to EPI is subjected to the replacements. As an example, see the derivations in Chap. 4.

3.6.3 Ramifications to J and K

With I now covariant, by proportionality (3.18) J will be covariant. Then, by (3.15), so is K. Then any solution to the extremization of K will be covariant. In this way, all aspects of the EPI approach become covariant, which of course had to be true (Goldstein, 1950).

3.7 Field dynamics from information

Suppose that the EPI principle Eqs. (3.16) and (3.18) yields a solution $\mathbf{q}$ for a given scenario. The amplitudes $\mathbf{q}$ are often regarded as a 'field'. Knowledge of the field implies complete knowledge of *the dynamics* (momentum, energy) of the phenomenon (Morse and Feshbach, 1953). This suggests that the EPI information quantities I and J somehow define the dynamics of the scenario. Exactly how this happens is shown next.

3.7.1 Transition to complex field functions

Field dynamical quantities are usually expressed in terms of generally *complex* field functions, rather than the purely real $\mathbf{q}$ that we use. Hence, we temporarily express information I in terms of complex amplitude functions, as follows.

Store the real $\mathbf{q}$ as the real and imaginary parts of generally complex field amplitude functions $\psi_1, \ldots, \psi_{N/2}$ according to Eq. (2.24). Although information I is defined in Eq. (2.19) in terms of the real amplitudes $\mathbf{q}$, it may be easily shown that I is equivalently expressed in terms of the complex ψ_n as

$$I = 4N \int d\mathbf{x} \sum_{n=1}^{N/2} \nabla\psi_n^* \cdot \nabla\psi_n \tag{3.38}$$

for general coordinates $\mathbf{x}$. The ψ_n are now our field functions. Like the $\mathbf{q}$, they represent a solution to an EPI problem.

3.7.2 Exercise

Derive Eq. (3.38) from Eq. (2.19), noticing that the imaginary cross-terms in the product *had* to cancel out because the resulting I must be real, by its definition Eq. (2.19).

For the specific space-time coordinates $\mathbf{x} \equiv (x_1, x_2, x_3, ict) \equiv (\mathbf{r}, ict)$ Eq. (3.38) becomes

$$I = \sum_{n=1}^{N/2} \iint d\mathbf{r}dt \, i_n(\mathbf{r}, t),$$

$$\tag{3.39}$$

$$\text{where } i_n(\mathbf{r}, t) \equiv 4Nc \left[(\nabla\psi_n)^* \cdot \nabla\psi_n - \frac{1}{c^2} \left(\frac{\partial\psi_n}{\partial t} \right)^* \frac{\partial\psi_n}{\partial t} \right].$$

The latter is the nth component Fisher information *density*, as at Eqs. (3.19). We do not use Eqs. (3.37) to make the theory gauge covariant and bring in field potentials. Instead, we work with a free-field case. This will not affect the argumentation of this section.

Eqs. (3.39) show that information component i_n depends purely upon the *rates of change* $\nabla\psi_n$, $\partial\psi_n/\partial t$ of a corresponding field component ψ_n. By contrast, the bound information component j_n (defined at Eq. (3.19)) always depends, instead, upon the component ψ_n *directly* (see, e.g., Eq. (4.22)). The result is that, functionally, the information components obey

$$i_n = i_n[\nabla\psi_n, \partial\psi_n/\partial t] \text{ and } j_n = j_n[\psi_n]. \tag{3.40}$$

3.7.3 Momentum and stress-energy

Now we turn to the physical quantities that specify the dynamics of the problem. These are the canonical momentum density μ_n, $n = 1, \ldots, N/2$ (*note*: notation p_n is reserved for probabilities) and the stress-energy tensor

$W_{\alpha\beta}$, for α, $\beta = 0$, 1, 2, 3 independently. These are normally defined in terms of the Lagrangian $\mathscr{L}$ for the problem as

$$\mu_n \equiv \frac{\partial \mathscr{L}}{\partial(\partial\psi_n/\partial t)} \quad \text{and} \quad W_{\alpha\beta} \equiv \sum_{n=1}^{N/2} \left(\frac{\partial\psi_n}{\partial x_\alpha}\right) \frac{\partial \mathscr{L}}{\partial(\partial\psi_n/\partial x_\beta)} - \mathscr{L}\delta_{\alpha\beta}, \qquad (3.41)$$

with $\delta_{\alpha\beta}$ the Kronecker delta. But by Eq. (3.16) the Lagrangian for our problem is actually $\mathscr{L} = \sum_n (i_n - j_n)$ so that, by Eqs. (3.19) and (3.40), Eqs. (3.41) become

$$\mu_n = C_1 \frac{\partial i_n}{\partial(\partial\psi_n/\partial t)},$$

$$\qquad\qquad\qquad\qquad\qquad\qquad\qquad\qquad\qquad\qquad\qquad (3.42)$$

$$W_{\alpha\beta} = C_2 \sum_{n=1}^{N/2} \left[\left(\frac{\partial\psi_n}{\partial x_\alpha}\right) \frac{\partial i_n}{\partial(\partial\psi_n/\partial x_\beta)} - (i_n - j_n)\delta_{\alpha\beta} \right].$$

Constants C_1 and C_2 were inserted so as to give correct units (their specific values are irrelevant to the argumentation). With these constants, Eqs. (3.42) may be regarded as defining equations for the μ_n and $W_{\alpha\beta}$ from the information viewpoint.

Eqs. (3.42) show how the dynamics of the problem are completely specified by the Fisher data information and the Fisher bound information, component by component. An interesting aspect of the results is that each momentum density μ_n only depends upon information I through a corresponding component i_n. Also, the stress-energy $W_{\alpha\beta}$ depends only upon the i_n, except for the diagonal elements $W_{\alpha\alpha}$ which depend upon the j_n as well.

In this way, Fisher information in its two basic forms I and J completely defines the dynamics of the measured phenomenon. In essence, the dynamics are manifestations of information.

3.8 An optical measurement device

There is a simple measurement device that clarifies and substantiates the axiomatic approach taken in Secs. 3.3.2 and 3.4.2. This follows the model shown in Fig. 3.3. It is the familiar optical 'localization experiment' given in Schiff (1955) and other books on introductory quantum mechanics. Usually it is employed heuristically to derive the Heisenberg uncertainty principle. Instead, we will use it – likewise heuristically – to demonstrate the following.

(i) *A unitary transformation of coordinates* naturally arises during the course of measurement of many (perhaps all) phenomena.
(ii) EPI, as defined by Eqs. (3.16) and (3.18), *is implied by* the unitary transformation and the perturbing effect of the measurement.

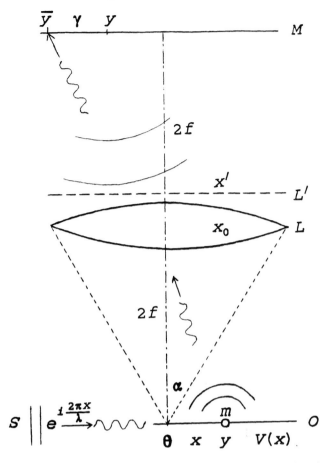

Fig. 3.3. A particle of mass m in object plane O is moving under the influence of a potential $V(x)$. The particle is at position $y = \theta + x$, where random displacement x is characteristic of quantum mechanics. A wave from monochromatic source S interacts with the particle. The wave traverses the lens L and Fraunhofer plane L' before registering in measurement space M at a data position $\bar{y} = y + \gamma$. Random displacement γ is characteristic of diffraction by the lens.

(iii) The well-known Von Neumann measurement Eq. (10.22), whereby the system wave function separates from the measurement function, naturally arises in the output image.

Hence, the localization experiment says a lot more than the Heisenberg uncertainty principle. The combination of points (i), (ii) and, then, the EPI principle, imply (as shown in Chap. 4 and Appendix D) that *an act of measurement gives rise to quantum mechanics.*

3.8.1 The measurement model

A particle in 'object space' (using optical terminology) is being measured for its X-coordinate position by the use of monochromatic light source S and a diffraction-limited, thin lens L. See Fig. 3.3. The particle has mass m and is moving under the influence of a potential function $V(x)$. The half-angle subtended by the lens at the particle is α. A monochromatic light source S illuminates object space with a plane wave $\exp(2\pi ix/\lambda)$.

The classical X-position of the particle is at the center of the field, located at a coordinate value θ. This is the 'ideal' position absent of any random phenomena. However, the phenomenon of quantum mechanics causes the particle to have a position fluctuation x from θ *just prior to its measurement*. Quantity x is an example of an 'intrinsic' fluctuation, as defined in Chap. 2. Consequently, the X-coordinate position of the particle at its input to the measuring system is

$$y \equiv \theta + x. \tag{3.43}$$

There is a probability amplitude on random values y given a fixed value of θ, called $\psi_y(y|\theta)$. By the shift invariance property Eq. (2.16), x is independent of the absolute position θ so that

$$\psi_y(y|\theta) = \psi_0(x), \quad x = y - \theta. \tag{3.44}$$

Fluctuations x follow the probability amplitude $\psi_0(x)$.

The optics are diffraction limited, and operate at equal object- and image distance, so that the magnification is unity. A scale is placed in 'image space' (at top of figure). Its origin is to the right (not shown), while that of object space O is equally off to the left (also not shown).

A wave from the source illuminates the particle. The wave and particle interact, causing the particle's intrinsic wave function $\psi_0(x)$ to be perturbed by an amount $\delta\psi(x)$, and effecting the information transition $J \to I$. These occur in object space. Also, the wave propagates through the lens, and continues on to image, or measurement, space. There, it gives rise to a light detection event (say, the sensitization of a photographic grain) at a random position

$$\bar{y} = y + \gamma = \theta + \eta, \quad \eta = x + \gamma. \tag{3.45}$$

We used Eq. (3.43) in the middle equality. Displacement γ is random, independent of y (by Eq. (2.16)), and the result of diffraction due to the finite aperture size of the lens (see below).

There is a probability amplitude on random values $\bar{y}$ in the presence of a given value of θ, called $\psi_{\bar{y}}(\bar{y}|\theta)$, which obeys

$$\psi_{\bar{y}}(\bar{y}|\theta) = \psi_{\bar{y}}(\theta + \eta|\theta) \equiv \psi_i(\eta). \tag{3.46}$$

The first equality is by Eq. (3.45), and the second is by shift-invariance (2.16) in image space. One of the aims of this analysis is to relate the image space amplitude $\psi_i(\eta)$ to the object space one $\psi_0(x)$ and a property of the lens.

To accomplish this, we have to establish the physical connection between a particle position y in object space and the corresponding coordinate $\bar{y}$ of the light detection event in image space.

3.8.2 Optical model

For simplicity, optical effects are treated classically, i.e., by the theory of coherent wave optics (Marathay, 1982). However, the particle positions are treated statistically, i.e., obeying a random phenomenon of some sort. This is akin to the 'semiclassical' treatment of radiation (Schiff, 1955) whereby the radiation is treated as obeying *classical* electromagnetic theory, while the particle with which it interacts is treated as obeying *quantum* mechanics. There is, however, an important distinction to be made: the fact that the particle obeys quantum mechanics is to be derived (by EPI), not assumed.

For further simplicity, only X-coordinate dependences are analyzed. Finally, only temporally stationary effects are considered. The presence of the time-independent potential $V(x)$ implies that energy is conserved, so that we may take the particle to be in a definite state of total energy. Then the time-dependence of the particle wave function separates out from the space coordinate. A like property holds for the optical waves, which are monochromatic. These separated time dependences can be ignored, for the purpose of the analysis to follow. The result is that all amplitude functions (optical and quantum mechanical) of the problem are assumed to be just functions of an appropriate X-coordinate.

At each particle position y, its interaction with a plane light wave from the source causes a spherical amplitude wave to be launched toward image space. Hence, the more probable the position y is for the particle, the higher will the *optical intensity* be at position y in the object intensity pattern. Call the optical amplitude value at a position y in *object space* $u(y|\theta)$. Then, on the basis of amplitudes, it must be that $u(y|\theta) \propto \psi_y(y|\theta)$. Define the units of optical intensity such that the area under the intensity curve $|u(y|\theta)|^2$ is unity. Then by normalization of the PDF $|\psi_y(y|\theta)|^2$ the proportionality constant may be taken to be unity, so that

$$u(y|\theta) = \psi_y(y|\theta). \tag{3.47}$$

By the Compton effect (Compton, 1923), the wavelength of the scattered light shifts according to the scattering angle. Assume that the particle mass is large enough and/or the range of scattering angles small enough, that the shift is negligible over the range of angles considered. Then the scattered light is effectively monochromatic.

Consider, next, formation of the image. Since the light is effectively monochromatic, the object-space amplitude function $u(y)$ propagates through the lens to image space according to the rules of *coherent image formation*. Then, as with Eq. (3.47), the output *optical* amplitude, suitably normalized, can be regarded as the data-space *probability* amplitude $\psi_{\bar{y}}(\bar{y}|\theta)$ for measurement events $\bar{y}$.

By this correspondence, we can use the known rules of coherent image formation to form the measurement space law $\psi_{\bar{y}}(\bar{y}|\theta)$. These are as follows.

Designate the coherent image of a perfectly localized 'point' source of radiation in object space as $w_{\bar{y}}(\bar{y}|y)$. The notation means the spread in positions $\bar{y}$ due to 'source' positions at y. Function $w_{\bar{y}}$ is the Green's function for the problem.

Coherent image formation obeys *linearity*, meaning that the output optical amplitude is a superposition of the input amplitude as weighted by the Green's function,

$$\psi_{\bar{y}}(\bar{y}|\theta) = \int dy \, \psi_y(y|\theta) w_{\bar{y}}(\bar{y}|y). \tag{3.48}$$

Assume, as is usual in image formation, a condition of *isoplanatism*: the point amplitude response keeps the same shape regardless of absolute 'source' position y,

$$w_{\bar{y}}(\bar{y}|y) = w(\bar{y} - y). \tag{3.49}$$

Optical amplitude w is commonly called the 'point amplitude response' of the lens. It measures the intrinsic spread due to diffraction in the coherent image of each object point.

Substituting Eqs. (3.44) and (3.49) into Eq. (3.48) gives

$$\psi_{\bar{y}}(\bar{y}|\theta) = \int dy \, \psi_0(y - \theta) w(\bar{y} - y) \tag{3.50a}$$

$$= \int dx \, \psi_0(x) w(\bar{y} - \theta - x) \tag{3.50b}$$

after an obvious change of integration variable. Using Eq. (3.46) on the left-hand side then gives

$$\psi_i(\eta) = \int dx \, \psi_0(x) w(\eta - x), \tag{3.51}$$

after again using Eq. (3.45). This relates the probability amplitude $\psi_i(\eta)$ on the *total fluctuation* η in image, or measurement, space to the probability amplitude $\psi_0(x)$ on intrinsic fluctuation x in object space and the point amplitude response $w(x)$ of the lens. The right-hand side of Eq. (3.51) takes the mathematical form of a 'convolution' of $\psi_0(x)$ with $w(x)$.

3.8.3 Von Neumann wave function reduction

We can alternatively express the measurement wave $\psi_{\bar{y}}(\bar{y}|\theta)$ as a coherent sum

$$\psi_{\bar{y}}(\bar{y}|\theta) \equiv \int dx \psi(x, \bar{y}|\theta) \tag{3.52}$$

where probability amplitude $\psi(x, \bar{y}|\theta)$ expresses the *joint state* of the particle under measurement and the instrument reading. Comparing Eqs. (3.50b) and (3.52) shows that

$$\psi(x, \bar{y}|\theta) = \psi_0(x)w(\bar{y} - \theta - x). \tag{3.53}$$

The joint state of particle and instrument separates into a product of the separate states. This is the celebrated Von Neumann separation effect after a measurement.

This equation also agrees with the result Eq. (10.26a) of using a rigorous four-dimensional approach to measurement. (Note that $\bar{x}$ in Eq. (10.26a) is, from Eq. (10.6), what we called $\bar{y} - \theta$ in Eq. (3.53).) The wave function is 'reduced', i.e., narrowed, by the product. This is explained further in Sec. 10.9.6.

The fact that there is such three-way agreement on measurement further validates the simple optical model.

3.8.4 Optical pupil space, particle momentum space

From Eqs. (3.44) and (3.47), the optical object $u(x)$ relates to the quantum mechanical probability amplitude $\psi_0(x)$ as

$$u(x) = \psi_0(x). \tag{3.54}$$

The intrinsic quantum mechanical fluctuations of the particle define the input amplitude wave to the optics. How does the latter propagate through the optics to the measurement plane?

Since the lens L in Fig. 3.3 is a thin, diffraction-limited lens, the plane just beyond the lens (line L' in figure) acts as the Fraunhofer, or Fourier, plane for amplitude functions in object space.

Hence, due to the input optical amplitude function $u(x)$, there is an optical disturbance $u(x')$ along the line L' that obeys

$$U(x') = \int_{|x'| \leq x_0} dx\, u(x) \exp\left[\frac{2\pi i x x'}{\lambda R}\right], \quad R \approx 2f. \tag{3.55}$$

Here x' is the pupil spatial coordinate and R is approximately the image distance. An irrelevant multiplicative constant has been ignored. Also, by the geometry of the figure, the finite pupil size $2x_0$ relates to the half-angle α subtended by the lens at the ideal object position θ by

$$x_0 \approx R \sin \alpha. \qquad (3.56)$$

Let us suppose that we want to apply EPI to this scenario. The aim is to find the law that the particle wave functions $\mathbf{q}$ should obey. The $\mathbf{q}$ are defined to be the successive, real and imaginary parts of a wave functions ψ (a case $N = 2$ in Eq. (2.24)). This is effectively assuming a particle without spin, as is evident from Chap. 4. Then there is one, complex wave function ψ to be determined. This is, in fact, what we have been calling $\psi_0(x)$.

As mentioned in Sec. 3.4.5, the calculation hinges critically upon the form of the functional $J[\psi]$. This, in turn, depends upon the choice of invariance principle. It turns out that the most efficient (in the sense of $\kappa = 1$) such principle is the expression of a unitary transformation of coordinates. By Sec. 3.3.2, the transformation is to define two coordinate spaces, defining I and J, respectively, which are connected by the information transition $J \to I$.

Ideally, the nature of the unitary transformation should not be imposed by the observer in some *ad hoc* manner. In the spirit of Wheeler's program (Sec. 0.1), *it should be defined by the internal processes of the measuring device.* The only transformations of coordinates that occur during operation of our measuring device are Fourier transformations. There are two such and, consistent with our aims, Fourier transformations are unitary (see Eq. (4.5)). One is the transition by which optical waves proceed from object space coordinates x to optical pupil coordinates x'. This is the statement of the Fourier transform Eq. (3.55). The other is the inverse Fourier relation to Eq. (3.55). This expresses the transition from wavefront coordinates x' to output measurement space coordinates η,

$$u_i(\eta) = \int_{-x_0}^{x_0} dx' \, U(x') \exp\left[-\frac{2\pi i x' \eta}{\lambda R}\right]. \qquad (3.57)$$

Hence, the lens causes a transition of coordinates

$$x \to x' \to \eta. \qquad (3.58)$$

This defines a flow of Fisher information from object space to lens space to measurement space. As will become apparent in Sec. 3.8.6, the photon is the carrier of the information.

This means that the lens is the 'transducer' of the flow of information that is essential to EPI. Recall that EPI requires a unitary transformation, or some other invariance principle, as its 'physical' input (Sec. 3.4.5). Then the lens itself should provide the invariance principle. In fact, the lens obeys the Fourier relation Eq. (3.55) connecting input space x with another, conjugate space x' (along line L' in the figure). It will become apparent later that a Fourier transformation is also a unitary one. Therefore, Eq. (3.55) defines, as well, the invariance principle that EPI needs; this is in the form of a unitary transformation.

For increased generality, rather than directly using the pupil coordinate x' as the conjugate coordinate, we use a coordinate proportional to x', call it μ, obeying

$$\mu \equiv \frac{hx'}{\lambda R}. \tag{3.59}$$

Parameter h is what we call 'Planck's parameter'. It is later found to be a constant; see Sec. 4.1.14.

Using definition (3.59) in Eq. (3.55) gives

$$U\left(\frac{\lambda R}{h}\mu\right) = \int dx \psi_0(x) \exp\left[\frac{i\mu x}{\hbar}\right] \equiv \sqrt{h}\phi_0(\mu). \tag{3.60}$$

Observing that the last equality expresses a Fourier transform relation, we can invert it to yield

$$\psi_0(x) = \frac{1}{\sqrt{2\pi\hbar}} \int d\mu \phi_0(\mu) \exp\left[-\frac{i\mu x}{\hbar}\right]. \tag{3.61}$$

Coordinate μ turns out to be the particle momentum (see Sec. 4.1.9). Thus, *the unitary conjugate space to x is momentum space*, and this can be pictured as being located in pupil amplitude space.

With coordinate μ that of momentum, function $\phi_0(\mu)$ becomes the *probability amplitude on momentum* (Sec. 4.1.12). Hence, the pupil amplitude function U is proportional to the probability amplitude on momentum for the particle in object space, and each pupil coordinate value corresponds 1:1 to a momentum value.

3.8.5 Confirming EPI Eq. (3.18) using unitarity

To review, the instrument defines a coordinate space, along line L' of the figure, for optical amplitudes that is a Fraunhofer, or Fourier, transformation away from x-space. This is obviously an optical property of the instrument. But this Fourier space has, as well, special physical significance *to the particle* in being its momentum space. Since a Fourier transformation is unitary, the optical Fraunhofer plane of the instrument defines a particle momentum space that is unitary to the particle position space x. This is the unitary transformation that EPI extracts from the information flow internal to the instrument.

It follows that the Fisher information I, expressed equivalently in momentum space, represents the 'bound' information for the particle:

$$I[\psi_0(x)] = \kappa J[\phi_0(\mu)], \; \kappa = 1. \tag{3.62}$$

This is Eq. (3.18) of the EPI principle.

The functional forms of $I[\psi(x)]$ and $J[\phi(x)]$ are as found in Appendix D, Eqs. (D1) and (D3), for the case $N = 2$ (a single complex wave function):

$$I[\psi] = \int dx \psi'^{*}(x)\psi'(x), \tag{3.63}$$

and

$$J[\phi] = \frac{1}{\hbar^2} \int d\mu \mu^2 \phi^{*}(\mu)\phi(\mu). \tag{3.64}$$

A superfluous factor of 8 has been dropped from both equations. Substituting in $\psi \equiv \psi_0$ and $\phi \equiv \phi_0$ gives the I and J of Eq. (3.62).

We can see that the form (3.64) for J follows trivially from the form (3.63) for I, as follows. Operating d/dx on Eq. (3.61) shows that the same unitary (Fourier) transform that connects $\psi(x)$ and $\phi(\mu)$ also connects $\psi'(x)$ and $-i\mu\phi(\mu)/\hbar$. Then Eq. (3.64) follows from Eq. (3.63) as merely a statement (Eq. (4.5)) of the measure-preserving property of the unitary transformation. This is also called Parseval's theorem in the particular case at hand of a *Fourier* unitary transformation.

Notice that Eq. (3.62) is interpreted as showing equal informations I and J *in object space*. Hence, the transition $J \to I$ that characterizes EPI takes place, for the instrument, *right at its input port*. We will show, below, that the EPI axiom 1 of $\delta I = \delta J$ is also obeyed, again, at the input. The result is that EPI is enacted as a physical process at the input. Hence, it is the *input law* $\psi_0(x)$ that is fixed by EPI as either the Schroedinger wave equation (shown in Appendix D), Klein–Gordon equation or Dirac equation (shown in Chap. 4).

What, then, is the need for the measurement in the first place? (1) It is the measurement *interaction*, again at the input, that causes perturbation $\delta\psi(x)$, an activity that causes perturbations δI and δJ and, hence, activates EPI in the first place. (2) With the input law $\psi_0(x)$ fixed by EPI, the output measurement law $\psi_i(\eta)$ is then fixed by the convolution Eq. (3.51). In this way, *EPI generates both the input and the output amplitude laws of the instrument.*

From the figure, some wave energy will miss the lens and, hence, not reach image space. This suggests a loss of information in transit from object- to image space. We next examine this possibility.

3.8.6 Output information

The output information is $I[\psi_i]$. This incorporates the possible degrading effects of instrumental noise, through function $\psi_i(\eta)$. According to Eq. (3.51), ψ_i is known if the point amplitude response of the lens $w(x)$ is known. By definition of $w(x)$, it may be found by applying Eq. (3.55) and then Eq. (3.57) to an input impulse $u_\Delta(x) = \delta(x)$. This is easily done, giving

$$w(\eta) = 2x_0 \, \text{sinc}\left(\frac{2\pi x_0}{\lambda R}\eta\right), \quad \text{sinc}\,(u) \equiv \frac{\sin u}{u}. \tag{3.65}$$

Use of the last equation in Eq. (3.51) gives

$$\psi_i(\eta) = 2x_0 \int dx \psi_0(x) \, \text{sinc}\left[\frac{2\pi x_0}{\lambda R}(\eta - x)\right]. \tag{3.66}$$

This is, again, a convolution. The output of a convolution tends to be smoother than either of its input functions. Hence, $\psi_i(\eta)$ tends to be a smoother function than $\psi_0(x)$.

The image information $I[\psi_i]$ may be found as follows. Taking the Fourier transform of Eq. (3.66) gives

$$\phi_i(\mu) = \phi_0(\mu) \, \text{Rect}\,(\mu/\mu_0), \tag{3.67}$$

where $\text{Rect}\,(x) \equiv 1$ for $|x| \leqslant 1$, or 0 otherwise .

This shows that momentum values at the image suffer cutoff at a value μ_0. By Eqs. (3.56) and (3.59), $\mu_0 = (h/\lambda)\sin\alpha$. The effect is due to the finite lens aperture size $2x_0$ or angle α. It is interesting that μ_0 is exactly the X-component of momentum of a 'particle' of total momentum h/λ. This, of course, agrees with the discrete, or photon, aspect of light. In general, photons will radiate from object space in all directions. Thus, some will miss the lens. These constitute a loss of information. This loss effect is further examined next.

Substituting Eq. (3.67) into Eq. (3.64) gives

$$J[\phi_i] = \frac{1}{\hbar^2} \int d\mu \mu^2 \phi_i^*(\mu)\phi_i(\mu) = \frac{1}{\hbar^2} \int_{-\mu_0}^{\mu_0} d\mu \mu^2 \phi_0^*(\mu)\phi_0(\mu) \leqslant J[\phi_0] \tag{3.68}$$

because of the finite limits. Then by Eq. (3.62),

$$I[\psi_i] \leqslant I[\psi_0]. \tag{3.69}$$

The image generally contains less Fisher information than does the object space.

This said, a special situation arises when the input particle has a maximum momentum value, call it $\mu_{\max}$. Then by correspondence (3.59) there is a corresponding pupil coordinate size

$$x'_{\max} = \frac{\lambda R \mu_{\max}}{h} \tag{3.70}$$

such that for larger coordinates x', the pupil amplitude $U(x') = 0$. By the particle picture (see above), there are no photons that radiate from object space in these directions. Hence, if the *physical* pupil size x_0 exceeds x'_{max}, then x_0 may be effectively replaced by ∞ in Eq. (3.57). Rederiving Eq. (3.66) under these circumstances gives the equation as evaluated in the limit as $x_0 \to \infty$. By Eq. (0.68) the sinc function in (3.66) approaches a δ function, giving

$$\psi_i(\eta) = \psi_0(\eta). \tag{3.71}$$

Therefore, $I[\psi_i] = I[\psi_0]$. There is now no loss of information in the image space.

Hence, in this scenario of a finite band of momentum values and a large enough aperture size, the EPI transition $J \to I$ can be pictured as ending in the output data space. In this case, EPI directly derives $\psi_i(\eta)$ as well as $\psi_0(x)$ (since they are now the same function). Convolution Eq. (3.51) need not be used.

We return to the general scenario, where all momentum values may be present. It is shown, next, that EPI Eq. (3.16) explicitly holds for this scenario.

3.8.7 *Confirming EPI Eq. (3.16) using unitarity*

We found that EPI Eq. (3.16) is implied by axiom 1 (cf. Sec. 3.4.1). In fact, axiom 1 may be *explicitly* shown to hold, for this measuring device, as follows.

Axiom 1 states that if informations I and J are perturbed by amounts δI and δJ, then the two perturbations are equal. This assumes that I and J are at extremized values due to a solution **q** (the single wave function $\psi_0(x)$ here) of EPI, and that the perturbations δI and δJ are caused by a perturbation $\delta \psi(x)$ of $\psi_0(x)$.

Hence, we have to first find a mechanism for the perturbation $\delta \psi(x)$. As we found in Sec. 3.8.5, *the optical source wave provides the perturbation, by interacting with the particle.*

The resulting perturbations δI and δJ are obviously related since $\psi_0(x)$ and $\phi_0(\mu)$ are Fourier transform mates. We next evaluate these changes, showing that they are equal. Subscripts 0 are dropped since they are now superfluous.

The variations in question obey, by definition Eq. (0.41),

$$\delta I[\psi] \equiv \left. \frac{\partial I[\psi]}{\partial \varepsilon} \right|_{\varepsilon=0} d\varepsilon, \tag{3.72a}$$

$$\delta J[\phi] \equiv \left. \frac{\partial J[\phi]}{\partial \varepsilon} \right|_{\varepsilon=0} d\varepsilon. \tag{3.72b}$$

Our aim, then, is to show that

$$\frac{\partial I[\psi]}{\partial \varepsilon}\bigg|_{\varepsilon=0} = \frac{\partial J[\phi]}{\partial \varepsilon}\bigg|_{\varepsilon=0}. \tag{3.73}$$

The easier perturbation to compute is δJ. From Eq. (3.64), this results from a perturbation of wave function $\phi(\mu)$. As at Eq. (0.3), let

$$\phi(\mu) = \phi_0 + \varepsilon \eta(\mu), \tag{3.74}$$

where $\eta(\mu)$ is any perturbing function of the EPI solution $\phi_0(\mu)$. Substituting Eq. (3.74) into Eq. (3.64) gives J as a function of ε,

$$J(\varepsilon) = \frac{1}{\hbar^2} \int d\mu \mu^2 [\phi_0^* + \varepsilon^* \eta^*][\phi_0 + \varepsilon \eta]. \tag{3.75}$$

Then directly

$$\frac{\partial J}{\partial \varepsilon} = \frac{1}{\hbar^2} \int d\mu \mu^2 [\phi_0^* + \varepsilon^* \eta^*] \eta. \tag{3.76}$$

Hence

$$\frac{\partial J}{\partial \varepsilon}\bigg|_{\varepsilon=0} = \frac{1}{\hbar^2} \int d\mu \mu^2 \phi_0^*(\mu) \eta(\mu). \tag{3.77}$$

Multiplication by $\delta \varepsilon$ gives δJ, as required.

We now proceed to $\delta I[\psi]$. The perturbation (3.74) in momentum space implies a perturbation in x space obeying

$$\psi(x) = \psi_0(x) + \varepsilon \alpha(x) \tag{3.78}$$

where, by Eq. (D2),

$$\psi_0(x) = \frac{1}{\sqrt{2\pi\hbar}} \int d\mu \phi_0(\mu) \exp(-i\mu x/\hbar) \text{ and}$$

$$\alpha(x) = \frac{1}{\sqrt{2\pi\hbar}} \int d\mu \eta(\mu) \exp(-i\mu x/\hbar). \tag{3.79}$$

(Note that $\varepsilon \alpha(x)$ is what we previously called $\delta \psi(x)$.) Use of Eq. (3.78) in Eq. (3.63) yields

$$I(\varepsilon) = \int dx [\psi_0'^* + \varepsilon^* \alpha'^*][\psi_0' + \varepsilon \alpha']. \tag{3.80}$$

Then by direct differentiation,

$$\frac{\partial I}{\partial \varepsilon}\bigg|_{\varepsilon=0} = \int dx \psi_0'^* \alpha'. \tag{3.81}$$

Differentiating Eqs. (3.79) gives

$$\psi_0'^*(x) = \frac{1}{\sqrt{2\pi\hbar}} \frac{i}{\hbar} \int d\mu \phi_0^*(\mu) \exp(-i\mu x/\hbar) \text{ and}$$

$$(3.82)$$

$$\alpha'(x) = \frac{1}{\sqrt{2\pi\hbar}} \frac{-i}{\hbar} \int d\mu' \eta(\mu')\mu' \exp(-i\mu' x/\hbar).$$

Substituting these equations into Eq. (3.81) and interchanging orders of integration gives

$$\frac{\partial I}{\partial \varepsilon}\bigg|_{\varepsilon=0} = \frac{1}{2\pi\hbar} \frac{i}{\hbar} \frac{-i}{\hbar} \int d\mu \phi_0^*(\mu)\mu \int d\mu' \eta(\mu')\mu' \int dx \exp\left[\frac{ix}{\hbar}(\mu - \mu')\right]. \quad (3.83)$$

The far-right integral is $2\pi\hbar\delta(\mu - \mu')$ by Eqs. (0.68) and (0.71). Then by use of the sifting property Eq. (0.63b), Eq. (3.83) implodes to

$$\frac{\partial I}{\partial \varepsilon}\bigg|_{\varepsilon=0} = \frac{1}{\hbar^2} \int d\mu \phi_0^*(\mu)\eta(\mu)\mu^2. \quad (3.84)$$

This is identical to Eq. (3.77), as we set out to show.

This result can be generalized to a case where *any unitary transformation* connects coordinates x and μ. Remarkably, the proof is much easier than the preceding. Observe the integrand of Eq. (3.77). As shown below Eq. (3.64), quantity $-i\mu\phi_0(\mu)/\hbar$ is a unitary (Fourier) transformation of quantity $\psi_0'(x)$ in Eq. (3.81). Also, $-i\mu\eta(\mu)/\hbar$ is the same unitary transformation of $\alpha'(x)$. Then by the measure-preserving property Eq. (4.5) of a unitary transformation, the right-hand sides of Eqs. (3.77) and (3.81) are equal.

Then, by Eqs. (3.72a,b), $\delta I[\psi] = \delta J[\phi]$.

3.8.8 Summary

In general, *the EPI procedure is directly implied by the existence of a unitary transformation* between object space and a physically meaningful conjugate space (Secs. 3.8.5, 3.8.7). In such a measurement scenario, EPI holds independent of the axiomatic approach and, hence, any assumption that the system PDF obey the Fokker–Planck equation or (even) the I-theorem. This results in a much wider scope of application for the EPI principle, e.g., to deriving the wave equations for particles with spin (as in Chap. 4). The situation may be further generalized, as shown two paragraphs below.

Both working equations of EPI, Eq. (3.16) and Eq. (3.18), were found to be physically obeyed during use of the device. These followed from the existence of a Fourier (unitary) transformation between X-coordinate space and momentum space. This unitary transformation, in turn, followed from the internal dynamics of the measurement device. The measurement gave rise, as well, to the transition $J \rightarrow I$ and the perturbation $\delta\psi$ that EPI requires to implement the variational principle Eq. (3.16). Hence, for the measurement system at hand, all tenets of EPI were found to hold physically. An interesting result is that *the nature of the measurement instrument defined the physics of the object under measurement.*

Was this a coincidence? In fact, the optical measuring device is a prototype for a wide class of measuring instruments. Any particle location scheme requires a 'probe' particle (or photon) for interaction with the measured particle. This will provide the required wave function perturbation for initiating EPI. Also, any linear measuring instrument has a Fourier transform plane, somewhere. (Even an 'optical' system without a lens has a Fourier plane at infinity.) As we found, these were the requirements for EPI to be enacted.

As an example of the use of a real probe particle, consider the case of a probe electron. Under the assumption that the probe particle interacts with the measured particle via *a weak* potential energy function $V_{12}(x)$, the probability rate for scattering of the electron in a given direction is the *Fourier transform of* $V_{12}(x)$. This is by use of the Born approximation (see, e.g., Eisberg, 1961, p. 527). Here, once again, the measuring instrument operates through the use of a Fourier plane that is conjugate to X-coordinate space. Moreover, as in the lens model of preceding sections, the space of the output Fourier transform is a momentum space. Hence, it again corresponds to the coordinate space along line L' of Fig. 3.3.

An EPI solution defines the wave function – here called $\psi_0(x)$ – for the object under measurement. This, in turn, forms the amplitude law $\psi_i(\eta)$ from which the observed data value $\bar{y}$ is sampled; an example is Eq. (3.66). The upshot is that *the measurement procedure elicits, or 'creates' in some sense, the probability law (the physics) from which the measurement is sampled.*

It should be mentioned that the existence of a unitary transformation implies the validity of the EPI approach, but does not, in itself, guarantee that the approach can be *carried through*. In a general scenario, the implementation of EPI requires its two functionals $I[\psi(\mathbf{x})]$ and $J[\phi(\boldsymbol{\mu})]$ to ultimately be expressed *in the same coordinate space* in order to define a useable Lagrangian. For example, in the application to quantum mechanics (Appendix D, Chap. 4), the functional $J[\phi(\boldsymbol{\mu})]$ is essentially the mean-square momentum, and this can be re-expressed as the mean kinetic energy, a mean that can be taken *in X-coordinate space*. Since functional $I[\psi(\mathbf{x})]$ is already expressed in X-coordinate space, the entire Lagrangian for the problem is now in X-space, allowing the EPI approach to be implemented. Conversely, if the functional $J[\phi(\boldsymbol{\mu})]$ could not be re-expressed in X-space, EPI could not have been implemented.

In general, the re-expression of a $J[\phi(\mu)]$ in X-space hinges on two effects: (i) that $I[\phi(\mu)]$ is a statistical average, and statistical averages can be re-expressed in various spaces; and (ii) the quantity being averaged – μ^2 – has an equivalent form in X-space, here, essentially the kinetic energy. Such a pair of effects also hold in the derivation of quantum gravitational effects in Chap. 11.

It is not obvious that the 'equivalence' effect (ii) holds in general, i.e., for any unitary transformation of physical coordinates. Effect (ii) is not satisfied, e.g., by merely re-transforming $I[\phi(\mu)]$ back to $I[\psi(x)]$ via the known unitary transformation. In this case the two functionals I and J become identical and, so, information $K = 0$ identically (for *all* choices of amplitude functions **q**). This is a mere tautology.

Instead, equivalence effect (ii) must be a distinct, *physical input* into the problem, as in Appendix D and Chap. 4 (Secs. 4.1.15, 4.1.16).

However, in some measurement scenarios there is not an obvious unitary transformation connecting X-space with another space. Such cases occur in Chaps. 5–9. But even in such cases there is still an invariance principle of some kind that is obeyed by the measured particle. An example is continuity of flow (see later chapters). In this situation the EPI procedure may not be directly deduced, as here, but rather is taken to rest upon the axiomatic approach of Secs. 3.3.2 and 3.4.2.

As will be seen, the EPI process is shaped, or constrained, by the form of the particular invariance principle for the scenario. This may be shown explicitly for phenomena that obey the knowledge 'game' of Fig. 3.2. All EPI solutions lie along the bottom row, and each phenomenon has a generally different solution point along that row. Each such solution point is defined by its invariance principle.

Obviously, there is an intimate connection between statistical unitary transformations and the EPI approach. More work needs to be done on exploring the various physical unitary transformations, and their implications via the EPI principle.

3.9 EPI as a state of knowledge

According to EPI, there is a hierarchy of *physical knowledge* present. At *the top* are:

(A) the Fisher *I*-theorem (Sec. 1.8.2), which states that I, like entropy H_B, is a physical entity that *monotonically* changes with time and, also, can be transferred, or can 'flow', from one system to another (Sec. 1.8.10);

(B) the concept of a level J of Fisher information that is intrinsic to, or 'bound' to, each phenomenon (Secs. 3.3.1, 3.4.5); and

(C) the invariance, or symmetry, principle (Sec. 3.4.5) governing each phenomenon.

The laws (A)–(C), which we call the 'top laws', exist prior to, or independent of, any explicit measurements. They can possibly be *verified* (or nullified) by measurement, but that's another matter.

At the second rung down the knowledge ladder are the *three axioms*:

(i) Conservation of information perturbation, Eq. (3.13), during a measurement;

(ii) Eq. (3.19) defining information densities $i_n(\mathbf{x})$, $j_n(\mathbf{x})$ on the microlevel; and

(iii) Eq. (3.20) governing the efficiency of information transition, on the microlevel, from phenomenon to intrinsic data.

At the third rung down the ladder is the EPI principle. This follows (as we found) from either the axioms or from the existence of a physically meaningful unitary transformation space.

Finally, at the fourth rung down the ladder, is the carrying through of EPI as a *calculation*. This requires the EPI principle, as augmented by top law (C). The output of the calculation is the law governing formation of the amplitudes $\mathbf{q}$ for that scenario. For example, in Chap. 4 it is the Klein–Gordon 'law' governing formation of the amplitude ψ.

The question of what should be regarded as the laws of physics is of interest. Should they, e.g., be the 'top' laws (A)–(C) mentioned above, or, as is conventionally assumed, the output laws, such as the Klein–Gordon equation? We can expect, and the chapters ahead will verify, that some invariance principles (C) do double (or more) duty in implying physical laws. For example, the continuity of flow condition is used by EPI to derive both Maxwell's equations (Chap. 5) and the Einstein field equations (Chap. 6). Therefore, there are more physical laws than there are invariance conditions (C) for their derivation. Clearly it is desirable to have to make the fewest assumptions about nature. On this basis, the EPI output laws can be regarded as subsidiary to the top laws. They are also subsidiary in being subject to a contingency situation – measurement – for their existence, as is clarified next.

3.10 EPI as a physical process

The *physical picture* that is provided by EPI should also be considered. We postulate that if real data are at hand, they must have been caused by a *physical process*. The EPI view is that an output law is part of an ongoing physical process that includes the measurement step as its activator. (In this sense, the measurement 'creates' the probability law from which it is sampled. Imagine that!)

The measurement must be a real one upon a real object, say, a particle. The measurement physically activates the three axioms (or the unitary transformation) and, subsequently, EPI as a continuation of the process. In the *absence* of a real measurement upon a real object, the process is not activated so that the output law does not *physically* occur. (This does not prevent us from computationally *using* the form of the output law, e.g., the Schroedinger wave equation, to predict future, or past, states of an unmeasured, hypothetical entity. We are here restricting attention to physical processes, not states of knowledge as in Sec. 3.9.)

The output law continues as a physical process until another measurement is made. This re-initializes the state of the particle; etc. This is a continuing physical process punctuated and refreshed by step-like jolts due to new measurements. The new measurements act as unpredictable, discontinuous, irreversible, instantaneous operations upon the object, somewhat like so many *deus ex machina* activities. Chap. 10 and Sec. 11.2.16 clarify these effects.

Since EPI output laws only physically occur as reactions to measurement they are subsidiary to the top laws (A)–(C), which exist as absolutes, i.e., whether or not measurements take place. On this basis, the *real* laws of physics are, again, the top laws.

Many of the preceding ideas were developed jointly with B. H. Soffer.

We mentioned, above, that the initiation of a measurement creates the probability law from which the data value will be sampled. That is, it locally creates the physics of the observed phenomenon. This view regards reality as being perpetuated by requests for knowledge. It adds a new, creative dimension to the nominally passive act of observation. A traditional view of reality called *logical positivism* holds that all statements other than those describing or predicting observations are meaningless. Creative observation goes one step further, stating that the observations are, themselves, meaningless except insofar as they *create local physics*.

Making a measurement is a quantitative way of asking a question. The idea of measurement begetting phenomenon seems to be the physical counterpart to the adage that a well-posed *mathematical* problem, or question, contains the seeds of its solution. It is interesting to consider whether asking a qualitative question, as well, leads in some sense to a physical phenomenon (partially addressed on pp. 250–2).

This view of measurement has some strange ramifications. For example, in the well-known Schroedinger's cat experiment, it is now *observation* of the cat that either kills it or endows it with life. Or, in the many-worlds theory of

Everett (1973), whereby each new observation occurs in a new world, the new world is now *created* by the observation (see also Sec. 11.2.16).

3.11 On applications of EPI

In each of the following chapters, the EPI principle is applied to a different measurement scenario. Each such scenario leads to the derivation of a different physical law. The ordering of the chapters is, in the main, arbitrary so that they may be read in any order. However, the chapters are grouped as to similarity of approach or of application.

The flow of operations in each chapter's derivation follows those in Fig. 3.4. A parameter θ is chosen to be measured. The measurement is to be carried through with an instrument that has a given 'instrument function' (Sec. 3.8, Chap. 10). The measurement is initiated. The measurement process interferes, and interacts, with the phenomenon governing the parameter. This results in the perturbation of all the probability amplitudes q describing the phenomenon in the input (object) space to the instrument.

The phenomenon is identified by a suitable invariance principle. The principle should, by Wheeler's proposal of Sec. 0.1, be identified by the *internal processes of the measuring instrument*. An example was the unitary transformation suggested by the optical device in Sec. 3.8. An alternative to a unitary transformation is a property of continuity of flow for *the sources*. This could likewise be implied by the operation of a measuring device that obeys continuity of flow. The invariance principle is the only physical input to the procedure and, ultimately, allows the bound information J to be solved for.

The continuity of flow and unitary transformation principles are, respectively, invariance principles of the non-equality and equality type. These are designated as types (a) and (b), respectively, in Sec. 3.4.6. Type (a) principles give rise to a unique EPI solution, while type (b) principles give rise to two distinct EPI solutions. It is interesting that type (b) scenarios only occur for quantum phenomena (Chaps. 4, 10, 11 and Appendix D). All other phenomena that are derived in this book are of type (a).

The perturbed probability amplitudes q perturb, in turn, the channel capacity I (through defining Eq. (2.19)) and information J (through Eq. (3.13)). This activates the steps (3.13)–(3.20) defining the EPI process.

The EPI solutions define the phenomenon in the *input space* to the measuring instrument. Solutions at the output, or measurement, space must be obtained by other means. As examples: the output solution is obtained by a simple convolution of the EPI solution with the instrument function (Eqs. (3.51), (10.26b)).

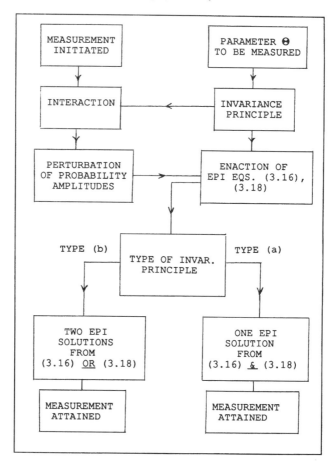

Fig. 3.4. EPI flow of operations.

We begin in Chap. 4 with the application to (relativistic) quantum mechanics since it is probably of major interest to more readers than any other application. Followups to this topic are Appendix D, where the Schroedinger wave equation is derived, and Chap. 10, where the wave equation *at* each measurement is found (counterpart to Feynman–Mensky measurement theory). Also, there is a very close resemblance between the development in Chap. 4 with that in Chap. 11 on quantum gravity. For example, each utilizes an equality-type invariance principle (type (b)) and, hence, should give two distinct output law solutions **q**. This correspondence and others are pointed out in Chap. 11.

Strong similarities also exist between the development in Chap. 5 of classical electromagnetic theory and that in Chap. 6 of the Einstein field equations of classical gravity theory. This is why the chapters follow one another. The correspondences are pointed out in Chap. 6.

Since EPI Eq. (3.16) is a variational principle, the general solution to an EPI problem is an Euler–Lagrange Eq. (0.34). This represents a second-order differential equation, whose solutions can be generally divided into two classes: *intrinsic* wave phenomena, for which the solution is a true wave equation, and non-wave phenomena. As will be seen, the latter is specified by a differential equation that can be immediately solved in final form for the distribution function.

For example, in Chap. 7 it is found that the EPI differential equation for classical particle velocity is soluble, in the form of the Maxwell–Boltzmann velocity distribution. Or, in Chap. 8 the differential equation for the power spectrum of a certain type of noise is immediately soluble as the ubiquitous $1/f$ power noise phenomenon. Likewise, in Chap. 9 a differential equation that is found for the probability amplitude describing the universal physical constants is immediately solved, giving a $1/x$ law for the PDF. Chaps. 8 and 9 are grouped together, as well, because they employ similar invariance principles in determining the physical information J for each.

The fact that many of the EPI differential equations are immediately soluble in this way extends the applicability of EPI to a wider scope of phenomena than simply wave phenomena. We believe it to be applicable to *all* phenomena.

4

Derivation of relativistic quantum mechanics

4.1 Derivation of Klein–Gordon equation

4.1.1 Goals

The EPI principle (3.16), (3.18) is perhaps best clarified by its application to quantum mechanics (Frieden, 1994; Frieden and Soffer, 1995). The Klein–Gordon and Dirac equations will be seen to follow as reactions to a real, four-space measurement. Other results will be the equivalence of mass and energy, and the constancy of the Planck constant h (see Sec. 3.4.14). To proceed with EPI, we need to first identify the four-parameters $\boldsymbol{\theta}$ (Sec. 3.5) to be measured. As we found in Sec. 3.8, their measurement perturbs the amplitudes $\mathbf{q}$ of the problem and starts EPI going as a process.

4.1.2 Choice of measured parameters

Let the observer measure, once, the space-time coordinates of a particle of mass m. (Then the ideal parameters $\boldsymbol{\theta}$ of the *intrinsic* scenario are N classical four-positions.) The single, real measurement perturbs (Sec. 3.8, Chap. 10) the N amplitude functions $q_n(\mathbf{x})$. Coordinates $\mathbf{x}$ are defined in terms of space-time coordinates x, y, z, t as

$$x_1 = ix, \ x_2 = iy, \ x_3 = iz, \ x_4 = ct$$

$$\boldsymbol{r} = (x, y, z), \ \mathbf{x} = (x_1, x_2, x_3, x_4).$$

(4.1)

This choice needs some explanation. In the case of mixed real and imaginary coordinates, I can be positive *or negative* (Appendix C). As it turns out, in this application if we make space coordinates real and the time coordinate imaginary then I becomes negative at the solution. Although this would permit the required wave equations to be derived by EPI, some awkwardness would arise in Sec. 4.1.17, where a negative resolution length would result. Conversely, we found that if we instead make the space coordinates purely imaginary and the time coordinate real, then *I* becomes positive. This is the sole reason for the coordinate choice (4.1).

4.1.3 Form of I

According to the plan of Sec. 3.6.2, replacements (3.37) must be imposed upon Eq. (2.19) for *I* in order for the latter to be expressed gauge covariantly. But, as also mentioned in Sec. 3.6.2, we can defer doing this until after *J*[**q**] is found. Hence, at this point in the derivation we stick with Eq. (2.19) as it stands, i.e., a field-free form.

4.1.4 Construction of complex amplitude functions

The use of complex amplitudes ψ_n in quantum mechanics is more a convenience than a necessity. One can instead work with the purely real amplitude functions that are the real and imaginary parts of the ψ_n. These are, in fact, the **q** obeying Eq. (2.24). In being the components of the ψ_n they are, in a sense, more fundamental than the ψ_n. However, for purposes of comparison with standard results, and, because they are indeed more convenient to use, we now regard the ψ_n as the new unknowns of the problem.

What are our basic quantities *p*, *I*, etc., in terms of these new complex amplitudes? We already found that probability $p(\mathbf{x})$ has the usual squared-modulus form (2.25) in terms of the ψ_n. We proceed now to *I*.

4.1.5 I in terms of the complex amplitudes ψ_n

We established above that we may defer the field replacements (3.37) in definition (2.19) of *I* until after the Euler–Lagrange solution to EPI is found. In essense, we first use a field-free scenario. By direct use of Eq. (4.1) in Eq. (2.19), we find that *I* obeys

$$4Nc \sum_{n=1}^{N/2} \iint dr\, dt \left[-(\nabla \psi_n)^* \cdot \nabla \psi_n + \left(\frac{1}{c^2}\right)\left(\frac{\partial \psi_n}{\partial t}\right)^* \frac{\partial \psi_n}{\partial t} \right] = I. \qquad (4.2)$$

(See the related Ex. 3.7.2.)

4.1.6 Finding J = J[ψ] by an invariance principle

By the general approach of Sec. 3.4.5, we need to find an invariance principle involving the ψ_n or I. We already found this, in one-dimensional form, for the position location problem of Sec. 3.8. This is the invariance of information I under a Fourier transformation from the space of coordinate x to that of a conjugate coordinate μ. As we found in Sec. 3.8, *the existence of such a (unitary) transformation guarantees the validity of the EPI approach.* We now suitably generalize the one-dimensional transformation to our four-dimensional problem.

4.1.7 Definition of Fourier coordinate space

Define new coordinates that are the Fourier conjugates to position-time,

$$\overset{old}{(ir,\ ct)} \overset{new}{\underset{F.T.}{\longleftrightarrow} (i\mu/\hbar,\ E/c\hbar)} \tag{4.3}$$

with

$$\psi_n(\mathbf{r},\ t) = \frac{1}{(2\pi\hbar)^2} \iint d\boldsymbol{\mu}\ dE\ \phi_n(\boldsymbol{\mu},\ E)\ e^{-i(\boldsymbol{\mu}\cdot\mathbf{r} - Et)/\hbar}. \tag{4.4}$$

Thus, ψ_n and the new functions ϕ_n are Fourier transform (F.T.) mates. (Note that the *subscripted* functions ϕ_n are distinct from the scalar potential ϕ defined in Eqs. (3.37).) The F.T. operation is unitary, obeying the measure-preserving requirement

$$\iint d\mathbf{r}\ dt\ \psi_m^* \psi_n = \iint d\boldsymbol{\mu}\ dE\ \phi_m^* \phi_n, \quad \text{all } m,\ n. \tag{4.5}$$

Exercise Show that Eq. (4.5) follows from direct substitution of Eq. (4.4) into its left-hand integral. *Hint:* Switch orders of integration so as to use the Fourier integral representation Eq. (0.70) of a Dirac delta function; then use the sifting property Eq. (0.69) of the delta function.

4.1.8 Nature of Planck's parameter

Equation (4.4) introduces a new parameter $\hbar$. At this point $\hbar$ is merely a parameter that is constant for a particular problem, but which is not necessarily the same constant over all quantum mechanical problems. That is, it is not necessarily a *universal* constant. EPI theory will later fix it as a universal constant (also see Sec. 3.4.14).

4.1.9 Indefinite nature of Fourier space coordinates

Quantities $\boldsymbol{\mu}$, E are simply regarded, at this point, as 'coordinates' of the Fourier space. These coordinates are not assumed by EPI to have any prior physical significance. For example, they may be related in any way. The proper relation will be *derived* later, as the famous equivalence (4.17) of energy, mass and momentum.

4.1.10 Correspondence between derivatives and product functions

By differentiating Eq. (4.4) we find the correspondences

$$(\nabla\psi_n, \partial\psi_n/\partial t) \overset{\text{F.T.}}{\longleftrightarrow} (-i\boldsymbol{\mu}\phi_n/\hbar, iE\phi_n/\hbar). \tag{4.6}$$

That is, the left-hand derivative functions in **x**-space are Fourier transforms of the corresponding product functions in conjugate space.

4.1.11 Use of Parseval's theorem

Since $\nabla\psi_n$ and $-i\boldsymbol{\mu}\phi_n/\hbar$ are seen to be Fourier mates, by Parseval's theorem their squared areas are equal,

$$\iint d\boldsymbol{r}dt(\nabla\psi_n)^* \cdot \nabla\psi_n = (1/\hbar^2)\iint d\boldsymbol{\mu}\,dE|\phi_n(\boldsymbol{\mu}, E)|^2\mu^2, \tag{4.7}$$

and

$$\iint d\boldsymbol{r}dt\left(\frac{\partial\psi_n}{\partial t}\right)^*\frac{\partial\psi_n}{\partial t} = \frac{1}{\hbar^2}\iint d\boldsymbol{\mu}\,dE|\phi_n(\boldsymbol{\mu}, E)|^2 E^2. \tag{4.8}$$

Using these two relations in Eq. (4.2) gives

$$I = \left(\frac{4Nc}{\hbar^2}\right)\sum_{n=1}^{N/2}\iint d\boldsymbol{\mu}\,dE|\phi_n(\boldsymbol{\mu}, E)|^2\left(-\mu^2 + \frac{E^2}{c^2}\right) \equiv J. \tag{4.9}$$

This is the invariance principle for the given scenario. The same value of I can be expressed in the new space $(\boldsymbol{\mu}, E)$, where it is called J (see Sec. 3.4.5). J is then the bound information for the scenario. Note also that $\kappa = 1$ here (see Sec. 3.4.2).

4.1.12 Physical properties of ϕ_n

Taking the special case $m = n$ in the unitarity Eq. (4.5) gives a Parseval's theorem

$$c \iint d\boldsymbol{r}\, dt |\psi_n|^2 = c \iint d\boldsymbol{\mu}\, dE |\phi_n|^2, \; n = 1, \ldots, N/2. \qquad (4.10)$$

Summing both sides over n, using correspondence (2.25) and, then, the normalization of p, gives

$$1 = \int d\boldsymbol{\mu}\, dE P(\boldsymbol{\mu}, E), \; P(\boldsymbol{\mu}, E) \equiv c \sum_{n=1}^{N/2} |\phi_n(\boldsymbol{\mu}, E)|^2. \qquad (4.11)$$

Thus, the new quantity P obeys $P \geqslant 0$ and normalization. This implies that P is a PDF in $(\boldsymbol{\mu}, E)$ space.

4.1.13 The bound information J

Following the lead of Sec. 4.1.6, Eqs. (4.9) and (4.11) give an expression for information J,

$$I \equiv J = \frac{4N}{\hbar^2} \iint d\boldsymbol{\mu}\, dE P(\boldsymbol{\mu}, E) \left(-\mu^2 + \frac{E^2}{c^2} \right). \qquad (4.12)$$

This is now an expectation

$$J = \left(\frac{4N}{\hbar^2} \right) \left\langle -\mu^2 + \frac{E^2}{c^2} \right\rangle. \qquad (4.13)$$

Hence J is an information that is expressed in terms of physical attributes of the problem (energy, momentum, etc.). Also, the equality (4.12) of I and J means that the constant κ in the condition Eq. (3.18) of EPI has the value unity in this scenario.

Result (4.13) has some important implications, discussed next.

4.1.14 Planck's parameter as a constant

By Sec. 3.4.14, J is regarded as a universal constant. Since the two factors in (4.13) are independent, each must be a constant. Then parameter $\hbar$ must be a universal constant.

4.1.15 Equivalence of matter and energy

Next, consider the second factor in Eq. (4.13). The character of the statistical fluctuations of E, $\boldsymbol{\mu}$ necessarily change from one set of boundary conditions to another. This would make J a variable, contrary to our aims, unless

$$-\mu^2 + \frac{E^2}{c^2} = constant \equiv A^2(m, c) \qquad (4.14)$$

where A is some function of the rest mass m and the speed of light c (the only other parameters of the free-field scenario). Solving for E gives

$$E^2 = c^2 \mu^2 + A^2(m, c)c^2. \tag{4.15}$$

By dimensional analysis, function $A(m, c)$ must obey a relation

$$A = mc, \tag{4.16}$$

where m is defined to be the mass of the particle. Eq. (4.15) then becomes the famous equivalence relation

$$E^2 = c^2 \mu^2 + m^2 c^4 \tag{4.17}$$

linking mass, momentum and energy. We take it *to define* 'coordinates' $\boldsymbol{\mu}$ and E as momentum and energy values, respectively.

4.1.16 Rest mass m as universal constant

Plugging Eq. (4.17) into Eq. (4.13) gives directly

$$I = 4N \left(\frac{mc}{\hbar} \right)^2 \equiv J. \tag{4.18}$$

Hence, the intrinsic information I in the four-position of a particle is proportional to the square of its intrinsic energy mc^2.

Since J is to be a universal constant, c is fixed as a constant in Sec. 3.5.5, and $\hbar$ has already been so fixed, we conclude from (4.18) that the rest mass m of the particle must be a universal constant. For example, if the particle is an electron, this condition fixes its rest mass as a universal constant.

4.1.17 Compton wavelength as an ultimate resolution length

By Sec. 2.3.2, I actually measures the *capacity* of the observed phenomenon to provide information about (in this case) four-length. Then I should translate into a figure for the ultimate fluctuation (resolution) length that is intrinsic to quantum mechanics; see also Sec. 3.4.15. Information (4.18) is identically

$$I = \frac{4N}{L^2}, \quad L \equiv \frac{\hbar}{mc}. \tag{4.19}$$

L is the 'reduced' (divided by 2π) Compton wavelength of the particle. By Eq. (2.9) and the efficiency of the intrinsic scenario,

$$I = 4N/e_{\min}^2. \tag{4.20}$$

We also made the reasonable assumption that each of the N estimates has the same accuracy. Combining Eqs. (4.19) and (4.20) gives

$$e_{\min} = L, \tag{4.21}$$

the reduced Compton length. This relates to a *resolution length* as follows.

Eq. (4.4) shows that, if the particle has a unique (fixed) momentum value $\boldsymbol{\mu} = m\mathbf{v}$, where $\mathbf{v}$ is the particle velocity, then $\psi_n(\mathbf{r}, t)$ is periodic, obeying $\psi_n(\mathbf{r} + \lambda_{DB}, t) = \psi_n(\mathbf{r}, t)$, $\lambda_{DB} = 2\pi\hbar/\mu = h/mv$. Wavelength λ_{DB} is called the DeBroglie wavelength and is, by its definition, a measure of the uncertainty in position for the particle. Comparison with L given in (4.19) shows that, since $v \leqslant c$, the Compton length $2\pi L$ is the *smallest* DeBroglie wavelength that is possible for a particle, regardless of speed. This is, then, the ultimate uncertainty in position for a particle. Then, by Eq. (4.21), the error $e_{\min}$ is the ultimate uncertainty divided by 2π, defining an ultimate resolution length. We recall, from Chaps. 1 and 2, that this result allows for the possibility of optimal data processing. Thus, *it represents one statement of the ultimate ability 'to know'.* Another statement of the limiting ability to know is the Heisenberg uncertainty principle; this is derived in Sec. 4.3.

4.1.18 Physical information

We can now proceed to form the physical information for the problem. Placing Eq. (4.17) in Eq. (4.12), using Eq. (4.11) and then Eq. (4.10), gives

$$J = (4Nm^2c^3/\hbar^2) \int\int d\boldsymbol{\mu} \, dE \sum_{n=1}^{N/2} \phi_n^* \phi_n$$

$$= (4Nm^2c^3/\hbar^2) \int\int d\mathbf{r} \, dt \sum_{n=1}^{N/2} \psi_n^* \psi_n. \tag{4.22}$$

The latter equality is again by use of Parseval's theorem. Then, by Eqs. (4.2) and (4.22), the physical information (3.16) is

$$K \equiv I - J = 4Nc \sum_{n=1}^{N/2} \int\int d\mathbf{r} \, dt$$

$$\times \left[-(\nabla\psi_n)^* \cdot \nabla\psi_n + \left(\frac{1}{c^2}\right) \left(\frac{\partial\psi_n}{\partial t}\right)^* \left(\frac{\partial\psi_n}{dt}\right) - \frac{m^2c^2}{\hbar^2} \psi_n^* \psi_n \right]. \tag{4.23}$$

4.1.19 Non-unique nature of the solutions

Recall that the EPI principle consists of two conditions: the extremum condition Eq. (3.16) and the zero-condition Eq. (3.18). As was discussed in Sec. (3.4.6), since our invariance principle (4.9) is directly of the form Eq. (3.18) the latter *by itself* gives a distinct solution **q**. Note that this **q** does not necessarily satisfy the extremum requirement (3.16) as well. Hence, in the following, we seek two generally different solutions **q** to the problem: those that satisfy the extremum condition Eq. (3.16) and those that satisfy the zero-condition Eq. (3.18). As we will see, these correspond to the classes of particles called bosons and fermions, respectively.

4.1.20 Klein–Gordon equation without fields

By EPI extremum condition (3.16), information K is to be extremized. This is through variation of the ψ, since the latter have replaced the purely real amplitudes **q** via Eq. (2.24). The variational solution is found in the usual way. The integrand of Eq. (4.23) is used as the Lagrangian $\mathscr{L}$ in the Euler–Lagrange Eq. (0.34), which is here

$$\frac{d}{dx}\left(\frac{\partial\mathscr{L}}{\partial\psi^*_{nx}}\right) + \frac{d}{dy}\left(\frac{\partial\mathscr{L}}{\partial\psi^*_{ny}}\right) + \frac{d}{dz}\left(\frac{\partial\mathscr{L}}{\partial\psi^*_{ny}}\right) + \frac{d}{dt}\left(\frac{\partial\mathscr{L}}{\partial\psi^*_{nt}}\right) = \frac{\partial\mathscr{L}}{\partial\psi^*_n}, \quad (4.24)$$

where

$$\psi^*_{nx} \equiv \partial\psi^*_n/\partial x, \quad (4.25)$$

etc., for y, z and t. After multiplying through by $c^2\hbar^2$, the result is

$$-c^2\hbar^2\nabla^2\psi_n + \hbar^2\frac{\partial^2\psi_n}{\partial t^2} + m^2c^4\psi_n = 0, \quad n = 1, \ldots, N/2. \quad (4.26)$$

This is the free field Klein–Gordon equation (Schiff, 1955). EPI gives this as the physical law that amplitudes ψ will obey.

4.1.21 Exercise

The derivation of this answer as described above is a straightforward exercise in use of the Euler–Lagrange approach. The motivated reader should try it out.

4.1.22 Value of N

Recall (Sec. 3.4.10) that the value of N is fixed as the number needed to satisfy the output physical law of EPI, here Eq. (4.26). Since this is the same second-order differential equation for each value of n, a value $N = 2$ suffices. Recalling definition (2.24) of the ψ this corresponds to a single complex wave function ψ.

4.1.23 Klein–Gordon equation with fields

Replacements (3.37) have been shown to make the theory gauge covariant
(Sec. 3.6.2). Making these replacements in Eqs. (4.23) and (4.26) give a field-
dependent information

$$K = 4Nc \iint dr\, dt \sum_{n=1}^{N/2} \left[-\left(\nabla + \frac{ieA}{c\hbar}\right)\psi_n^* \cdot \left(\nabla - \frac{ieA}{c\hbar}\right)\psi_n \right]$$

$$+ \left[\frac{1}{c^2}\left(\frac{\partial}{\partial t} - \frac{ie\phi}{\hbar}\right)\psi_n^*\left(\frac{\partial}{\partial t} + \frac{ie\phi}{\hbar}\right)\psi_n - \frac{m^2 c^2}{\hbar}|\psi_n|^2\right] \quad (4.27)$$

and a field-dependent solution

$$-c^2\hbar^2\left(\nabla - \frac{ieA}{c\hbar}\right)\cdot\left(\nabla - \frac{ieA}{c\hbar}\right)\psi_n + \hbar^2\left(\frac{\partial}{\partial t} + \frac{ie\phi}{\hbar}\right)^2 \psi_n + m^2 c^4 \psi_n = 0.$$

$$(4.28)$$

As with the free-field solution (4.26), value $N = 2$ suffices to define a general
solution to (4.28). This represents, again, one complex wave function ψ.

Eq. (4.28) defines the Klein–Gordon equation with fields A and ϕ generally
present (Schiff, 1955). If the rest mass m is finite, then (4.28) defines the
motion of particles with spin zero, called π mesons (Eisberg, 1961). These
particles belong to a wider class of particles, called 'bosons', which have
generally integral values of the spin. Or, if the mass m and charge e are both
zero, (4.28) describes 'particles' of spin 1 – photons – in a restricted sense
(see Sec. 5.1.19).

4.1.24 Schroedinger wave equation limit

The non-relativistic limit of (4.28) can be readily taken. Following Schiff
(1955), we take the limit in Appendix G. The result, Eq. (G8), is the well-known
Schroedinger wave equation (SWE). Hence, the SWE is arrived at by EPI via the
Klein–Gordon, equation, i.e., indirectly. Rigorous use of EPI cannot derive the
SWE directly because EPI is a Lorentz covariant approach and can only derive
covariant phenomena. The SWE is, of course, not covariant.

4.1.25 SWE from an approximate EPI approach

Nevertheless, the SWE can follow from an approximate use of EPI. See
Appendix D. This use abandons Lorentz covariance in ignoring the time
coordinate t. Although it produces the correct stationary SWE result, because

the approach is approximate it does not derive many of the other effects that were previously found: (a) the time-dependence of the SWE is not found; (b) the mass–energy relation (4.17) has to be assumed, rather than derived as here; and (c) the constancy of h and m are not proved. Evidently, the covariance requirement is powerful enough that, to ignore it, means losing many benefits.

4.1.26 The question of normalization

Eqs. (4.10) and (4.11), with $N = 2$, express a condition of normalization over space and time,

$$c \iint d\mathbf{r}\, dt |\psi(\mathbf{r},\, t)|^2 = 1. \tag{4.29}$$

This can be recast as

$$1 = c \int dt \int d\mathbf{r} |\psi(\mathbf{r},\, t)|^2 = c \int_{t_1}^{t_2} dt p_t(t) \tag{4.30a}$$

$$\text{where } p_t(t) \equiv \int d\mathbf{r} |\psi(\mathbf{r},\, t)|^2. \tag{4.30b}$$

We have assumed the fluctuations in detection time values to be bounded by the given numbers t_1, t_2. The difference $(t_2 - t_1)$ represents a finite total time interval for detection of the particle. At the observer's discretion it can be made small or large.

Now, normalization in quantum theory is conventionally taken over space $\mathbf{r}$ alone. That is, the right-hand side of Eq. (4.30b) is taken to be a constant (unity) in time. For example, such is the case for a ψ obeying the *non-relativistic* Schroedinger wave Eq. (G8) (Schiff, 1955, pp. 22, 23). However, it is known that a solution ψ to the *Klein–Gordon* equation (4.28) *does not necessarily* have a constant normalization integral (4.30b) (Morse and Feshbach, 1953, p. 256). *This is an inconsistency in the standard theory.* (Also see, in this regard, the Overview section.) Does our four-dimensional theory give constant normalization?

Yes, since our ψ has a four-dimensional argument and obeys *four-dimensional* normalization Eq. (4.29) the latter can only result in a constant. Also Eq. (4.30b), which is the conventional normalization equation, becomes merely a marginal PDF $p_t(t)$ in EPI theory. If this is time-varying, there is no inconsistency. In fact, this PDF has an interesting interpretation.

The PDF $p_t(t)$ represents, by its definition, the probability that the particle is measured somewhere (anywhere) within measurement space at the time $(t,\, t + dt)$. A small $p_t(t)$, for example, signifies that the particle has a low probability of being measured. This is effectively the same as saying that the particle has a low probability of *existing*, a property usually called *annihilation*. Such a situation can occur, e.g., if the particle is absorbed or otherwise exits the volume. Conversely, a high value for $p_t(t)$ connotes *creation* of the particle. The particle has either been re-emitted or has re-entered the volume at the time $(t, t + dt)$.

In cases where $p_t(t)$ turns out to be a constant, Eq. (4.30a) gives the constant as $p_t(t) = [c(t_2 - t_1)]^{-1}$. This states that the probability of detecting the particle is the same at all times, and varies inversely with the total detection time interval. If the constant $p_t(t)$ is small, then there is a constant tendency toward annihilation of the particle, etc. for the other extreme.

Do our four-dimensionally normalized wave functions obey different kinematics than do the standard, three-dimensionally normalized Klein–Gordon wave functions? Since the four-dimensional wave functions $\psi(\mathbf{r}, t)$ obey the standard Klein–Gordon equation (4.28) the answer is obviously no. The energy values and wave functions must be the standard answers. The only difference between the two solutions is the nature of the normalization factor as previously discussed. The same will be true of solutions to the Dirac equation, derived below.

In summary, a benefit of the four-dimensional approach is that it overcomes the problem of a variable three-dimensional normalization integral that occurs in the standard theory. In doing so, it allows for annihilation and creation events among Klein–Gordon solutions.

4.2 Derivation of Dirac equation

4.2.1 *Alternative solutions to EPI*

We have previously satisfied the extremum condition (3.16) of EPI. This gave rise to the Klein–Gordon equation as the solution. As we noted, this describes particles with zero spin. The other 'half' of EPI is condition (3.18). Because we had information $J = I$, with $\kappa = 1$, condition (3.18) becomes simply

$$I[\psi] - J[\psi] = 0. \qquad (4.31)$$

We now want to solve this condition for the complex amplitudes ψ. This corresponds to a case (b) as discussed in Sec. 3.4.6. The solution to (4.31) will not necessarily be coincident with the one that extremized Eq. (4.27).

In fact this is a desirable effect. The first half of EPI, condition (3.16), gave rise to the equation defining a particle with zero spin (a boson). There is, then, a possibility that the second half, condition (3.18), will correspondingly give rise to the equation defining the 'complement' to the zero-spin particle, namely a particle with a spin of $1/2$ (a fermion). Even further multiplicity of solutions ψ is necessary if the wave equations obeyed by all the elemental particle spin states are to be found. Quite possibly each 'root' ψ of (4.31) (with the ψ now, more generally, *multiply* subscripted) will define a different such particle as specified by its spin.

Results, below, agree with this conjecture: the lowest-component $(N = 4)$ solution to (4.31) defines the wave equation for the massless neutrino of spin $1/2$; and the next lowest $(N = 8)$ solution defines wave equations for the spin $1/2$ particles with finite mass, the electron and the positron.

4.2.2 Free-field case

By Eqs. (4.23) and (4.31), we require the solution ψ to obey

$$4Nc\iint dr\,dt \sum_{n=1}^{N/2}\left[-(\nabla\psi_n)^*\cdot\nabla\psi_n + \lambda^2\left(\frac{\partial\psi_n}{\partial t}\right)^*\frac{\partial\psi_n}{\partial t} - \eta^2|\psi_n|^2\right] = 0, \quad (4.32)$$

where we have introduced two parameters

$$\lambda = 1/c,\ \eta = mc/\hbar. \quad (4.33)$$

It may be noted that the ensuing derivation will not follow Dirac's historic procedure. Thus, we do not start from an *ad hoc* Hamiltonian operator equation where derivatives represent momentum or energy. The EPI formalism is a stand-alone procedure, not needing such assumptions. Instead, we simply seek the roots of Eq. (4.32). This will be implemented by a factorization procedure that uses Dirac's matrices (Dirac, 1928; 1947).

4.2.3 Dirac matrices introduced

Regard a sequence of matrices $[\alpha_x]$, $[\alpha_y]$, $[\alpha_z]$, $[\beta]$, each dimensioned $(N/2) \times (N/2)$, with constant elements that are to be determined. Length N is, as usual, left undetermined until it is fixed by sufficiency of solution at the end. For convenience of notation, define a vector of matrices

$$[\alpha] \equiv ([\alpha_x]\ [\alpha_y]\ [\alpha_z])^T, \quad (4.34)$$

where T denotes the transpose. Thus the dimensions of $[\alpha]$ are $(3N/2) \times (N/2)$. Also, regard ψ as a vector (see Eq. (2.24))

$$\psi = (\psi_1, \ldots, \psi_{N/2})^T. \quad (4.35)$$

Hence, ψ is a dimension $(N/2) \times 1$ vector.

Define the inner product of $[\alpha]$ with $\nabla\psi$ as

$$[\alpha]\cdot\nabla\psi \equiv [\alpha]^T\nabla\psi \equiv [\alpha_x]\partial\psi/\partial x + [\alpha_y]\partial\psi/\partial y + [\alpha_z]\partial\psi/\partial z. \quad (4.36)$$

Notice that each right-hand term is of dimension $(N/2) \times 1$.

4.2.4 Factorization vectors introduced

Introduce two 'helper' vectors of dimension $(N/2) \times 1$,

$$\mathbf{v}_1 \equiv i[\alpha] \cdot \nabla \psi - [\beta] \eta \psi + i\lambda \partial \psi / \partial t, \tag{4.37a}$$

$$\mathbf{v}_2 \equiv i[\alpha^*] \cdot \nabla \psi^* + [\beta^*] \eta \psi^* - i\lambda \partial \psi^* / \partial t. \tag{4.37b}$$

Exercise From these definitions, if $\psi_1(r, t)$ is a solution of $\mathbf{v}_1 = 0$ then

$$\psi_2^*(r, t) \equiv \psi_1^*(r, -t) \tag{4.38}$$

is a solution of $\mathbf{v}_2 = 0$. Show this. *Hint:* Take minus the complex conjugate of the equation $\mathbf{v}_1[\psi_1(r, t)] = 0$ and evaluate it at $t = -t$. The result should be $\mathbf{v}_2[\psi_2^*(r, t)] = 0$ for a ψ_2^* defined by Eq. (4.38).

4.2.5 Essential property

Vectors $\mathbf{v}_1, \mathbf{v}_2$ have an important property of factorization. If matrices $[\alpha_x], [\alpha_y], [\alpha_z], [\beta]$ are Hermitian and anticommute with one another, then

$$\iint dr\, dt \mathbf{v}_1 \cdot \mathbf{v}_2 = \int dr\, dt \sum_{n=1}^{N/2} \left[-(\nabla \psi_n)^* \cdot \nabla \psi_n + \lambda^2 \left(\frac{\partial \psi_n}{\partial t} \right)^* \frac{\partial \psi_n}{\partial t} - \eta^2 |\psi_n|^2 \right]$$

$$+ i \iint dr\, dt (S_4 + S_5). \tag{4.39}$$

Except for the presence of new functions S_4 and S_5, the helper vectors $\mathbf{v}_1, \mathbf{v}_2$ *factor the Klein–Gordon information form*. This is shown in Appendix E.

4.2.6 Resulting EPI solutions

Comparing result (4.39) with requirement (4.32) of EPI, the solution is a pair of vectors $\mathbf{v}_1, \mathbf{v}_2$ obeying

$$\iint dr\, dt\, [\mathbf{v}_1 \cdot \mathbf{v}_2 - i(S_4 + S_5)] = 0. \tag{4.40}$$

A microlevel solution Eq. (3.20) to this problem makes the integrand zero. It is shown in Appendix E that this is satisfied by either of $\mathbf{v}_1$ or $\mathbf{v}_2$ being zero. By definition (4.37a), the result of setting $\mathbf{v}_1 = 0$ is

$$\mathbf{v}_1 \equiv i[\alpha] \cdot \nabla \psi - [\beta] \eta \psi + i\lambda \frac{\partial \psi}{\partial t} = 0. \tag{4.41}$$

By result (4.38), setting $v_2 = 0$ gives the conjugate solution ψ^* to Eq. (4.41) as evaluated at negative time.

4.2.7 Dirac equations (free field)

Eq. (4.41) is the free-field Dirac equation (Morse and Feshbach, 1953). It is obeyed by a particle with spin 1/2 (see below), the electron. The fact that $\psi^*(r, -t)$ obeys the condition $v_2 = 0$, with v_2 defined at Eq. (4.37b), means that the conjugate (anti-) particle to the electron, the positron, moves backwards in time. This is the well-known Feynman interpretation of the positron.

In summary, use of the two solutions $v_1 = 0$, $v_2 = 0$ in EPI gives two Dirac equations, one for the electron and one for the positron.

4.2.8 Dirac equations (with fields)

The replacements (3.37) make the theory covariant. Making these replacements in Eq. (4.41) gives a field-dependent solution

$$i[\alpha]\cdot\left(\nabla - \frac{ieA}{c\hbar}\right)\psi - \eta[\beta]\psi + i\lambda\left(\frac{\partial}{\partial t} + \frac{ie\phi}{\hbar}\right)\psi = 0. \qquad (4.42)$$

This is the Dirac equation including the effects of electromagnetic fields. It describes the probability amplitudes ψ of a particle with spin. The spin is embedded in matrices $[\alpha]$, as shown next.

4.2.9 Dimension N = 8 case, resulting spin

As usual, the dimension N of the amplitude functions ψ has been left arbitrary up to the solution step. At this point we have to find the smallest value of N that is sufficient to describe the solution. The particle is assumed to have generally non-zero rest mass so that $\eta \neq 0$ in Eqs. (4.37). The solution (4.41) is expressed in terms of the matrices $[\alpha]$, $[\beta]$ of the theory. It is required that they be Hermitian and mutually anticommute (Appendix E). The smallest N that allows these properties to be obeyed is value $N = 8$, i.e., 4 complex wave functions ψ_n (Schiff, 1955, p. 326). This describes a spin-1/2 particle, the electron. Explicit representations of the matrices are

$$[\alpha_x] = \begin{bmatrix} 0 & \sigma_x \\ \sigma_x & 0 \end{bmatrix} \quad [\alpha_y] = \begin{bmatrix} 0 & \sigma_y \\ \sigma_y & 0 \end{bmatrix}$$

$$[\alpha_z] = \begin{bmatrix} 0 & \sigma_z \\ \sigma_z & 0 \end{bmatrix} \quad [\beta] = \begin{bmatrix} 1 & 0 \\ 0 & -1 \end{bmatrix}. \tag{4.43}$$

The 'elements' of these matrices are themselves 2×2 matrices

$$[\sigma_x] = \begin{bmatrix} 0 & 1 \\ 1 & 0 \end{bmatrix} \quad [\sigma_y] = \begin{bmatrix} 0 & -i \\ i & 0 \end{bmatrix}$$

$$[\sigma_z] = \begin{bmatrix} 1 & 0 \\ 0 & -1 \end{bmatrix} \quad [1] = \begin{bmatrix} 1 & 0 \\ 0 & 1 \end{bmatrix} \tag{4.44}$$

with [0] a matrix of all zeroes. Matrices $[\sigma_x]$, $[\sigma_y]$, $[\sigma_z]$ are called the Pauli spin matrices (Pauli, 1927).

4.2.10 Dimension N = 4 case

Consider, next, the case of a particle with zero rest mass so that, by (4.33), $\eta = 0$. Now the Klein–Gordon information (4.23) does not contain the end term in mass m. As before, this is to be factored. We may again use the approach of Appendix E to achieve the factorization. Obviously the problem is a special case $m = 0$ of our previous solution Eqs. (4.41), (4.38). Hence, the resulting requirements are again those of anticommutation, but now only for the matrices $[\alpha_x]$, $[\alpha_y]$, $[\alpha_z]$. Matrix $[\beta]$ no longer enters into the problem since the coefficient η is zero in factor (4.41).

Setting the factorization vector $\mathbf{v}_1$ in (4.41) equal to zero now gives a requirement

$$i[\alpha] \cdot \nabla \psi + i\lambda \frac{\partial \psi}{\partial t} = 0. \tag{4.45}$$

In fact, the three Pauli spin matrices $[\sigma_x]$, $[\sigma_y]$, $[\sigma_z]$ satisfy our requirement of anticommutation, as may easily be verified from their definitions (4.44). Hence, we identify

$$[\alpha_x] = [\sigma_x], [\alpha_y] = [\sigma_y], [\alpha_z] = [\sigma_z]. \tag{4.46}$$

Since the dimension of these matrices is 2×2, we see that now we have $N = 4$ as the solution dimension (i.e., $N/2 = 2$ complex wave functions ψ_1, ψ_2). Eq. (4.45) is now Weyl's equation (Roman, 1960) describing the massless neutrino of spin $1/2$.

Hence, leaving the dimensionality N general until the end of the problem and, then, fixing it by sufficiency of solution has merit. As shown in this section and the preceding one, in this measurement scenario it results in the wave equations for some well known elementary particles.

4.2.11 Non-relativistic limit

This limit can be directly taken in Eq. (4.42) (Schiff, 1955, pp. 329–30). As is well known, it gives the Schroedinger wave Eq. (G8) plus a term involving interaction of the particle spin with the magnetic field H. This term does not disappear unless the particle has zero spin.

4.3 Uncertainty principles

4.3.1 For position–momentum measurements

The Heisenberg uncertainty principle states that, at a given time t, a particle's position and momentum intrinsically fluctuate by amounts x and μ from ideal (classical) values θ_x and θ_μ with variances ϵ_x^2 and ϵ_μ^2 obeying

$$\epsilon_x^2 \epsilon_\mu^2 \geq (\hbar/2)^2. \tag{4.47}$$

This is conventionally derived from the Fourier transform relation (4.4) connecting position and momentum spaces; see, e.g., Bracewell (1965).

The relation may be shown, as well, to arise out of the use of Fisher information. Specifically, it will be seen to be the expression of the Cramer–Rao inequality (1.1) for X-measurements in the *intrinsic* scenario (Sec. 2.1.1).

Consider an intrinsic measurement scenario where the position and momentum of a particle are measured at the same time. Like all measurements, the X-measurements must obey the Cramer–Rao inequality (1.1),

$$e_x^2 I_x \geq 1, \quad e_x^2 \equiv \langle (\hat{\theta}_x(y) - \theta_x)^2 \rangle. \tag{4.48}$$

The general estimator function is $\hat{\theta}_x(y)$. The question is, what is information I_x for this scenario? (See also Stam, 1959.)

As at Eq. (2.24), define complex amplitude functions $\psi_n(x)$ as those whose real and imaginary components are our real amplitude functions $\mathbf{q}$. By Eq. (4.4), represent each $\psi_n(x)$ as the Fourier transform of a corresponding function $\phi_n(\mu)$ of momentum fluctuations μ,

$$\psi_n(x) = \frac{1}{\sqrt{2\pi\hbar}} \int d\mu \phi_n(\mu) \exp(-i\mu x/\hbar). \tag{4.49}$$

This is a case $N = 2$ of but one complex wave function. Then the total information is the one-dimensional version of Eq. (4.2) due to $N = 2$ intrinsic data. Also, by direct evaluation of Eq. (D1) using $\psi = |\psi| \exp(iS)$ we get $I = I_x + \langle (dS/dx)^2 \rangle$, so that $I_x \leq I$ or

$$2I_x \leq 8 \int dx \psi'^* \psi', \quad \psi \equiv \psi_1, \quad \psi' \equiv d\psi/dx. \qquad (4.50)$$

The factor 2 on the far left-side follows from the additivity of information, Eq. (1.61), for the $N = 2$ data. Substituting Eq. (4.49) into (4.50) gives, as the information *per data value*,

$$I_x \leq \frac{4}{\hbar^2} \int d\mu \mu^2 |\phi(\mu)|^2, \quad \phi \equiv \phi_1. \qquad (4.51)$$

By Eqs. (4.11), $|\phi(\mu)|^2$ is just the marginal PDF $P(\mu)$, so that the integral in (4.51) is a mean value,

$$I_x \leq \frac{4}{\hbar^2} \langle \mu^2 \rangle \equiv \frac{4}{\hbar^2} \epsilon_\mu^2. \qquad (4.52)$$

The far-right equality follows because the μ are fluctuations from the mean momentum. Using Eq. (4.52) in Eq. (4.48) gives directly

$$e_x^2 \epsilon_\mu^2 \geq (\hbar/2)^2. \qquad (4.53)$$

This is of the form (4.47), the Heisenberg principle for the X-coordinate measurement, although we have allowed for a difference of interpretation by representing the X-position errors differently, ϵ_x^2 vs. e_x^2. The two principles are not quite the same, as is discussed below.

4.3.2 For time–energy measurements

A Heisenberg principle is also obeyed by simultaneous measurements of time and energy,

$$e_t^2 \epsilon_E^2 \geq (\hbar/2)^2, \qquad (4.54)$$

where e_t^2 is the mean-square time fluctuation and ϵ_E^2 is the mean-square energy fluctuation. This principle may be derived from the Cramer–Rao inequality for the intrinsic time measurements, by analogous steps to the preceding.

4.3.3 Exercise

Carry through the derivation.

4.3.4 Discussion

We observed that there are (at least) two routes to the Heisenberg principle: (a) based upon Fourier complementarity (4.49) or (b) arising out of the

Cramer–Rao inequality. There are conceptual differences between approaches (a) and (b).

The taking of data means a process of randomly sampling from 'prior' probability laws $p(x)$ and $P(\mu)$ (In the usual statistical sense, 'prior' means prior to any measurements.) However, the Fourier approach (a) does not assume the actual taking of data. The spreads ϵ_x^2, ϵ_μ^2 that it defines in Eq. (4.47) are just parameters that measure the widths of the prior laws. Hence, the approach (a) derives a Heisenberg principle (4.47) that holds independent of, and *prior to*, any measurement. This, of course, violates the spirit of EPI, according to which physical laws follow *in reaction to* measurement.

By contrast, the EPI-based approach (b) states that, if an X-coordinate *is measured* (with an ideal detector; see Sec. 2.1.1), then there is a resulting uncertainty in momentum that obeys reciprocity (Eq. 4.53)) with the coordinate uncertainty. Since this interpretation of the Heisenberg principle is measurement-based, it agrees with the spirit of EPI.

In a nutshell, our disagreement with the usual Heisenberg interpretation lies in the presumed nature of the fluctuations. The conventional interpretation is that these are intrinsic to the phenomenon and independent of measurement. Our view is that they are intrinsic to the phenomena (Sec. 2.1.1), and *arise out of* measurement, as do all phenomena.

Of course if the detector is not ideal, and contributes noise of its own to the measurements, both interpretations need to be modified (Arthurs and Goodman, 1988; Martens and de Muynck, 1991; Caves and Milburn, 1987).

The second difference between approaches (a) and (b) lies in the nature of their predictions. Eq. (4.47) states that the spreads in positions will obey the principle. By contrast, Eq. (4.53) states that the spreads in *any functions* $\hat{\theta}_x(y)$ of the data positions obey the principle. The latter includes the former, then, as a particular case (where $\hat{\theta}_x(y) = y$). In this sense, version (4.53) is the more general of the two.

4.3.5 Uncertainty principles expressed by entropies

The uncertainty principle (4.53) has been seen to derive from the fact that Fisher information I measures the spread in momentum values. Another measure of the spread in momentum is H, its Shannon entropy, here denoted as

$$H(\mu) = -\int d\mu \, P(\mu) \ln P(\mu), \quad P(\mu) = \sum_n |\phi_n(\mu)|^2. \qquad (4.55)$$

(See Secs. 1.3, 1.7.) The entropy associated with space x is, correspondingly,

$$H(x) = -\int dx p(x) \ln p(x), \quad p(x) = \sum_n |\psi_n(x)|^2. \tag{4.56}$$

Then, can the complementarity of width of the two PDFs $p(x)$ and $P(\mu)$ be expressed in terms of these two entropies? The answer is yes.

Hirschman's inequality (Hirschman, 1957; Beckner, 1975) states that

$$H(x) + H(\mu) \geqslant \ln(\pi e \hbar/2). \tag{4.57}$$

Parameter e is here the Naperian base. This states that it is impossible for $p(x)$ and $P(\mu)$ to both be arbitrarily narrow functions. It is interesting to evaluate (4.57) for the case where the PDFs are Gaussian. Since the entropy for a Gaussian obeys $H = \ln \sigma + \ln \sqrt{2\pi e}$, Eq. (4.57) gives

$$\ln \sigma_\mu + \ln \sqrt{2\pi e} + \ln \sigma_x + \ln \sqrt{2\pi e} \geqslant \ln(\pi e \hbar) \text{ or } \sigma_\mu^2 \sigma_x^2 \geqslant (\hbar/2)^2, \tag{4.58}$$

again the Heisenberg principle (4.47). A further similarity between the two principles is that the amplitude functions that attain a minimum Heisenberg product also attain a minimum 'Hirschman sum'. This is the Gaussian case cited above.

4.4 Overview

Of central importance to the derivations is the existence of the unitary transformation space $(i\boldsymbol{\mu}/\hbar, E/c\hbar)$ (see Eq. (4.3)). As we found in Sec. 3.8, the existence of such a conjugate space to the measurement space $(i\boldsymbol{r}, ct)$ guarantees the validity of the EPI approach for the given problem.

Particles are known to follow either Bose–Einstein or Fermi–Dirac statistics (Jauch and Rohrlich, 1955). The former are characterized by integral spin, the latter by $1/2$-integral spin (an odd number times $1/2$). What we have found is that these basic particle types derive, respectively, from the two requirements (3.16) and (3.18) of EPI. Conversely, this tends to confirm these two requirements as being physically meaningful.

The solutions for spin 0 and spin $1/2$ were found, in particular. The derivations were for the particular case where the parameters $\boldsymbol{\theta}$ are simple vectors (Sec. 4.1.2). We expect that higher-spin EPI solutions will result similarly when the parameters are made to be general tensors instead of vectors.

That parameter $\kappa = 1$ for this scenario is of interest. Recall that κ represents, from Eq. (3.18), the ratio of the intrinsic Fisher information I to the information J that is bound to the phenomenon. That $\kappa = 1$ here and in Appendix D suggests, by Sec. 3.4.5, that quantum mechanics is an accurate theory. The intrinsic information I contains as much information as the phenomenon can supply. As we saw at Eq. (4.9), this arose out of the unitary nature of the invariance principle that was employed. Many other phenomena, specifically, non-quantum ones, will not obey such information efficiency. See later chapters.

We want to emphasize that the derivations do not rely on the well-known association of gradient operators with momentum or energy. No Hamiltonian operators are used. All gradients in the derivations arise consistently from *within* the approach: ultimately from the definition (2.19) of Fisher information.

In the case of a conservative, scalar force field $\phi(r)$, with $A(r, t) = 0$, the Klein–Gordon solution Eq. (4.28) may be regarded, alternatively, as a solution to the knowledge acquisition game of Sec. 3.4.12 (see *Exercise* at end). On the other hand, since the Dirac solution Eq. (4.42) resulted from the zero-principle Eq. (3.18) of EPI, and not from the extremization principle Eq. (3.16), it does not follow from a knowledge acquisition game. This is a strange distinction between the two solutions and, consequently, between the nature of fermions and bosons.

The *mathematical* route to the *Klein–Gordon equation* that is taken in Sec. (4.1) is via the conventional Lagrangian for the problem. The unique contribution of the EPI approach is in *deriving* the terms of the Lagrangian from physical and information considerations. However, even mathematically, the route taken in Sec. (4.2) to the *Dirac equation* is unique. It is not a Lagrangian variational problem, but rather finds the Dirac equations for the electron and the positron as the two roots of an equation.

A further advantage of EPI over conventional theory is worth mentioning. Assume, for simplicity, a case $N = 2$ of a single complex wave function $\psi(r, t)$. We give an argument of Schiff (1955). By multiplying the Klein–Gordon Eq. (4.26) on the left by ψ^*, multiplying the complex conjugate of (4.26) on the left by ψ, and subtracting the results, one obtains a pseudo conservation of flow equation

$$\frac{\partial}{\partial t}\overline{P}(r, t) + \nabla\cdot\overline{S}(r, t) = 0, \tag{4.59}$$

where

$$\overline{P}(r, t) = \frac{i\hbar}{2mc^2}\left(\psi^*\frac{\partial\psi}{\partial t} - \psi\frac{\partial\psi^*}{\partial t}\right) \tag{4.60}$$

and $\overline{S}(r, t)$ is another form which doesn't concern us here. In (4.59), one conventionally identifies the quantity that is differentiated $\partial/\partial t$ as a probability density. Then quantity $\overline{P}(r, t)$ obeying (4.60) would be our PDF $p(r, t)$. But, obviously (4.60) is of a generally different form than our presumed form (2.25), here

$$p(\mathbf{r},\ t) = \psi^* \psi. \tag{4.61}$$

(The two agree in the nonrelativistic limit.)

Which of the two forms is correct? *The form (4.60) can go negative* (Schiff, 1955, p. 319). Since a PDF must always obey positivity, this rules out the use of (4.60) as a PDF. Also, an alternative choice $|\psi(\mathbf{r},\ t)|^2$ of $p(\mathbf{r},\ t)$, which does not have this problem, instead suffers from a generally *variable* three-dimensional normalization integral $d\mathbf{r}$ (Morse and Feshbach, 1953, p. 256). These problems are distinct flaws in the conventional theory, which is premised upon three-dimensional normalization $d\mathbf{r}$.

Fortunately, as we found, the problem of variable normalization can be overcome by the *four-dimensionality* of the EPI approach. No matter how variable the integral is after three integrations, after the fourth it must be a constant! This allowed us, then, to accept the squared modulus $|\psi(\mathbf{r},\ t)|^2$ as the form for $p(\mathbf{r},\ t)$. This is the standard EPI choice (2.25), and is mathematically consistent, since it obeys positivity.

In summary of the last two paragraphs, the covariant EPI approach leads to a mathematically consistent definition of the PDF for the problem whereas the usual, non-covariant approach leads to serious inconsistencies.

These benefits ultimately follow from the way time is treated by EPI theory: on an equal footing with space (cf. Secs. 3.1.2, 3.5.8). Thus, the time 'coordinate' t and space 'coordinate' $\mathbf{r}$ for an amplitude function $\psi(\mathbf{r},\ t)$ are both regarded as random fluctuations from ideal values. Also, just as space is not presumed to 'flow' in standard treatments of quantum mechanics, time is likewise not interpreted to 'flow' in EPI theory. As with the Einstein field equations of general relativity (Chap. 6), the Klein–Gordon and Dirac equations merely provide an 'arena' for the occurrence of random space and random time fluctuations.

Also, we mentioned in Sec. 4.1.2 that either choice of coordinates $(i\mathbf{r},\ ct)$ or $(\mathbf{r},\ ict)$ would suffice in leading to the end products of the approach – the Klein–Gordon and Dirac equations. Such arbitrariness implies that either 'space-like' or 'time-like' coordinates (Jackson, 1975) describe equally well the phenomena of quantum mechanics. Again, space and time have an equal footing.

Quantum mechanics is well-known to obey additivity of amplitudes rather than additivity of probabilities. This leads to non-classical behavior, such as in the famous two-slit experiment (Schiff, 1955, pp. 5, 6) whereby the net probability at the receiving plane is not merely the sum of the two contributions from the individual slits, but also includes a cross-term contribution due to the interference of amplitudes from the slits. This is often taken to be the signature effect that distinguishes quantum- from classical mechanics. How does EPI account for this effect?

The effect originates in Eqs. (1.23) and (1.24), whereby the PDF $p(x)$ *is defined* as the square of an amplitude function $q(x)$ and, resultingly, the information I is expressed directly in terms of $q(x)$. When this I is used in EPI principle Eqs. (3.16), (3.18) the result is a differential equation in $q(x)$ [$\psi(\mathbf{r}, t)$ here], not in $p(x)$ directly. Solutions $q(x)$ to this equation are often in the form of superposition integrals. Then, squaring the superposition to get $p(x)$ brings in all cross-terms of the superposition, e.g., the two-slit cross-term mentioned above.

We noted below Eq. (1.23) that such use of probability amplitudes is actually classical in origin, tracing to work of Fisher. Hence, oddly enough, this signature quantum mechanical effect originates in classical statistics. It is perhaps not surprising, then, that this cross-term effect occurs in classical statistical physics as well (see Eqs. (7.56) and (7.59)).

We have shown that the wave equations of quantum mechanics follow from the EPI principle. These wave equations hold at the input space to the measuring apparatus (see Secs. 3.8.1, 3.8.5). One might ask, then, what wave equation is obeyed at the *output* of the apparatus? This topic is taken up in Chap. 10.

5

Classical electrodynamics

5.1 Derivation of vector wave equation

5.1.1 Goals

The aim of this chapter is to derive Maxwell's equations in vacuum via the EPI principle. This is done in two steps: by (1) deriving the vector wave equation; and then (2) showing that this implies the Maxwell equations. The latter is an easy task. The main task is problem (1).

We follow the EPI procedure of Sec. 3.4. This requires identifying the parameters to be measured (Sec. 3.4.4), forming the corresponding expression (2.19) for I, and then finding J by an appropriate invariance principle (Sec. 3.4.5). These steps are followed next.

5.1.2 Choice of ideal parameter θ

The four components of the electromagnetic potential **A** are defined as

$$\mathbf{A} \equiv (A_n, \, n = 1, \, \ldots, \, 4) \equiv (A_1, \, A_2, \, A_3, \, \phi) \equiv (\mathbf{A}, \, \phi) \qquad (5.1)$$

in terms of the vector *three*-potential A and the scalar potential ϕ. Note that Eq. (5.1) does not define the so-called *four-potential* of electromagnetic theory. In our notation (3.34c), this would require an imaginary i to multiply either ϕ or A in the equation. We do not use a four-potential (and four-current) in this chapter, as it would merely add unnecessary complication to the approach that is taken. Vector **A** and the main output of the approach, Eq. (5.51), can easily be placed in four-vector form anyhow; see Sec. 5.1.23.

This agrees with the fact that the potential **A** and vector **q** are proportional for this problem, and **q** was found (Sec. 2.4.4) to not generally be a four-vector, i.e., not be covariant. Instead, *each component A_n and q_n will be covariant*, i.e., will obey a covariant differential equation; again, as implied in Sec. 2.4.4.

On the other hand, the Fisher coordinates **x** will remain a four-vector, as is generally required by EPI (Sec. 3.5.8).

The magnetic and electric field quantities *B* and *E* are defined in terms of **A** by Eqs. (5.85) and (5.86) (Jackson, 1975). Vector **A** obeys the Lorentz condition, Eq. (5.84). Fields *E* and *B* are direct observables.

Consider an experiment whose aim is to determine one of the fields *E* or *B* at an ideal four-position $\boldsymbol{\theta}$. The measuring instrument that is used for this purpose acts, as in Sec. 3.8, to perturb the amplitude functions **q** of the problem at its input space. (Note that the **q** are not identified with specifically electromagnetic quantities until Sec. 5.1.21.) Also, in the input space, random errors in position **x** occur that define the 'intrinsic data' of the problem

$$\mathbf{y} = \boldsymbol{\theta} + \mathbf{x}. \tag{5.2}$$

Thus, *E* or *B* is measured, but *precisely where* is unknown. The **x** are the Fisher variables of the problem.

The Fisher variables **x** must be related to the space *r* and time *t* fluctuations of the measurement location. We choose to use

$$\mathbf{x} \equiv (x_k, \; k = 1, 2, 3, 4) \equiv (\boldsymbol{r}, \; ict), \; \boldsymbol{r} \equiv (x, y, z). \tag{5.3}$$

These are as in Eq. (4.1) defining the measurement problem in quantum mechanics, except for the location of the factor *i*. Actually, the coordinates of Eq. (4.1) would work as well here; we simply prefer working with imaginary time rather than imaginary space.

Eq. (5.2) states that input positional accuracy is limited only by fluctuations that characterize the electromagnetic field. The most fundamental of these are the 'vacuum fluctuations', which give rise to an uncertainty in position of amount

$$\Delta x \sim (\hbar / m_0 \omega_0)^{1/2}. \tag{5.4}$$

Quantities m_0 and ω_0 are the mass and frequency of an oscillator defining the electromagnetic field (see Misner *et al.*, 1973, p. 1191). (Note: At this point Eq. (5.4) is purely motivational; it is derived at Eq. (5.47) below.)

5.1.3 Information I

By Eq. (2.11), the intrinsic data $\mathbf{y}_n$ are assumed to be collected independently, and there is a PDF

$$p_n(\mathbf{y}_n | \boldsymbol{\theta}_n) = p_n(\mathbf{x}) \equiv q_n^2(\boldsymbol{r}, t) \tag{5.5}$$

describing each (*n*th) four-measurement. The Fisher information *I* in the *position* measurements (see preceding) is to be found. The result is Eq. (2.19) which, with our choice (5.3) of coordinates becomes

$$I = 4c \iint d\boldsymbol{r}\, dt \sum_{n=1}^{N} \left[\nabla q_n \cdot \nabla q_n - \frac{1}{c^2} \left(\frac{\partial q_n}{\partial t} \right)^2 \right].$$ (5.6)

Note that this is very similar to the quantum mechanical answer (4.2) for I. Here we simply don't pack the amplitude functions $\mathbf{q}$ to form new, *complex* amplitudes.

5.1.4 Invariance principles

The purpose of the invariance principle for a scenario is to define the bound information J. In Chap. 4, this was accomplished by a principle of unitary transformation between coordinate–time space and momentum–energy space. Here, there is no unitary transformation that connects $(\mathbf{x}, t)$ space with another physically meaningful coordinate space. Since information J is to specifically relate to (be 'bound' to) the phenomenon, we instead seek as the invariance principle a defining property of the source (see Sec. 3.11). Assuming that the system is a closed one, the most basic such principle is a property of *continuity of charge and current flow*. This is

$$\frac{\partial \rho}{\partial t} + \nabla \cdot \boldsymbol{j} = 0, \ \nabla \equiv (\partial/\partial x_1, \ldots, \partial/\partial x_3) \equiv (\partial/\partial x, \partial/\partial y, \partial/\partial z)$$

(5.7)

$$\rho = \rho(\boldsymbol{r}, t), \ \boldsymbol{j} = \boldsymbol{j}(\boldsymbol{r}, t),$$

with the current/area $\boldsymbol{j}$ and the charge/volume ρ as the sources.

As was mentioned in Secs. 3.5.7 and 3.6.3, the information K should obey gauge- and coordinate covariance. Use of the (covariant) Lorentz condition

$$\frac{1}{c} \frac{\partial q_4}{\partial t} + \sum_{n=1}^{3} \frac{\partial q_n}{\partial x_n} = 0,$$ (5.8)

helps to achieve these aims. Eq. (5.8) may be regarded as an auxiliary condition that is supplemental to the invariance principle (5.7). Alternatively, since (5.8) is itself the mathematical expression of an equation of continuity of flow, we can regard the problem to be physically defined by a single *joint* condition of continuity of flow, (5.7) and (5.8).

5.1.5 Fixing N = 4

The Lorentz condition (5.8) is the only physical input into EPI that functionally involves the unknown amplitudes $\mathbf{q}$. Since the highest index n that enters into Eq. (5.8) is value 4, we take this to mean that the sufficient number of amplitudes needed to express the theory is $N = 4$. This step could have been taken at the end of the derivation, but is more convenient to fix here once and for all.

5.1.6 On finding the bound information J

In the quantum mechanical scenario of Chap. 4, the invariance principle in use permitted J to be directly expressed as I (see Eq. (4.9)). Then we immediately knew that $\kappa = 1$ and were able to solve (3.16) and (3.18) for their distinct solutions.

By comparison, principles (5.7) and (5.8) do not directly express I in terms of a physically dependent information J. Therefore, at this point we do not know either J or the efficiency constant κ. They have to be solved for, based upon use of the conditions (5.7), (5.8). This corresponds to a case (a) described in Sec. 3.4.6, where both EPI conditions (3.16), (3.18) must be solved *simultaneously* for a common solution $\mathbf{q}$. We do this below.

5.1.7 General form for J

The bound information J is a scalar *functional* of all physical aspects of the problem, i.e. quantities $\mathbf{q}(\mathbf{x})$, $j(\mathbf{x})$ and $\rho(\mathbf{x})$ at all $\mathbf{x}$. It can be represented generally as an inner product

$$J = 4c \int\int dr\, dt \sum_{n=1}^{4} E_n J_n, \quad E_n = const_n, \quad J_n = J_n(\mathbf{q}, j, \rho). \quad (5.9)$$

The constants E_n and functions $J_n(\mathbf{q}, j, \rho)$ are to be found.

5.1.8 EPI variational solution

As mentioned in the plan above, we first find the solution for $\mathbf{q}$ to problem (3.16). Subtracting Eq. (5.9) from (5.6) gives the information Lagrangian

$$\mathscr{L} = 4c \sum_n \left[\nabla q_n \cdot \nabla q_n - \frac{1}{c^2} \left(\frac{\partial q_n}{\partial t} \right)^2 - E_n J_n(\mathbf{q}, j, \rho) \right]. \quad (5.10)$$

This is used in the Euler–Lagrange Eq. (0.34) for the problem,

$$\sum_{k=1}^{3} \frac{d}{dx_k} \left(\frac{\partial \mathscr{L}}{\partial q_{nk}} \right) + \frac{1}{c} \frac{d}{dt} \left(\frac{\partial \mathscr{L}}{\partial q_{n4}} \right) = \frac{\partial \mathscr{L}}{\partial q_n},$$

$$n = 1, \ldots, 4; \quad q_{nk} \equiv \frac{\partial q_n}{\partial x_k}, \quad q_{n4} \equiv \frac{\partial q_n}{\partial t}.$$

$$(5.11)$$

The result is directly

$$\Box q_n = -\frac{1}{2} \sum_m E_m \frac{\partial J_m}{\partial q_n}, \tag{5.12a}$$

$$\text{where } \Box \equiv \nabla^2 - \frac{1}{c^2} \frac{\partial^2}{\partial t^2}, \quad \nabla^2 \equiv \sum_{k=1}^{3} \frac{\partial^2}{\partial x_k^2}. \tag{5.12b}$$

$\Box$ is called the d'Alembertian operator and ∇^2 is called the Laplacian operator.

5.1.9 *Alternative form for I*

The expression (5.6) for *I* may be integrated by parts. First note the elementary result

$$\int dx \left(\frac{dq}{dx}\right)^2 = \frac{dq}{dx} q \bigg|_{-\infty}^{\infty} - \int dx q \frac{d^2q}{dx^2} = 0 - \int dx q \frac{d^2q}{dx^2} \tag{5.13}$$

for a probability amplitude q. The zero occurs because we will assume Dirichlet or Neumann conditions to be obeyed by the potential **A** at the boundaries to the observation space (Eq. (5.55)); and each component of **A** will be made proportional to a corresponding component of **q** (Eq. (5.48)). Integrating by parts in this manner for every coordinate (r, t), information (5.6) becomes

$$I = -4c \int\int dr\, dt \sum_n q_n \Box q_n. \tag{5.14}$$

This is used in the following.

5.1.10 *EPI zero-root solution*

The second EPI problem is to find the roots **q** of (3.18),
$$I[\mathbf{q}] - \kappa J[\mathbf{q}] = 0, \tag{5.15}$$
I given by (5.14) and *J* given by (5.9). The problem is thus

$$-4c \int\int dr\, dt \sum_n (q_n \Box q_n + \kappa E_n J_n(\mathbf{q}, \mathbf{j}, \rho)) = 0. \tag{5.16}$$

Its solution is the microscale Eq. (3.20), which is here
$$q_n \Box q_n = -\kappa E_n J_n. \tag{5.17}$$

5.1.11 *Common solution q*

From solutions (5.12a) and (5.17), it must be that

$$\tfrac{1}{2} q_n \sum_m E_m \frac{\partial J_m}{\partial q_n} = \kappa E_n J_n. \tag{5.18}$$

This allows us to specify more precisely the form for unknown functions $J_n(\mathbf{q}, \boldsymbol{j}, \rho)$, as follows.

Notice that the right-hand side of (5.18) has been 'sifted' out of the sum on the left-hand side. Then it must be that

$$\tfrac{1}{2} q_n \frac{\partial J_m}{\partial q_n} = \kappa J_m \delta_{mn}, \tag{5.19}$$

where δ_{mn} is the Kronecker delta function. Now consider two cases.

Case 1: $m \neq n$. Eq. (5.19) gives $\partial J_m/\partial q_n = 0$. This implies that

$$J_m(\mathbf{q}, \boldsymbol{j}, \rho) = J_m(q_m, \boldsymbol{j}, \rho). \tag{5.20}$$

Case 2: $m = n$. Eq. (5.19) gives $q_n \partial J_n/\partial q_n = 2\kappa J_n$. But since, by (5.20), the $\mathbf{q}$ dependence of each J_n is only through the single q_n, it follows that $\partial J_n/\partial q_n = dJ_n/dq_n$, the full derivative. Using this in Eq. (5.19) allows the latter to be integrated,

$$\ln J_n = 2\kappa \ln q_n + D_n(\boldsymbol{j}, \rho), \quad \text{or} \quad J_n = q_n^{2\kappa} G_n(\boldsymbol{j}, \rho). \tag{5.21}$$

Quantities D_n and G_n are integration 'constants' after the integration in q_n, and hence can still functionally depend upon $(\boldsymbol{j}, \rho)$ as indicated.

5.1.12 Resulting wave equation

Using form (5.21) for J_n in wave equation (5.12a) gives

$$\Box q_n = -\tfrac{1}{2} E_n \cdot 2\kappa q_n^{2\kappa-1} \cdot G_n(\boldsymbol{j}, \rho), \quad \text{or} \quad \Box q_n = q_n^{b} F_n(\boldsymbol{j}, \rho), \tag{5.22a}$$

$$\text{where } F_n \equiv -\kappa E_n G_n, \ n = 1 - 4, \text{ and } b \equiv 2\kappa - 1. \tag{5.22b}$$

The parameter b and the new functions $F_n(\boldsymbol{j}, \rho)$ need to be found.

5.1.13 Where we stand so far

It is interesting to note that the information approach has allowed us to proceed as far as a wave equation, Eq. (5.22a), without the need for any more specifically electromagnetic a requirement than an unspecified dependence F_n upon the electromagnetic sources $\boldsymbol{j}$ and ρ. To evaluate this dependence will, finally, require use of the invariance principles (5.7) and (5.8). This is done in the following sections. Note that principles (5.7) and (5.8) are actually quite weak as statements of specifically electromagnetic phenomena. That their use can lead to Maxwell's equations seems remarkable. The key is to combine them with EPI, as will become apparent.

5.1.14 Use of Lorentz and conservation of flow conditions

We proceed in the following sections to find the parameter b and the functions $F_n(\boldsymbol{j}, \rho)$ in Eq. (5.22a). In this section we find a set of conditions (5.25a,b) that the unknowns $F_n(\boldsymbol{j}, \rho)$ must obey. This is accomplished through use of the two invariance conditions (5.7), (5.8).

Start by operating upon Eq. (5.22a) for $n = 1$ with $\partial/\partial x_1$. Save the resulting equation. Then operate upon Eq. (5.22a) for $n = 2$ with $\partial/\partial x_2$; *etc.*, through $n = 4$ and operation $(1/c)\partial/\partial t$. Add the resulting equations. (Of course this would all be indicated more directly, albeit with some sacrifice of simplicity, if we used tensor notation. However, for the sake of clarity we defer use of tensor notation until the chapter on general relativity, where its use is mandatory.) Because the operations $\partial/\partial x_1$, $\partial/\partial x_2$, *etc.*, commute with $\square$, the left-hand side becomes

$$\square \left(\sum_{k=1}^{3} \frac{\partial q_k}{\partial x_k} + \frac{1}{c} \frac{\partial q_4}{\partial t} \right) = 0 \tag{5.23a}$$

since the quantity within parentheses is the Lorentz form (5.8).

Then operating in the same way on the *right-hand* side of Eq. (5.22a) should likewise give zero. Operating $\partial/\partial x_1$ in this way for $n = 1$ gives

$$q_1^b \left(\sum_{m=1}^{3} \frac{\partial F_1}{\partial j_m} \frac{\partial j_m}{\partial x_1} + \frac{\partial F_1}{\partial \rho} \frac{\partial \rho}{\partial x_1} \right) \tag{5.23b}$$

plus another term which is zero if $b = 0$ (as will be the case). Or, operating in the same way for $n = 2, 3, 4$ gives analogous equations. Adding them gives

$$\sum_{n=1}^{3} q_n^b \left(\sum_{m=1}^{3} \frac{\partial F_n}{\partial j_m} \frac{\partial j_m}{\partial x_n} + \frac{\partial F_n}{\partial \rho} \frac{\partial \rho}{\partial x_n} \right) + q_4^b \left(\sum_{m=1}^{3} \frac{\partial F_4}{\partial j_m} \frac{1}{c} \frac{\partial j_m}{\partial t} + \frac{\partial F_4}{\partial \rho} \frac{1}{c} \frac{\partial \rho}{\partial t} \right) \equiv S \equiv 0 \tag{5.24}$$

by our requirement.

We show next that this is satisfied if

$$q_n^b \frac{\partial F_n}{\partial j_m} = B\delta_{mn}, \quad q_n^b \frac{\partial F_n}{\partial \rho} = 0; \quad m, n = 1, 2, 3; \tag{5.25a}$$

$$q_4^b \frac{\partial F_4}{\partial j_m} = 0, \quad m = 1, 2, 3; \quad q_4^b \frac{\partial F_4}{\partial \rho} = cB; \tag{5.25b}$$

$$\text{where } B \equiv B(\boldsymbol{j}, \rho), \tag{5.25c}$$

a new *scalar* function of the sources. Substituting the identities (5.25a) into (5.24) and using the sifting property of the Kronecker delta makes the first sum over n collapse to

$$B \sum_{n=1}^{3} \frac{\partial j_n}{\partial x_n} \equiv B \nabla \cdot \boldsymbol{j}. \tag{5.26}$$

Operator ∇ is the usual 3-D gradient operator (5.7). Similarly, substituting the identities (5.25b) makes the second sum collapse to

$$B \frac{\partial \rho}{\partial t}. \tag{5.27}$$

Hence, the total sum S in Eq. (5.24) becomes

$$S = B \left(\nabla \cdot \boldsymbol{j} + \frac{\partial \rho}{\partial t} \right) \equiv 0 \tag{5.28}$$

by the equation of flow (5.7). This is what we set out to show. Therefore, Eq. (5.24) is satisfied by the solution Eqs. (5.25a–c).

5.1.15 Finding exponent b

Condition (5.25a) permits us to find the exponent b. Since $F_n = F_n(\boldsymbol{j}, \rho)$, the left-hand side of the first relation (5.25a) only varies in $\boldsymbol{q}$ as q_n^b. This means that the right-hand side's dependence upon $\boldsymbol{q}$ is most generally of this form as well. But by Eq. (5.25c) function B does not depend upon the $\boldsymbol{q}$. Hence, the only way the $\boldsymbol{q}$-dependence in the first relation (5.25a) could balance is by having

$$b = 0. \tag{5.29}$$

5.1.16 Finding function B(j, ρ)

This unknown function may be found from requirements (5.25a) *under the condition* (5.29). The second requirement (5.25a) implies that

$$F_n(\boldsymbol{j}, \rho) = F_n(\boldsymbol{j}), \ n = 1, 2, 3. \tag{5.30a}$$

The *first* requirement (5.25a) for $m \neq n$ shows, using Eq. (5.29), that

$$F_n(\boldsymbol{j}) = F_n(j_n). \tag{5.30b}$$

Then, the first requirement (5.25a) gives, for $m = n$,

$$\frac{\partial F_n(j_n)}{\partial j_n} = B(\boldsymbol{j}, \rho), \ n = 1, 2, 3. \tag{5.31}$$

Since the left-hand side doesn't depend upon ρ, necessarily

$$B(\boldsymbol{j}, \rho) = B(\boldsymbol{j}). \tag{5.32}$$

On the other hand, the first Eq. (5.25b) implies that $F_4(\boldsymbol{j}, \rho) = F_4(\rho)$ alone. Then the left-hand side of the *second* Eq. (5.25b) has no $\boldsymbol{j}$-dependence, implying that on the right-hand side $B(\boldsymbol{j}, \rho) = B(\rho)$ alone.

The only way that the latter and Eq. (5.32) can simultaneously be true is if

$$B(j, \rho) = B \equiv Const. \tag{5.33}$$

5.1.17 Finding functions $F_n(j, \rho)$

Since $b = 0$, the first condition (5.25b) shows that

$$F_4(j, \rho) = F_4(\rho). \tag{5.34}$$

Then the second condition (5.25b), combined with results (5.29) and (5.33), becomes a simple differential equation

$$\frac{dF_4(\rho)}{d\rho} = cB = Const. \tag{5.35}$$

This has the elementary solution

$$F_4(\rho) = cB\rho + C_4, \; C_4 = Const. \tag{5.36}$$

Likewise, the first Eq. (5.25a) becomes, for index $m = n$ and using Eqs. (5.29) and (5.30a,b), a simple differential equation

$$\frac{dF_n(j_n)}{dj_n} = B = Const., \; n = 1, 2, 3. \tag{5.37}$$

The solution is

$$F_n(j_n) = Bj_n + C_n, \; C_n = Const., \; n = 1, 2, 3. \tag{5.38}$$

In summary, all functions F_n are linear in their corresponding currents or charge density. Constants B, C_n remain to be found.

5.1.18 Efficiency parameter κ

By Eqs. (5.22b) and (5.29), for this application of EPI the efficiency parameter

$$\kappa = 1/2. \tag{5.39}$$

The precise meaning of this effect is still uncertain. However, by the reasoning of Sec. 3.4.5, the implication is that classical electromagnetics is an approximation: only 50% of the bound or phenomenological information J is utilized in the intrinsic information I. This, of course, agrees with the fact that the classical approach *is* an approximation. An EPI approach that allowed for quantization of the field might yield a value of $\kappa = 1$ as, e.g., we obtained in the quantum approaches of Chap. 4.

5.1.19 Constants C from photon case

Substitution of results (5.29), (5.36) and (5.38) into the wave equation (5.22a) gives

$$\Box\mathbf{q} = B\mathbf{J_s} + \mathbf{C}, \quad \mathbf{q} \equiv (q_1, \ldots, q_4)$$

$$\mathbf{C} \equiv (C_1, \ldots, C_4), \quad \mathbf{J_s} \equiv (\mathbf{j}, c\rho). \tag{5.40}$$

All quantities are four-component vectors as indicated (but not relativistic four-vectors; see below Eq. (5.1)). This equation would take the form of the classical vector wave equation for the electromagnetic potentials $(A_1, A_2, A_3, \phi) \equiv \mathbf{A}$ if they were substituted for the amplitudes $\mathbf{q}$ (much more on this in Sec. 5.1.21 *et seq.*). However, the vector wave equation does not have an additive constant $\mathbf{C}$ on its right-hand side. In fact, $\mathbf{C}$ must be $\mathbf{0}$, as is shown in the following.

We have not yet defined the physical origin of fluctuations $\mathbf{x}$. By comparison with Eq. (5.40), consider the Klein–Gordon equation (4.28) for a particle with mass $m = 0$ and charge $e = 0$,

$$\Box\psi = 0, \quad \psi = Q_1 + iQ_2, \tag{5.41}$$

where Q_1, Q_2 are real amplitude functions. This is a d'Alembert's equation. It can be taken to describe the probability amplitude ψ for the electromagnetic quantum (Eisberg, 1961, p. 697), i.e., the photon. This is in the following sense.

It had been conjectured (see, e.g. Akhiezer and Berestetskii, 1965, pp. 10, 11) that a wave function $\psi(\mathbf{r}, t)$ for localization of a photon does not exist. However, it was recently shown (Bialynicki-Birula, 1996) that a wave function obeying Eq. (5.41) represents the local occurrence of photon energy over finite, but small (the order of a wavelength) regions of space. Thus, ψ represents the probability amplitude for a 'coarse-grained' (Cook, 1982) space of possible photon locations. See also the related work (Sipe, 1995) and (Deutsch and Garrison, 1991). It is interesting to note, as well, the corresponding *particle* case-limitation mentioned in Sec. 4.1.17, whereby particle localization can be no finer than the Compton length. Hence, in practice, both photons and particles can only be defined over coarse-grained spaces.

Compare Eq. (5.41) with Eq. (5.40) *under the same charge-free* conditions $\mathbf{J_s} = 0$,

$$\Box\mathbf{q} = \mathbf{C}. \tag{5.42}$$

From their common origin as electromagnetic phenomena, Eqs. (5.41) and (5.42) must be describing the same physical effect: by (5.41) a propagating photon. We see that the two equations are equivalent if $\mathbf{q}$ is linear in the components Q_1, Q_2 of ψ and if $\mathbf{C} = \mathbf{0}$,

$$q_n = K_{n1}Q_1 + K_{n2}Q_2, \; n = 1 - 4, \text{ and } \mathbf{C} = \mathbf{0}. \qquad (5.43)$$

This confirms our assertion that $\mathbf{C} = \mathbf{0}$. Note that if $\mathbf{C}$ were not zero, then it would represent an artificial electromagnetic source in (5.42) – one that exists in the absence of real sources $\mathbf{J_s}$. Such a source has not been observed on the macroscopic level.

5.1.20 *Plane wave solutions*

We can conclude, then, that in the absence of real sources the fluctuations $\mathbf{x}$ in the measured field positions are those in the (coarse-grained) positions of photons. These trace from the d'Alembert Eq. (5.41) governing photon positions. A solution ψ to (5.41) in the case of definite energy E and momentum $\boldsymbol{\mu}$ is in the form of a plane wave

$$\psi = \exp\left[\frac{i}{\hbar}(\boldsymbol{\mu} \cdot \mathbf{r} - Et)\right], \; E = c\mu. \qquad (5.44)$$

Assume the wave to be travelling along the x direction. Then (5.44) shows periodic behavior in x with a wavelength

$$\Delta x = \frac{2\pi\hbar}{\mu}. \qquad (5.45)$$

The equivalent classical oscillator of mass m_0 has an energy

$$E_0 = \frac{\mu^2}{2m_0} \equiv \hbar\omega_0, \text{ so that } \mu = (2\hbar m_0 \omega_0)^{1/2}. \qquad (5.46)$$

Using the latter in Eq. (5.45) gives an uncertainty

$$\Delta x = \pi\sqrt{2}(\hbar/m_0\omega_0)^{1/2}. \qquad (5.47)$$

This confirms Eq. (5.4) for the expected vacuum fluctuation.

5.1.21 *A correspondence between probability amplitudes and potentials*

We found in Sec. 5.1.19 that $\mathbf{C} = \mathbf{0}$. Then the EPI output Eq. (5.40) takes the form of a wave equation in the amplitudes $\mathbf{q}$ due to an *electromagnetic source* $\mathbf{J_s}$. This and the photon plane wave correspondences of Secs. 5.1.19 and 5.1.20 imply that the amplitudes $\mathbf{q}$ in (5.40) are *linear* in the electromagnetic potentials (A_1, A_2, A_3, ϕ),

$$q_n \equiv aA_n, \; n = 1, 2, 3, \; q_4 \equiv a\phi, \; a = Const. \qquad (5.48)$$

Using $\mathbf{C} = \mathbf{0}$ and the latter in Eq. (5.40) gives

$$\Box \mathbf{A} = B\mathbf{J_s}. \tag{5.49}$$

We absorbed the constant a into the constant B. This is the vector wave equation with an, as yet, unspecified constant multiplier B.

Regarding uniqueness, note that we could have added an arbitrary function $\mathbf{f}$ to the right-hand sides of definitions (5.48). Provided $\Box \mathbf{f} = 0$, Eq. (5.49) would still result. However, for purposes of defining $\mathbf{q}$ uniquely, we do not do this. Function $\mathbf{f}$ amounts to an arbitrary choice of gauge function for potential $\mathbf{A}$. We later show that the solution $\mathbf{A}$ to Eq. (5.49) obeys Eq. (5.57), where it is uniquely defined by given sources $\mathbf{J_s}$. By Eq. (5.48), $\mathbf{q}$ is then unique.

5.1.22 The constant B

Here we find the value of the constant in Eq. (5.49). Since B does not depend upon either the sources $\mathbf{J_s}$ or the potentials $\mathbf{A}$ it must be a constant. By Sec. 3.4.14 it must also be a *universal* constant. Next use dimensional analysis. From the known units for fields $\boldsymbol{B}$ and $\boldsymbol{E}$, by Eqs. (5.85) and (5.86) those for $\mathbf{A}$ are charge/length. Then the units of $\Box A$ are charge/length^3. Also, those of $\mathbf{J_s}$ are charge/length^2-time. Hence, to balance units in (5.49) B must have units of time/length. Then the universal constant is the reciprocal of a velocity. From the correspondence Eq. (5.48), this must be the velocity of propagation of electromagnetic disturbances, i.e., c. With the particular use of m.k.s. units as well, we then get

$$B = -\frac{4\pi}{c}. \tag{5.50}$$

5.1.23 Vector wave equation

Then Eq. (5.49) becomes

$$\Box \mathbf{A} = -\frac{4\pi}{c} \mathbf{J_s}. \tag{5.51}$$

This is the vector wave equation in the Lorentz gauge. Its derivation was the main goal of this chapter.

It is more usual to express Eq. (5.51) in terms of four-vectors rather than the four-component vectors $\mathbf{A}$ and $\mathbf{J_s}$ that we use. If the fourth component of Eq. (5.51) is merely multiplied by the imaginary number i then (5.51) takes the usual form.

5.1.24 Probability law on fluctuations

By the use of Eqs. (5.5) and (5.48) in Eq. (2.23), we get

$$p(\mathbf{x}) = a^2 \sum_{n=1}^{4} A_n^2, \ p(\mathbf{x}) \geq 0 \tag{5.52}$$

as a PDF on the position errors $\mathbf{x}$. A factor of $1/4$ is absorbed into the constant a. The important property of positivity for the PDF is noted. To be useful, a PDF should be uniquely defined by given physical conditions. Also, it should be normalizable or, if not, at least useable for computing averages. We examine these questions, in turn.

5.1.25 Uniqueness property

Is the PDF $p(\mathbf{x})$ defined by Eq. (5.52) unique? By this equation, the uniqueness of p depends upon the uniqueness of $\mathbf{A}$ (up to a $\pm$ sign). The latter is defined by solution $\mathbf{A}$ to Eq. (5.51). It is known that solutions to this problem under stated boundary value conditions and initial conditions are unique (Morse and Feshbach, 1953, p. 834). Typical boundary conditions are as shown in Fig. 5.1.

Suppose that the solution $\mathbf{A}(r, t)$ is to be determined at field positions r that are within a volume V that contains the source $\mathbf{J_s}$. Field- and source-coordinates are denoted as (r, t) and (r', t'), respectively. Volume V is enclosed by a given surface σ. Denote field positions r lying within V as $r \subset V$. We consider cases of both finite and infinite V. The solution is given by (Morse and Feshbach, 1953, p. 837)

$$\mathbf{A}(r, t)_{r \subset V} = \mathbf{A}_0(r, t) + \frac{1}{c} \int_0^t dt' \int_V dr' \, G\mathbf{J_s}(r', t')$$

$$+ \frac{1}{4\pi} \int_0^{t^+} dt' \oint d\boldsymbol{\sigma} \cdot (G \nabla \mathbf{A} - \mathbf{A} \nabla G), \ G \equiv G(r, t; r', t'),$$

$$\tag{5.53a}$$

where

$$\mathbf{A}_0(r, t) = -\frac{1}{4\pi c^2} \int_V dr' \left[\left(\frac{\partial G}{\partial t'} \right) \mathbf{A}(r', t') - G \frac{\partial \mathbf{A}(r', t')}{\partial t'} \right]_{t'=0}. \tag{5.53b}$$

In Eq. (5.53a), the indicated gradients $\nabla \mathbf{A}$ and ∇G are with respect to the primed coordinates (r', t'). Also, the notation $\oint d\boldsymbol{\sigma}$ denotes an integral over the surface $\boldsymbol{\sigma}$ enclosing the source, and $d\boldsymbol{\sigma}$ is a vector whose direction is the outward-pointing normal to the local element $d\boldsymbol{\sigma}$ of the surface. Finally, the new function G is called a 'Green's function'. It is the scalar solution to

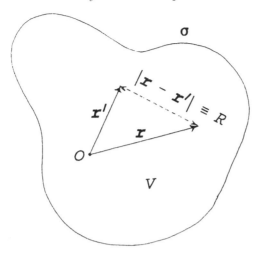

Fig. 5.1. Geometry for the electromagnetic problem. Boundary surface σ encloses volume V. With an origin at point O, position r locates a field point, and r' locates a source point, within V. Quantities r and r' are also random errors in position measurements, by the EPI model.

Eq. (5.51) with the vector source $\mathbf{J}_s$ replaced by a scalar impulse source $\delta(r - r')\delta(t - t')$.

We next show that, due to the isolation requirement (Secs. 1.2.1, 1.3.2, 1.8.5, 3.2) of EPI, under typical electromagnetic boundary conditions only the second right-hand term in Eq. (5.53a) contributes to $\mathbf{A}$.

The integral (5.53b) represents the contribution of initial conditions in $\mathbf{A}$ and G to $\mathbf{A}$ as evaluated at later times t. The source is assumed to be turned 'on' at a time $t' = t_1 \gg 0$ and 'off' at a time $t' = t_2 > t_1$. Thus, the integral (5.53b) is evaluated at a time $t' = 0$ in the *infinitely distant past* to t. Since, by correspondence (5.48) $\mathbf{A}$ represents a probability amplitude, by the isolation requirement of EPI it must be that $\mathbf{A}(r', 0) = 0$: the probability for the infinitely past event vanishes. This is isolation in the temporal sense. Also, from the viewpoint of normalization, a PDF must go to zero as its argument approaches $\pm\infty$. So must its first derivative. The upshot is that both $\mathbf{A}(r', t')$ and $\partial\mathbf{A}(r', t')/\partial t'$ are zero at $t' = 0$, and Eq. (5.53b) obeys

$$\mathbf{A}_0(r, t) = 0. \tag{5.54}$$

The initial potential is zero for this problem.

We next turn to evaluation of the third right-hand term in Eq. (5.53a), a surface integral. Typically, the potential $\mathbf{A}$ or its first derivatives $\nabla\mathbf{A}$ are zero on the surface,

$$\mathbf{A}\bigg|_{\sigma} = 0 \text{ or } \nabla\mathbf{A}\bigg|_{\sigma} = 0. \tag{5.55}$$

These are called zero 'Dirichlet' or 'Neumann' boundary conditions, respectively. We can choose the Green's function to obey the same boundary condition as does the potential function, so that

$$G\bigg|_{\sigma} = 0 \text{ or } \nabla G\bigg|_{\sigma} = 0. \tag{5.56}$$

Substituting Eqs. (5.54)–(5.56) into Eq. (5.53a) gives

$$\mathbf{A}(\mathbf{r}, t)_{r \subset V} = \frac{1}{c}\int_{0}^{t} dt' \int_{V} d\mathbf{r}' G(\mathbf{r}, t; \mathbf{r}', t')\mathbf{J}_{s}(\mathbf{r}', t'). \tag{5.57}$$

This is what we set out to show.

The integral (5.57) is well-defined for both continuous and impulsive (point, line or surface) sources. The sifting property (0.69) of the delta function would merely turn the right-hand integral in (5.57) into a sum over the sources.

We next turn to evaluation of the right-hand integral. As it turns out, this depends upon whether the bounding surface σ is at a finite or infinite distance from the source. The former is a 'closed space problem'; the latter is an 'open space problem'.

Consider the closed space problem. In view of the limited time interval $t_1 \le t' \le t_2$ for which the source is 'on' (i.e., is non-zero), and the fact that the output potential $\mathbf{A}$ is proportional to the source $\mathbf{J}_s$, potential $\mathbf{A}$ is effectively zero outside some time interval. Hence, we limit our attention to time fluctuation values t obeying $\bar{t}_1 \le t \le \bar{t}_2$. Denote these as $t \subset T$. Since the space of positions $\mathbf{r}$ is closed as well (Eq. (5.53a)), we see that we have a four-dimensional closed space problem

$$\mathbf{r} \subset V, t \subset T. \tag{5.58}$$

From Eq. (5.57) we see that the solution $\mathbf{A}$ is defined by the Green's function G for the problem. For the boundary conditions mentioned above, the solution is known to be (Morse and Feshbach, 1953, p. 850)

$$G(\mathbf{r}, t; \mathbf{r}', t') = 4\pi c^2 \sum_{n} \frac{\sin[\omega_n(t - t')]}{\omega_n} u(t - t')\psi_n^*(\mathbf{r}')\psi_n(\mathbf{r}), \tag{5.59}$$

where $u(t - t') = 1$ for $t > t'$ or 0 for $t < t'$ (the Heaviside function). Also, $\omega_n \equiv ck_n$ where the k_n are eigenvalues, and the functions $\psi_n(\mathbf{r})$ are eigenfunctions, of an associated problem. The ψ_n and k_n are to (a) satisfy the Helmholtz equation

$$\nabla^2\psi_n + k_n^2\psi_n = 0, \psi_n \equiv \psi_n(\mathbf{r}), \tag{5.60}$$

within the volume V, subject to the same Dirichlet or Neumann zero conditions on the surface σ as $\mathbf{A}$, and (b) form an orthonormal set within the volume V,

$$\int_V d\mathbf{r}\, \psi_n^* \psi_m = \delta_{nm}. \tag{5.61}$$

Obviously the shape of the surface σ is decisive in defining the orthonormal functions $\psi_n(\mathbf{r})$ of the problem.

Substituting Eq. (5.59) into Eq. (5.57) gives the unique solution to the closed space problem. We next briefly turn to the open space problem.

In the open space problem, the boundary σ recedes toward infinity. As before assume the source $\mathbf{J_s}$ to extend over a limited space – call it V_0 – and to be 'on' for a limited amount of time. Assume the same initial conditions as before, so that Eq. (5.54) holds. Also, the surface integral in Eq. (5.53a) vanishes once again, this time because of the isolation requirement of EPI (Secs. 1.2.1, 1.8.5, 3.2, etc.). No spatial 'events' $\mathbf{r} \to \infty$ are allowed so that, in its role as a *probability* amplitude, $\mathbf{A}$ must obey

$$\lim_{r \to \infty} \mathbf{A}(\mathbf{r},\, t) = 0, \quad \lim_{r \to \infty} \nabla \mathbf{A}(\mathbf{r},\, t) = 0. \tag{5.62}$$

That is, there is zero probability for such events.

Hence, the solution for $\mathbf{A}$ obeys Eq. (5.57) once more. This requires knowledge of the open space Green's function G. Assume a situation of 'causality', that is, that the effect of a source located a finite distance $|\mathbf{r} - \mathbf{r}'|$ away that is turned 'on' at a time t' be felt a *later* time $t > t'$. Then the Green's function is (Morse and Feshbach, 1953, p. 838)

$$G(\mathbf{r},\, t;\, \mathbf{r}',\, t') = \frac{\delta[(R/c) - (t - t')]}{R}, \quad R \equiv |\mathbf{r} - \mathbf{r}'|, \tag{5.63}$$

where $t > t'$. Substituting this into Eq. (5.57) gives

$$\mathbf{A}(\mathbf{r},\, t) = \frac{1}{4\pi c} \int_{V_0} d\mathbf{r}' \left[\frac{\mathbf{J_s}(\mathbf{r}',\, t - R/c)}{R} \right]. \tag{5.64}$$

This is the well-known 'retarded potential' solution, whereby the potential at a time t arises from source values that occurred at an earlier time $t - R/c$. This, in turn, follows from the fact that the velocity of propagation of the disturbance is the finite value c. The solution for $\mathbf{A}$ is once more unique.

It may be noted that the solution (5.63) for the Green's function of the open space problem only depends upon the difference of all coordinates, i.e., $(\mathbf{r} - \mathbf{r}')$, $(t - t')$. The resulting Eq. (5.57) is then in a covariant form (Jackson, 1975, p. 611). This agrees with Einstein's precept that the laws of physics should be covariant (Sec. 3.5.8). However, the Green's function for the closed space problem is Eq. (5.59), and this does not seem to be expressible as a function of differences of coordinates for *arbitrary* surface shapes σ. Hence, only certain closed space scenarios, and the open space scenario, admit of covariant solutions. This seems, at first, curious since the progenitor of these solutions – the wave Eq. (5.51) – *is* a covariant law. It also leads us to ask whether the potential **A** and its wave equation are lacking in some way as valid physical quantities.

In fact, other well-known covariant laws suffer the same fate. For example, the Klein–Gordon Eq. (4.28) of relativistic quantum mechanics is covariant. This agrees with Einstein's precept. But, the *particular solution* for which the particle is in a definite state of rest energy has a separate time factor $\exp(-imc^2t/\hbar)$, as in Eq. (G3). Such a solution is not covariant: it ultimately gives rise to the non-relativistic, non-covariant Schroedinger wave equation (Appendix G).

One concludes that the laws of physics (electromagnetic wave equation, Klein–Gordon equation, etc.) must be covariant, as Einstein required, but their solutions under particular conditions need not be. If one knows a non-covariant solution **A** in one frame of reference and wants it in another, this can be implemented by use of the Lorentz transformation rules (3.29), (3.33).

5.1.26 *On covariant boundary conditions*

We assumed above Eq. (5.54) that both $\mathbf{A}(r, t)$ and $\partial\mathbf{A}(r, t)/\partial t$ are zero at the initial time $t = 0$. These are 'Cauchy conditions' on the time coordinate. By contrast, we assumed either Dirichlet or Neumann conditions on the space coordinates (see below Eq. (5.54)). Hence, the combined space-time boundary conditions are not covariant, since they treat space and time differently. As we discussed above, the result is a non-covariant form (5.59) for the resulting Green's function for the problem. This led to a non-covariant solution **A**. It is interesting to see what happens if we depart from Cauchy conditions on the time, utilizing instead the *same* Dirichlet or Neumann conditions that were satisfied by the space coordinates. The resulting boundary conditions would now be covariant. However, the resulting solution **A** under such conditions is not unique! This is shown in Appendix H.

Now, as we noted above Eq. (5.54), the Cauchy conditions on the time are actually *consistent with* the EPI assumption of an isolated system. Hence, in the EPI formulation, particular solutions to the wave equation cannot be covariant (although the wave equation itself, as derived by EPI in preceding sections, *is covariant*). The non-covariance problem can be avoided, as we saw, by use of the Lorentz transformation.

Does, then, the EPI approach violate the postulates of relativity? As we noted above, the Einstein requirement of covariance applies to laws of physics

– here, the wave equation – and not to particular solutions of them. On this level, EPI does not violate the postulates of relativity.

5.1.27 Normalization for a closed space

Normalization requires

$$\int_T\int_V dt\, d\mathbf{r}\, p(\mathbf{r},\, t) \propto \int_T\int_V dt\, d\mathbf{r}\, |\mathbf{A}|^2 = const., \qquad (5.65)$$

based upon representation (5.52) for p. We consider cases of finite and, then, infinite volume V for the position coordinate, i.e., closed and open spaces. The time coordinate t is always within a closed temporal space T (see Eq. (5.58)), as in quantum mechanics (Sec. 4.1.26).

For the closed space case, we first establish an identity. The Green's function G in the expansion Eq. (5.59) for G is real, so that the eigenfunctions $\psi_n(\mathbf{r})$ in that expansion can be chosen to be purely real. This is equivalent to representing a real function by a series of sin and cos functions, rather than by complex exponentials. The choice is arbitrary. Using Eq. (5.59), the orthonormality property (5.61) and the sifting property (0.62b) of the Kronecker delta function δ_{mn}, gives

$$\int_V d\mathbf{r}\, G(\mathbf{r},\, t;\, \mathbf{r}',\, t')G(\mathbf{r},\, t;\, \mathbf{r}'',\, t'') = \sum_n \int_{t_1}^{t_2} dt\, \frac{\sin[\omega_n(t-t')]\sin[\omega_n(t-t'')]}{\omega_n^2}$$

$$\times u(t-t')u(t-t'')\psi_n(\mathbf{r}')\psi_n(\mathbf{r}'').$$

$$(5.66)$$

An unimportant multiplier has been ignored. This is the identity we sought.

For simplicity of presentation we find the normalization integral (5.65) for *any one component* A_m^2 of $|\mathbf{A}|^2$, with subscript m suppressed for convenience. This component arises from the corresponding component J_s of $\mathbf{J_s}$ in Eq. (5.57). Using the latter and identity (5.66) in Eq. (5.65) gives, after rearranging orders of integration,

$$\int_T\int_V dt\, d\mathbf{r}\, p(\mathbf{r},\, t) =$$

$$\sum_n \omega_n^{-2} \int_{t_1}^{t_2} dt\, \left[\int_0^t dt' \int_V d\mathbf{r}'\, \sin[\omega_n(t-t')]J_s(\mathbf{r}',\, t')\psi_n(\mathbf{r}') \right]^2. \qquad (5.67)$$

The orthonormality property (5.61) was also used, and the Heaviside functions u were replaced by unity because their arguments are only positive here.

At this point, the analysis branches, depending upon whether the source component J_s contains singularities; this includes poles or impulsive lines (of any shape) or surfaces. Suppose, first that there are no singularities. Then use of the Schwarz inequality on the right-hand side of Eq. (5.67) gives

$$\int_T dt \int_V d\mathbf{r}\, p(\mathbf{r},\, t) \leq \sum_n \omega_n^{-2} \int_{t_1}^{t_2} dt\, U_n(\omega_n,\, t) W_n(t), \text{ where} \qquad (5.68a)$$

$$U_n(\omega_n,\, t) \equiv \int_0^t dt' \int_V d\mathbf{r}'\, \sin^2\left[\omega_n(t - t')\right] J_s^2(\mathbf{r}',\, t') \text{ and} \qquad (5.68b)$$

$$W_n(t) \equiv \int_0^t dt' \int_V d\mathbf{r}'\, \psi_n^2(\mathbf{r}'). \qquad (5.68c)$$

Factor $W_n(t)$ may be directly evaluated using orthonormality property (5.61), as

$$W_n(t) = t. \qquad (5.69)$$

Also, we may again apply the Schwarz inequality, now to Eq. (5.68b), giving

$$U_n(\omega_n,\, t) \leq \int_0^t dt' \int_V d\mathbf{r}'\, \sin^4\left[\omega_n(t - t')\right] \int_0^t dt' \int_V d\mathbf{r}'\, J_s^4(\mathbf{r}',\, t'). \qquad (5.70)$$

The first double integral on the right-hand side may be evaluated analytically. The second double integral – call it $J(t)$ – is finite since volume V is finite and function J_s is assumed to be non-singular. Substituting these results and Eq. (5.69) into Eq. (5.68a) gives

$$\int_T dt \int_V d\mathbf{r}\, p(\mathbf{r},\, t) \leq V \sum_n \frac{1}{\omega_n^2} \int_{t_1}^{t_2} dt\, t^2 [\tfrac{3}{8} - \tfrac{1}{2}\operatorname{sinc}(2\omega_n t) + \tfrac{1}{8}\operatorname{sinc}(4\omega_n t)] J(t).$$

$$(5.71)$$

It is useful to examine convergence properties in the limit of ω_n large. In this limit the sinc functions contribute effectively zero to Eq. (5.71), and we get

$$\int_T dt \int_V d\mathbf{r}\, p(\mathbf{r},\, t) \leq \tfrac{3}{8} V \left[\int_{t_1}^{t_2} dt\, t^2 J(t)\right] \sum_n \frac{1}{\omega_n^2}. \qquad (5.72)$$

The right-hand integral is finite for a well-behaved function $J(t)$ over the given interval. Also, we had $\omega_n = ck_n$, where k_n is the nth eigenvalue. For geometries that are one-dimensional (Morse and Feshbach, 1953, p. 712) or rectangular these are proportional to n, with $n = 1, 2, \ldots$. Observing the summation

$$\sum_{n=1}^{\infty} \frac{1}{n^2} = \zeta(2) \approx 1.645, \qquad (5.73)$$

the Riemann zeta function of argument 2, we see that the normalization integral (5.72) converges for large n as the Riemann zeta function. At smaller n, the contributions to the integral now include terms $\operatorname{sinc}(2\omega_n t)$, and sinc $(4\omega_n t)$ and, so, depart from terms of the Riemann function, but are still finite.

Hence, the normalization integral for the single component A_m^2 in the sum (5.65) is finite. Since the assumed component m was arbitrary (and suppressed) in the preceding analysis, each of them gives a finite contribution, so that their total contribution is finite. Hence, the total normalization integral (5.65) is finite.

The preceding was for the case of no impulses in the source component J_s. The same kind of analysis can be gone through for impulsive sources, including a finite number of point sources, line sources (of whatever shape) and surface sources. We indicate the approach for the case of a surface source. The other cases can be treated analogously.

Suppose the equation of the surface is

$$f(x', y') = z' \text{ for } z_1 \leqslant z' \leqslant z_2 \tag{5.74}$$

where z_1, z_2 are finite because V is finite. Then the surface source function can be represented as

$$J_s(\mathbf{r}', t') = J_0(x', y', t')\delta[z' - f(x', y')] \tag{5.75}$$

where J_0 is a well-behaved function of its arguments, containing no singularities. Using this in Eq. (5.67) gives

$$\int_T \int_V dt\, d\mathbf{r}\, p(\mathbf{r}, t)$$

$$= \sum_n \omega_n^{-2} \int_{t_1}^{t_2} dt \left[\int_0^t dt' \int_A d\boldsymbol{\rho}' \sin[\omega_n(t - t')]J_0(\boldsymbol{\rho}', t')\psi_n(\boldsymbol{\rho}', f(\boldsymbol{\rho}'))\right]^2$$

$$\boldsymbol{\rho}' \equiv (x', y'),\, d\boldsymbol{\rho}' \equiv dx'\, dy',\, A \equiv (x_1 \leqslant x \leqslant x_2),\, (y_1 \leqslant y \leqslant y_2). \tag{5.76}$$

The sifting property (0.63b) of the Dirac delta function was used.

Equation (5.76) is again of the form (5.67). Then new functions U_n and W_n may be formed as in Eqs. (5.68a,b,c). U_n is bounded above, by essentially the same arguments as were made below Eq. (5.70), where J_0 now replaces J_s. The new W_n is

$$W_n(t) = \int_0^t dt' \int_A d\boldsymbol{\rho}'\, \psi_n^2(\boldsymbol{\rho}', f(\boldsymbol{\rho}')) = t \int_V d\mathbf{r}'\, \psi_n^2(\mathbf{r}')\delta[z' - f(x', y')]. \tag{5.77}$$

In the last step we did the integration dt' and 'anti-sifted' with the delta function. Since it obeys orthonormality property (5.61), $\psi_n^2(\mathbf{r}')$ may be regarded as a PDF on a fluctuation $\mathbf{r}'$. This is consistent, as well, with the EPI interpretation that a source 'position' $\mathbf{r}'$ represents, in fact, a random fluctuation in position. The last integral (5.77) then represents the probability density for the event that a point r' that is randomly sampled from the given PDF will lie somewhere on the surface $f(x', y')$. For any well-behaved PDF this is a finite number. As examples, if the ψ_n are products of trigonometric functions or Bessel functions, then they obey $[\psi_n(r')]^2 \leq 1$ and, so, can only lead to a finite integral in (5.77). Once again, the normalization integral (5.65) is finite.

5.1.28 *Normalization for an open space*

The potential **A** for this case was shown to obey Eq. (5.64), the retarded potential solution. Then the normalization integral becomes

$$\int_T \int_V dt\, dr\, p(r,\, t) = \frac{1}{(4\pi c)^2} \int_{t_1}^{t_2} dt \int_{-\infty}^{\infty} dr \left[\int_{V_0} dr' \frac{J_s(r',\, t - R/c)}{R} \right]^2,$$

(5.78)

$$R \equiv |r - r'|.$$

The integration region for coordinate r is of interest. The finite integration limits (t_1, t_2) on t mean that we are detecting the potential **A** over a finite time interval. We also have assumed that the source J_s is turned 'on' for a finite time interval. Because of this latter property, and because the source extends over a finite space V_0 as well, there is a largest and a smallest value for r that will contribute non-zero values of J_s to the normalization integral. For r values outside this range, R is so large that the time factor $(t - R/c)$ for J_s lies outside the 'on' interval. Denoting the resulting integration region for r as V_1, Eq. (5.78) now reads

$$\int_T \int_V dt\, dr\, p(r,\, t) = \frac{1}{(4\pi c)^2} \int_{t_1}^{t_2} dt \int_{V_1} dr \left[\int_{V_0} dr' \frac{J_s(r',\, t - R/c)}{R} \right]^2. \quad (5.79)$$

Because of the finite integration regions all around, it is evident that finiteness for the overall integral hinges on finiteness for the squared innermost integral,

$$Q \equiv \left[\int_{V_0} dr' J_s(r',\, t - R/c) \frac{1}{R} \right]^2. \quad (5.80)$$

We wrote the integral this way to suggest use of the Schwarz inequality,

$$Q \leq \int_{V_0} dr' J_s^2 \left(r',\, t - \frac{|r - r'|}{c} \right) \int_{V_0} dr' \frac{1}{|r - r'|^2}. \quad (5.81)$$

The indicated integrals are for fixed values of t and r.

Assume that the source function J_s is 'square-integrable', i.e.,

$$\int_{V_0} dr' J_s^2\left(r', t - \frac{|r - r'|}{c}\right) \le C, \tag{5.82}$$

a finite constant, for any values of t and r. This means that J_s should be a well-behaved function, without any poles whose squared areas are unbounded. A wide class of source functions obey Eq. (5.82).

An exception to (5.82) would be impulsive sources (points, lines or surfaces), since the area under a squared delta function is infinity. The latter situation can be avoided by substitution of the impulsive sources into Eq. (5.79), performing the integration dr using the sifting property, *and then squaring* as indicated. The Schwarz inequality can then be used on the resulting *summation* (rather than integral) Q for this case. Assuming a finite number of such sources, the sums must likewise be finite.

Also, the second integral in (5.81) can be evaluated, by change of variable to $R \equiv r - r'$ (r fixed), as

$$\int_{V_0} dr' \frac{1}{|r - r'|^2} \le \int_{R_0} dR \frac{1}{R^2} = 4\pi \int_{R_1}^{R_2} dR \frac{R^2}{R^2} = 4\pi(R_2 - R_1), \tag{5.83}$$

a finite number. Due to the change of variable, the integration region for the new coordinate R – call it region R_0 – is not spherical, but has some generally irregular shape. By comparison, the region R_0 for coordinates R is defined to be the smallest radially symmetric region, defined by inner and outer radii R_1 and R_2, that *includes within it* the region R_0. Thus, part of this radial region has points of zero yield in place of the values $1/R^2$ at the remaining ones. This gives rise to the inequality sign in Eq. (5.83).

Since both right-hand integrals in Eq. (5.81) are bounded from above, Q is bounded from above. As we saw (below Eq. (5.79)), this means that the normalization integral (5.78) is likewise bounded. For an open space problem, the integral of the square of the four-potential $\mathbf{A}$ over four-space is finite, and the corresponding PDF $p(r, t)$ given by Eq. (5.52) is normalizeable.

This result could also have been proved by a different approach. Consider the case where the temporal and spatial parts of the source function $J_s(r', t')$ separate into $J_1(r')J_2(t')$ (as in the case of a harmonic source). Then the integral in Eq. (5.80) takes on the form of a convolution of $J_1(r')$ with the function $J_2(t - |r'|/c)/|r'|$. Use of Young's inequality (Stein and Weiss, 1971, p. 31) then establishes the upper bound to the integral.

We found in Sec. 5.1.27 that the PDF is normalizeable over the closed space as well. Hence, the PDF is normalizeable for either class of observation space.

Finally, by Sec. (5.1.25) the PDF is uniquely defined under suitable conditions of physical isolation for the measurement scenario. Thus, the proposed

PDF has the basic properties of uniqueness and normalizeability that are required of a probability density.

I would like to acknowledge the help of Prof. William Faris in establishing normalizeability for the open-space case.

5.2 Maxwell's equations

As mentioned at the outset, Maxwell's equations follow in straightforward manner from the vector wave Eq. (5.51). This is shown as follows.

5.2.1 Lorentz condition on potential A

By Eqs. (5.8) and (5.48), the four-potential A obeys the Lorentz condition

$$\sum_{n=1}^{3} \frac{\partial A_n}{\partial x_n} + \frac{1}{c} \frac{\partial \phi}{\partial t} = 0. \tag{5.84}$$

5.2.2 Definitions of fields

The magnetic field B and electric field E are defined in terms of the three-potential $A \equiv (A_1, A_2, A_3)$ and scalar potential ϕ in the usual way:

$$B = \nabla \times A, \; B \equiv (B_x, B_y, B_z) \tag{5.85}$$

and

$$E \equiv -\nabla \phi - \frac{1}{c} \frac{\partial A}{\partial t}, \; E \equiv (E_x, E_y, E_z). \tag{5.86}$$

5.2.3 The equations

The four Maxwell's equations derive as follows:

$$\nabla \cdot B = 0 \tag{5.87}$$

by definition (5.85) and since $\nabla \cdot (\nabla \times A) = 0$ is a vector identity.

$$\frac{1}{c} \frac{\partial B}{\partial t} \equiv \frac{1}{c} \nabla \times \frac{\partial A}{\partial t} = \nabla \times (-\nabla \phi - E) = -\nabla \times E, \tag{5.88}$$

using definitions (5.85), (5.86) and the fact that $\nabla \times \nabla \phi \equiv 0$, respectively.

$$\nabla \times B \equiv \nabla \times (\nabla \times A) \equiv \nabla(\nabla \cdot A) - \nabla^2 A$$

$$= -\frac{1}{c}\frac{\partial}{\partial t}\nabla\phi - \frac{1}{c^2}\frac{\partial^2 A}{\partial t^2} + \frac{4\pi}{c}j$$

$$= \frac{1}{c}\frac{\partial}{\partial t}\left(-\nabla\phi - \frac{1}{c}\frac{\partial A}{\partial t}\right) + \frac{4\pi}{c}j \qquad (5.89)$$

$$= \frac{1}{c}\frac{\partial E}{\partial t} + \frac{4\pi}{c}j.$$

Here the first line is by definition (5.85) and a vector identity; the second line is by wave Eq. (5.51) and the Lorentz condition (5.84); the third line is a rearrangement of the second; and the fourth line is by definition (5.86). Finally,

$$\nabla \cdot E = \nabla \cdot \left(-\nabla\phi - \frac{1}{c}\frac{\partial A}{\partial t}\right)$$

$$= -\nabla^2\phi - \frac{1}{c}\frac{\partial}{\partial t}(\nabla \cdot A) \qquad (5.90)$$

$$= -\nabla^2\phi + \frac{1}{c^2}\frac{\partial^2\phi}{\partial t^2}$$

$$= -\Box\phi = 4\pi\rho.$$

Here the first line is by definition (5.86); the second line is a rearrangement of the first; the third line is by the Lorentz condition (5.84); and the fourth line is by the wave Eq. (5.51).

5.3 Overview

Maxwell's equations may be derived in two steps: (1) by deriving the vector wave Eq. (5.51) under the Lorentz condition, and then (2) showing that (5.51) together with the definitions (5.85) and (5.86) of the fields imply Maxwell's Eqs. (5.87)–(5.90).

In order to accomplish step (1), we use EPI. The measurement that initiates EPI is of the electric field E or the magnetic field B at a four-position $\mathbf{y}$. The latter is in error by intrinsic noise of amount $\mathbf{x} = (x, y, z, ict)$. As usual in EPI derivations, the 'noise' is the physical effect to be quantified. The four-measurement is assumed to be made in the presence of an electromagnetic four-source $\mathbf{J_s}$, so that the noise is electromagnetic in origin. A typical spread Δx in x-coordinate is given by Eq. (5.47). The aim of the EPI approach is to find the probability amplitudes $(q_n(\mathbf{x}), n = 1, \ldots, N) = \mathbf{q}$. These are the degrees of freedom of the measured effect.

The invariance principles that are used to find information J are the Lorentz condition (5.8) and conservation of flow (5.7) for the sources. Condition (5.8) implies that the minimum value of N consistent with the givens of the problem is value 4. With information J determined, the four amplitudes $\mathbf{q}$ are found to obey a general vector wave Eq. (5.40) with, however, a specifically *electromagnetic* source term $\mathbf{J_s}$. The latter implies that the wave equation is the electromagnetic one, which requires the *probability* amplitudes $\mathbf{q}$ to be linear in corresponding electromagnetic *potentials* $\mathbf{A}$; see Eq. (5.48). The resulting vector wave equation in the potentials, Eq. (5.51), is the most important step of the overall derivation.

The vector wave Eq. (5.40) that is initially derived contains an additive constant source vector $\mathbf{C}$. If it were non-zero, then in perfect vacuum (no real sources) there would still be artificial, non-zero source present. Since there is no experimental evidence for such a vacuum source this implies that $\mathbf{C} = 0$. The constant $\mathbf{C}$ arises in electromagnetic theory precisely as Einstein's additive field source Λ does (see next chapter).

Another constant of the theory that must be defined is B [Eq. (5.33)]. It must be a universal constant, have units of 1/velocity, and define electromagnetic wave propagation. Hence it is identified as being inversely proportional to the speed of light c, Eq. (5.50).

With the vector wave Eq. (5.51) derived, it is a simple matter to derive from it the Maxwell equations in the Lorentz gauge. This is shown in Sec. 5.2.

An interesting outcome of the approach is a new PDF $p(\mathbf{x})$ defining four-space localization within an electromagnetic field. This is given by Eq. (5.52), and follows from the connection Eq. (5.48) between probability amplitudes $\mathbf{q}$, on one hand, and *field amplitudes* $\mathbf{A}$ on the other. A PDF should be unique and normalizeable. The PDF $p(\mathbf{x})$ is shown to have these properties in Secs. (5.1.25)–(5.1.28). These are obeyed for open or closed space geometries, and in the presence of continuous or impulsive sources.

An enigma is that completely covariant boundary conditions led to a non-unique solution to the wave equation (Sec. 5.1.26). To attain uniqueness, non-covariant conditions were needed: Cauchy conditions for the time coordinate and either Dirichlet or Neumann conditions for the space coordinates (Sec. 5.1.25). However, as discussed in Secs. 5.1.25 and 5.1.26, the Einstein requirement of covariance is for the *laws* (here, differential equations) of physics, and not necessarily their solutions due to particular boundary conditions.

The use of statistics, as in this chapter, to derive the Maxwell equations, is not unique. DeGroot and Suttorp (1972) derive the macroscopic Maxwell's equations from an assumption of Maxwell–Lorentz field equations operating at the microscopic level. The link between the two equations is provided by methods of statistical mechanics, i.e., another statistical approach. Their assumption of the Maxwell–Lorentz field equations is, however, a rather strong one. These already have the mathematical form of the Maxwell equations. By comparison, the only physical inputs that EPI required were the two invariance conditions Eqs. (5.7) and (5.8), weak by comparison with an assumption of the Maxwell–Lorentz field equations. EPI seems to have the ability to derive a physical theory from a weaker set of assumptions than by the use of any other conventional approach (when it exists).

EPI regards Maxwell's equations as *fundamentally* statistical entities. Where have the statistics crept in? It was in admitting that any measurement within an electromagnetic field must suffer from random errors in position. Then the link Eq. (5.48) between the probability amplitude **q** *for the errors* and the electro-magnetic four-potential **A** makes the connection to electromagnetic theory. This connection is unconventional.

Classically, the potential **A** is merely a convenient mathematical tool to be used for computing the *observables* of the problem – the field strengths *B* and *E*. **A** has no physical meaning *per se*, except in the rarely used Proca theory of electromagnetism (Jackson, 1975, p. 598). By contrast, EPI regards **A** as a physical entity: it is a probability amplitude. This interpretation gives some resolution to the problem of what could possibly propagate through the emptiness of a vacuum: a field of probabilities, according to EPI.

In viewing **A** as a probability amplitude, EPI regards classical electromag-netism as defining a 'quantum mechanics' of its own (Sec. 3.4.16). In this interpretation, the general vector wave Eq. (5.51) becomes roughly a 'Klein–Gordon' equation for positional fluctuations. In this spirit, the open space solution (5.64) plus the initial 'wave' A_0 given by Eq. (5.53b) becomes a kind of scattering equation, much resembling that of quantum theory (Schiff, 1955, p. 101). Also, in the case of no sources Eq. (5.51) becomes a d'Alembert equation, where **A** represents the *probability amplitude* on coarse-grained photon positions (Sec. 5.1.19).

But if so, where can the discrete effects that are the hallmarks of quantum mechanics enter in? In fact, solutions for the vector potential **A** are often in the form of discrete eigenfunctions and eigenvalues: we need look no farther than the combination of Eqs. (5.57) and (5.59). Such solutions also, of course, occur in solution of the microwave or laser cavity resonator. In other words, discrete electromagnetic solutions arise out of boundary-value conditions, just as they do in ordinary quantum mechanics.

More generally, EPI regards all of its output differential equations as defining the physics of probability amplitudes (Sec. 3.4.16). Thus, even the classical field equation of general relativity defines a metric function that is, alternatively, a probability amplitude (on the four-position of gravitons). This is taken up next.

6

The Einstein field equation of general relativity

6.1 Motivation

As is well-known, gravitational effects are caused by a local distortion of space and time. The Einstein field equation relates this distortion to the local density of momentum and energy. This phenomenon seems, outwardly, quite different from the electromagnetic one just covered. However, luckily, this is not the case. As viewed by EPI they are very nearly the same.

This chapter on gravity is placed right after the one on electromagnetic theory because the two phenomena derive by practically identical steps. Each mathematical operation of the EPI derivation of the electromagnetic wave equation has a 1:1 counterpart in derivation of the weak-field Einstein field equation. For clarity, we will point out the correspondences as they occur. The main mathematical difference is one of *dimensionality*: quantities in the electromagnetic derivation are vectors, i.e., singly subscripted, whereas corresponding quantities in the gravitational derivation are doubly subscripted tensors. This does not amount to much added difficulty, however.

A further similarity between the two problems is in the simplicity of the final step: in Chap. 5 this was to show that Maxwell's equations follow from the wave equation. Likewise, here, we will proceed from the weak-field approximation to the general-field result. In fact, the argumentation in this final step is even simpler here than it was in getting Maxwell's equations.

Gravitational theory uses tensor quantities and manipulations. We introduce these in the section to follow, restricting attention to those that are necessary to the EPI derivation that follows. A fuller education in tensor algebra may be found, e.g., in Misner *et al.* (1973) or Landau and Lifshitz (1951).

161

6.2 Tensor manipulations: an introduction

It may be noted that the inner product of the coordinates **x** chosen in Eq. (5.3),

$$\mathbf{x} \equiv (x_0,\, x_1,\, x_2,\, x_3) \equiv (ict,\, x,\, y,\, z) \tag{6.1}$$

obeys

$$\mathbf{x \cdot x} = -(ct)^2 + x^2 + y^2 + z^2. \tag{6.2}$$

The imaginary unit i in coordinates (6.1) is bothersome to many people, since it places the time t in a different number space from the three space coordinates. The i may be avoided by supplementing subscripted variables (6.1) with superscripted ones, as follows.

6.2.1 Contravariant-, covariant coordinates

Define two sets of coordinates,

$$x^\nu \equiv (x^0,\, x^1,\, x^2,\, x^3) \equiv (x^0,\, x^j),\, j = 1,\, 2,\, 3 \tag{6.3}$$

and

$$x_\nu \equiv (x_0,\, x_1,\, x_2,\, x_3) \equiv (x_0,\, x_j),\, x_0 = ct,\, j = 1,\, 2,\, 3. \tag{6.4}$$

The former are called contravariant coordinates and the latter are called covariant coordinates. We now note an index convention that is used in Eqs. (6.3) and (6.4). A *Greek* index such as ν is implied to take on all four values 0–3. A *Latin* index such as j only takes on the three values 1–3 corresponding to the three space coordinates. The inner product is defined in terms of the two sets of coordinates as

$$\mathbf{x \cdot x} \equiv \sum_{\nu=0}^{3} x_\nu x^\nu. \tag{6.5}$$

This will allow us to express one set in terms of the other.

6.2.2 Einstein summation convention

Before doing this, we simplify the notation by using the *Einstein summation convention*. By this convention, any repetition of an index in an expression means that that index is to be summed over. Since index ν is repeated in Eq. (6.5), the latter becomes

$$\mathbf{x \cdot x} = x_\nu x^\nu. \tag{6.6}$$

This summation convention will be assumed in all equations to follow, except for those that state it to be 'turned off'.

6.2.3 Raising and lowering operators

Since Eqs. (6.2) and (6.5) must have identical right-hand sides, it must be that

$$x^\mu = \eta^{\mu\nu} x_\nu \text{ where } \eta^{\mu\nu} \equiv diag(-1, 1, 1, 1) = \eta^{\nu\mu}. \qquad (6.7)$$

Since matrix $\eta^{\mu\nu}$ is diagonal it is symmetric, as indicated. The matrix is called the 'flat field metric', for reasons that are found below. Eq. (6.7) shows that $\eta^{\mu\nu}$ acts as a 'raising' operator, transforming a subscripted variable x_ν to a super-scripted variable x^μ, that is, transforming a covariant coordinate to a contra-variant one.

Likewise, there is a 'lowering' operator $\eta_{\mu\nu}$ that causes

$$x_\mu = \eta_{\mu\nu} x^\nu, \ \eta_{\mu\nu} = \eta^{\mu\nu}. \qquad (6.8)$$

Combining the raising and lowering operators gives a delta function. By (6.7) and (6.8) $x_\alpha = \eta_{\alpha\nu}(\eta^{\nu\beta} x_\beta)$ so that

$$\eta_{\alpha\nu}\eta^{\nu\beta} = \delta^\beta_\alpha \text{ where } \delta^\beta_\alpha = 0 \text{ for } \alpha \neq \beta \text{ and } \delta^\alpha_\alpha = 1 \text{ (no summation).} \quad (6.9)$$

6.2.4 Metric tensor g

The inner product form (6.6) holds for differentials as well. In this way we can define a four-length ds obeying

$$(ds)^2 = g_{\mu\nu} dx^\mu dx^\nu \text{ where } g_{\mu\nu} = g_{\mu\nu}(x^\alpha) \qquad (6.10)$$

functionally. In contrast to matrix $\eta_{\mu\nu}$, matrix $g_{\mu\nu}$ is not generally diagonal. From the form of Eq. (6.10), $g_{\mu\nu}$ defines the coefficients of the components of a squared differential length for a given coordinate system in use. As an example, let spherical coordinates t, r, θ, ϕ define a four-space. The line element in this coordinate system, in vacuum, is

$$(ds)^2 = -\left(1 - \frac{2M}{r}\right)dt^2 + \frac{dr^2}{1 - 2M/r} + r^2 d\theta^2 + r^2 \sin^2\theta d\phi^2. \qquad (6.11)$$

From this, the metric tensor coefficients $g_{\mu\nu}$ may be 'picked off'.

6.2.5 Exercise

Find the metric tensor $g_{\mu\nu}$ in this way.

6.2.6 Flat field metric

As a more simple case, the geometry of space-time of special relativity incurs a differential length

$$(ds)^2 = -(dx^0)^2 + (dx^1)^2 + (dx^2)^2 + (dx^3)^2. \tag{6.12}$$

Comparison with Eq. (6.10) shows that here $g_{\mu\nu} = \eta_{\mu\nu}$. This space is 'flat', in that $g_{\mu\nu}(x^\alpha) = g_{\mu\nu} = Const.$ Hence, $\eta_{\mu\nu}$ is called the 'flat field' metric.

6.2.7 Derivative notation

Since we have two sets of coordinates x^ν and x_ν there are two different kinds of derivatives,

$$\frac{\partial f_{\alpha\beta}}{\partial x^\nu} \equiv f_{\alpha\beta,\nu} \quad \text{and} \quad \frac{\partial f_{\alpha\beta}}{\partial x_\nu} \equiv f_{\alpha\beta},^\nu. \tag{6.13}$$

Hence, a comma denotes differentiation with respect to the coordinate whose index follows it. The $\eta_{\mu\nu}$ and $\eta^{\mu\nu}$ operators lower and raise derivative indices as well. For example, the chain rule of differentiation gives

$$f_{\alpha\beta,\nu} \equiv \frac{\partial f_{\alpha\beta}}{\partial x^\nu} = \frac{\partial f_{\alpha\beta}}{\partial x_\mu} \frac{\partial x_\mu}{\partial x^\nu} = f_{\alpha\beta},^\mu \eta_{\mu\nu} \tag{6.14}$$

by Eq. (6.8). Hence, $\eta_{\mu\nu}$ serves to lower a raised derivative index. Likewise,

$$f_{\alpha\beta},^\nu = f_{\alpha\beta,\mu} \eta^{\mu\nu}. \tag{6.15}$$

Operator $\eta^{\mu\nu}$ serves to raise a lower derivative index.

6.2.8 Fisher information

As an example of the simplifications that the comma notation and preceding operations allow, consider the squared gradient with respect to coordinates (6.1). This becomes

$$\nabla q_n \cdot \nabla q_n \equiv \sum_{\lambda=0}^{3} \left(\frac{\partial q_n}{\partial x_\lambda}\right)^2 = q_{n,\lambda} q_n,^\lambda \tag{6.16}$$

after conversion to covariant coordinates (6.4), and use of the summation convention, lowering operation (6.8) and comma notation (6.13). This result allows the Fisher information Eq. (2.19) to be expressed as (2.20),

$$I = 4 \int d^4 x \, q_{n,\lambda} q_n,^\lambda \quad \text{where } d^4 x \equiv dx^0 \cdots dx^3. \tag{6.17}$$

Regarding covariance properties, the salient feature of this equation is its covariance in the derivative indices λ. When (6.17) is used in the EPI principle, the output wave equation is generally covariant. Of course the EPI output equation of this chapter, the Einstein field equation, is manifestly covariant.

6.2.9 Exercise: tensor Euler–Lagrange equations

Suppose that the Lagrangian $\mathscr{L}[q_v, q_{v,\lambda}]$, $q_v = q_v(x^\lambda)$ is known. Show that the Euler–Lagrange Eq. (0.34) for determination of q_v is, in tensor notation,

$$\frac{\partial}{\partial x^\lambda}\left[\frac{\partial \mathscr{L}}{\partial(q_{v,\lambda})}\right] = \frac{\partial \mathscr{L}}{\partial q_v}. \tag{6.18a}$$

6.2.10 Exercise: tensor form of d'Alembertian

Using Eqs. (6.7) and (5.12b), show that

$$q_v{}^\lambda{}_{,\lambda} = \eta^{\lambda\alpha}q_{v,\alpha,\lambda} = \Box q_v. \tag{6.18b}$$

6.3 Derivation of the weak-field wave equation

6.3.1 Goals

The aim of this chapter is to derive the Einstein field equation. This is done in two steps: by (1) deriving the weak-field equation; and then (2) showing that this implies the field equation for a *general* field. The latter is an easy matter. The main task is problem (1).

We follow the EPI procedure of Sec. 3.4. This requires identifying the parameters to be measured, forming the corresponding expression for I, and then finding J by an appropriate invariance principle.

This derivation follows in almost 1:1 fashion that of the electromagnetic wave equation in Chap. 5. Most sections in this chapter have a direct counter-part in Chap. 5: Sec. 6.3.1 corresponds to Sec. 5.1.1, Sec. 6.3.2 corresponds to Sec. 5.1.2, etc.

The main departure operationally is to replace singly subscripted quantities of the electromagnetic theory, such as vector potential A_n, by doubly sub-scripted quantities such as the weak-field metric $\bar{h}_{\mu v}$. This double subscripting is naturally present at all levels of the EPI derivation, as will be seen.

6.3.2 Choice of measured parameters θ

The gravitational metric $g_{\mu v}$ can be measured (Misner *et al.*, 1973, p. 324). Consider such a measurement procedure, where the gravitational field is so weak that there are only small local departures from flat space. Then the metric $g_{\mu v}$ may be expanded as

$$g_{\mu\nu} = \eta_{\mu\nu} + h_{\mu\nu}, \quad |h_{\mu\nu}| \ll 1. \tag{6.19}$$

The $h_{\mu\nu}$ are small metric perturbations. It is customary to work with an associated metric of perturbations

$$\overline{h}_{\mu\nu} \equiv h_{\mu\nu} - \tfrac{1}{2}\eta_{\mu\nu}h, \quad h \equiv \eta^{\mu\nu}h_{\mu\nu} \tag{6.20}$$

that is linear in $h_{\mu\nu}$. We call $\overline{h}_{\mu\nu}$ the 'weak field' metric.

It has been suggested (Lawrie, 1990) that $\overline{h}_{\mu\nu}$ can be interpreted as the wavefunction $q_{\mu\nu}$ of a graviton. This idea is confirmed later in this chapter; see Secs. 6.3.20–6.3.26. (Also see the corresponding Secs. 5.1.19–5.1.24 that allow the source-free electromagnetic potential **A** to be interpreted as the probability amplitude **q** for a photon.)

Consider, then, an experiment consisting of measurements of the weak-field metric. The latter is to be found at ideal (target) four-positions $\theta_{\mu\nu}$. The double subscripting is, of course, an arbitrary naming (or ordering) of the positions.

The measuring instrument perturbs the amplitude functions **q** of the problem, which initiates the EPI process. Also, at the input space to the instrument, fluctuations $x_{\mu\nu}$ occur that define the intrinsic data (Sec. 2.1.1) of the problem

$$y_{\mu\nu} = \theta_{\mu\nu} + x_{\mu\nu}. \tag{6.21}$$

By our notation convention (Sec. 6.2.1) this amounts to 16 measurements although, as we will see, $N = 10$ measurements suffice. This means that the position coordinates $x_{\mu\nu}$ are the Fisher variables (and not the metric values $g_{\mu\nu}$ themselves).

Eq. (6.21) states that input positional accuracy is limited only by fluctuations $x_{\mu\nu}$ that are characteristic of the gravitational metric. As is well known, these fluctuations occur on the scale of the Planck length

$$\Delta x \sim (\hbar G/c^3)^{1/2} \tag{6.22}$$

(cf. Eq. (5.4)). G is the gravitational constant. At this point Eq. (6.22) is purely motivational. It is derived at Eq. (6.64) below.

6.3.3 Information I

By Eq. (2.11) the four-data data values $\mathbf{y}_{\mu\nu}$ are assumed to be independent, and there is a PDF

$$p_{\mu\nu}(y_{\mu\nu}|\theta_{\mu\nu}) = p_{\mu\nu}(x^\alpha) \equiv q^2_{\mu\nu}(x^\alpha) \tag{6.23}$$

for the $(\mu\nu)$th measurement. The information is then Eq. (6.17), where the probability amplitudes **q** are now doubly subscripted,

$$I = 4 \int d^4x\, q_{\mu\nu,\lambda} q_{\mu\nu}{}^{,\lambda}. \tag{6.24}$$

6.3.4 Invariance principles

As in the electromagnetic problem, we will use two equations of continuity of flow for the invariance principle required by EPI.

The first equation of flow governs the sources that are driving the probability amplitudes. In Chap. 5, the general source was the vector current $\mathbf{J} = J_\mu$. Here the source is the 'stress-energy tensor' $T_{\mu\nu}$, defined as follows.

Consider a two-dimensional area A_k at rest in the observer's frame, with the normal to A_k pointing in the direction $\hat{e}_k$. Due to moving matter in the vicinity of the frame there will be an amount of momentum P^μ crossing the surface area A_k in the direction $\hat{e}_k$ during a time Δt. The (μk) component of the stress-energy tensor is defined as

$$T^{\mu k} = \frac{P^\mu}{A_k \Delta t} = T^{k\mu}. \tag{6.25}$$

The four-momentum P^μ mentioned here requires definition. A particle of mass m has a momentum

$$P^\mu \equiv mu^\mu \tag{6.26}$$

where u^μ is the particle's four-velocity $u^\mu \equiv (dt/d\tau, u^j)$ with τ its 'proper' time (time as seen by the particle). The final component, T^{00}, is the total mass-energy measured in the observer's frame.

In general, four-momentum is conserved. This is expressed as a conservation of flow theorem (cf. Eq. (5.7))

$$T_{\mu\nu}{}^{,\nu} = 0. \tag{6.27}$$

The second equation of flow is a Lorentz condition (usually obeyed by $\bar{h}_{\mu\nu}$) for the probability amplitudes $q_{\mu\nu}$,

$$q^{\mu\nu}{}_{,\nu} = 0. \tag{6.28}$$

(cf. Eq. (5.8)). We will later show that $q_{\mu\nu}$ and $\bar{h}_{\mu\nu}$ are proportional, justifying this condition.

A final invariance principle is a demand that the weak-field equations be 'form invariant', i.e., remain invariant in form to Lorentz and gauge transformations. Regarding *constant* tensors of order two, the only ones that obey such invariance are the flat-field metric and the Kronecker delta,

$$\eta_{\mu\nu}, \delta^\mu_\nu. \tag{6.29}$$

6.3.5 Fixing N

The Lorentz condition (6.28) is the only physical input into EPI that functionally involves the unknown amplitudes $q_{\mu\nu}$. Since each of indices μ, ν goes from 0 to 3 it might appear that there are $N = 16$ independent amplitudes in this theory. However, because of the symmetry (6.25) in the stress-energy tensor there are only $N = 10$ independent stress-energy values. We anticipate, at this point, that the $q_{\mu\nu}$ will obey symmetry as well so that $N = 10$ also defines the number of independent amplitudes that are sufficient for defining the theory. This step could have been taken at the end of the derivation, but it is more convenient to fix here once and for all.

6.3.6 On finding the bound information J

The invariance principles (6.27) and (6.28) do not explicitly express information I in terms of a physically parameterized information J. This means that we cannot at this point know J or the efficiency parameter κ. They have to be solved for, based upon *use* of the two invariance principles in the two EPI conditions (3.16), (3.18). The latter must be solved *simultaneously* for a common solution $q_{\mu\nu}$ (Sec. 3.4.6). We do this in the following sections.

6.3.7 General form for J

At this point the bound information J is an unknown scalar functional of the $q_{\mu\nu}$ and sources $T_{\mu\nu}$. It can therefore be represented as an inner product

$$ J \equiv 4 \int d^4 x A^{\alpha\beta} J_{\alpha\beta}, \quad A^{\alpha\beta} = const., \quad J_{\alpha\beta} = J_{\alpha\beta}(q_{\mu\nu}, T_{\mu\nu}). \tag{6.30} $$

The notation $J_{\alpha\beta}(q_{\mu\nu}, T_{\mu\nu})$ means that each $J_{\alpha\beta}$ depends in an unknown way upon *all* $q_{\mu\nu}$ and $T_{\mu\nu}$.

6.3.8 EPI variational solution

As planned above, we must first find the solution for $q_{\mu\nu}$ to EPI problem (3.16). Using the I form (6.17) for, now, a doubly subscripted $q_{\mu\nu}$ and using (6.30) for J gives an information K Lagrangian

$$ \mathcal{L} = q_{\mu\nu,\lambda} q_{\mu\nu}{}^{,\lambda} - A^{\mu\nu} J_{\mu\nu}. \tag{6.31} $$

We ignored a common factor 4. This is used in the Euler–Lagrange equation for the problem, which by Eq. (6.18a) is

$$ \frac{\partial}{\partial x^\lambda} \left[\frac{\partial \mathcal{L}}{\partial(q_{\mu\nu,\lambda})} \right] = \frac{\partial \mathcal{L}}{\partial q_{\mu\nu}}. \tag{6.32} $$

By Eq. (6.31) and lowering operation (6.8),

$$\frac{\partial \mathscr{L}}{\partial q_{\mu\nu,\lambda}} = 2\eta^{\lambda\alpha} q_{\mu\nu,\alpha} \text{ and } \frac{\partial \mathscr{L}}{\partial q_{\mu\nu}} = -A^{\alpha\beta} \frac{\partial J_{\alpha\beta}}{\partial q_{\mu\nu}}. \tag{6.33}$$

Using these in Eq. (6.32) gives a formal solution

$$2\eta^{\lambda\alpha} q_{\mu\nu,\alpha,\lambda} = -A^{\alpha\beta} \frac{\partial J_{\alpha\beta}}{\partial q_{\mu\nu}} \text{ or } \Box q_{\mu\nu} = -\tfrac{1}{2} A^{\alpha\beta} \frac{\partial J_{\alpha\beta}}{\partial q_{\mu\nu}} \tag{6.34}$$

by identity (6.18b) (cf. Eq. (5.12a)). This expresses $q_{\mu\nu}$ in terms of the unknown information components $J_{\mu\nu}$.

6.3.9 Alternative form for I

Once again use form (6.17) for I with singly subscripted amplitudes replaced by doubly subscripted ones. This form may be integrated by parts. First note the elementary result (5.13). There, the zero held because we assumed that either $q = 0$ or $dq/dx = 0$ (Dirichlet or Neumann conditions) hold on the boundaries of measurement space. Assume such conditions to hold here as well. Then integrating by parts Eq. (6.17) gives an information

$$I = 4 \int d^4 x q_{\mu\nu,\lambda} q_{\mu\nu}{}^{,\lambda} = 4 \sum_\lambda q_{\mu\nu}{}^{,\lambda} q_{\mu\nu} \bigg|_{-\infty}^{+\infty} - 4 \int d^4 x q_{\mu\nu} q_{\mu\nu}{}^{,\lambda}{}_{,\lambda}$$

$$\tag{6.35}$$

$$= \qquad 0 \qquad - 4 \int d^4 x q_{\mu\nu} \Box q_{\mu\nu}$$

by identity (6.18b).

6.3.10 EPI zero-root solution

The second EPI problem is to find the roots $q_{\mu\nu}$ of Eq. (3.18),

$$I[q_{\mu\nu}] - \kappa J[q_{\mu\nu}] = 0, \tag{6.36}$$

I given by (6.35) and J given by (6.30). The problem is then

$$I - \kappa J = -4 \int d^4 x (q_{\mu\nu} \Box q_{\mu\nu} + \kappa A^{\mu\nu} J_{\mu\nu}) = 0. \tag{6.37}$$

The solution is the microscale Eq. (3.20), which is here (cf. Eq. (5.17))

$$q_{\mu\nu} \Box q_{\mu\nu} = -k A^{\mu\nu} J_{\mu\nu} \text{ (no summation).} \tag{6.38}$$

6.3.11 Common solution $q_{\mu\nu}$

By Sec. 6.3.6, Eqs. (6.34) and (6.38) are to be simultaneously solved. Eliminating $\Box q_{\mu\nu}$ between them gives

$$\tfrac{1}{2} A^{\alpha\beta} q_{\mu\nu} \frac{\partial J_{\alpha\beta}}{\partial q_{\mu\nu}} = \kappa A^{\mu\nu} J_{\mu\nu} \text{ (no summation on } \mu\nu). \tag{6.39}$$

Noting that the right-hand side of (6.39) has been sifted out of the left-hand sum over $\alpha\beta$, it must be that

$$\tfrac{1}{2} q_{\mu\nu} \frac{\partial J_{\alpha\beta}}{\partial q_{\mu\nu}} = \kappa \delta^{\mu}_{\alpha} \delta^{\nu}_{\beta} J_{\mu\nu} \text{ (no summation).} \tag{6.40}$$

where δ^{μ}_{ν} is the Kronecker delta (6.9). Then the case $\alpha\beta \neq \mu\nu$ gives $\partial J_{\alpha\beta}/\partial q_{\mu\nu} = 0$, assuming non-trivial solutions $q_{\mu\nu} \neq 0$. This implies that

$$J_{\alpha\beta}(q_{\mu\nu}, T_{\mu\nu}) = J_{\alpha\beta}(q_{\alpha\beta}, T_{\mu\nu}) \text{ (no summation)} \tag{6.41}$$

That is, for the particular $\alpha\beta$, function $J_{\alpha\beta}$ depends on $q_{\alpha\beta}$ alone among all the q, but upon all the T.

Next, the opposite case $\alpha\beta = \mu\nu$ in Eq. (6.40) gives a condition

$$q_{\mu\nu} \frac{\partial J_{\mu\nu}}{\partial q_{\mu\nu}} = 2\kappa J_{\mu\nu}(q_{\mu\nu}, T_{\alpha\beta}) \text{ (no summation).} \tag{6.42}$$

Because of the particular dependence (6.41), the left-hand side partial derivative in (6.42) is actually an ordinary derivative. This allows the equation to be integrated with respect to $q_{\mu\nu}$. The result is

$$J_{\mu\nu} = q_{\mu\nu}^{2\kappa} B_{\mu\nu}(T_{\alpha\beta}) \text{ (no summation)} \tag{6.43}$$

(cf. Eq. (5.21)). Functions $B_{\mu\nu}(T_{\alpha\beta})$ arise as the constants (in $q_{\mu\nu}$) of integration and are to be found, along with parameter κ.

6.3.12 Resulting wave equation

Using the result (6.43) in either wave equation (6.34) or (6.38) gives an answer for $q_{\mu\nu}$ of the same form,

$$\Box q_{\mu\nu} = q_{\mu\nu}^{b} F_{\mu\nu}(T_{\alpha\beta}) \text{ (no summation) where } b \equiv 2\kappa - 1. \tag{6.44}$$

The parameter b and the new functions $F_{\mu\nu}(T_{\alpha\beta})$ need to be found. In general, each $F_{\mu\nu}$ depends upon all the $T_{\alpha\beta}$.

6.3.13 Where we stand so far

At this point we know that the amplitude functions $q_{\mu\nu}$ obey a wave equation (6.44), but with an unknown dependence upon the right-hand source functions

$F_{\mu\nu}(T_{\alpha\beta})$. Hence, the EPI approach has taken us quite a way toward solution to the problem, even without the use of any specifically gravitational inputs of information. But we can proceed no further without one. Hence, the invariance principles (6.27) and (6.28) will now be used. These are actually quite weak as statements of specifically gravitational phenomena. (Notice that they are practically identical to the electromagnetic invariance principles (5.7) and (5.8).) Hence, the fact that their use in EPI will lead to the weak-field equation is a definite plus for the theory.

6.3.14 Use of Lorentz- and conservation of matter-energy flow conditions

We proceed in the following sections to find the exponent b and functions $F_{\mu\nu}$. These will be defined by the conservation law (6.27), Lorentz condition (6.28) and invariance properties (6.29).

We start by operating with $\eta^{\nu\alpha}\partial/\partial x^{\alpha}$ on Eq. (6.44), with implied summation over ν and α. The left-hand side becomes

$$\eta^{\nu\alpha}\Box q_{\mu\nu,\alpha} = \eta^{\nu\alpha}\eta_{\mu\beta}\eta_{\nu\gamma}\Box q^{\beta\gamma}{}_{,\alpha} = \eta_{\mu\beta}\Box q^{\beta\alpha}{}_{,\alpha}. \qquad (6.45)$$

(The middle equality used lowering operations (6.8). The second equality used the delta function definition (6.9).) This vanishes, by the Lorentz condition (6.28).

Operating in the same way on the *right-hand* side of Eq. (6.44) should likewise give zero. By the chain rule of differentiation, it gives

$$\eta^{\nu\alpha}\left[q_{\mu\nu}^{b}\frac{\partial F_{\mu\nu}(T_{\alpha\beta})}{\partial T_{\beta\gamma}}\right]T_{\beta\gamma,\alpha} \qquad (6.46)$$

plus another term which is zero if $b = 0$ (as will be the case).

Observing the resemblance of the derivative $T_{\beta\gamma,\alpha}$ in (6.46) to the term $T_{\mu\nu}{}^{,\nu}$ in conservation equation (6.27), we see that (6.46) can be made to vanish if

$$q_{\mu\nu}^{b}\frac{\partial F_{\mu\nu}(T_{\alpha\beta})}{\partial T_{\beta\gamma}} = \tfrac{1}{2}(B_{\mu}^{\beta}\delta_{\nu}^{\gamma} + B_{\mu}^{\gamma}\delta_{\nu}^{\beta}), \text{ where } B_{\lambda}^{\gamma} \equiv B_{\lambda}^{\gamma}(T_{\alpha\beta}, q_{\alpha\beta}). \qquad (6.47)$$

The B_{λ}^{γ} are new, unknown, functions of all the $q_{\alpha\beta}$ and $T_{\alpha\beta}$. As a check, one may substitute the first right-hand term of (6.47) into (6.46) to obtain

$$B_{\mu}^{\beta}\eta^{\nu\alpha}\delta_{\nu}^{\gamma}T_{\beta\gamma,\alpha} = B_{\mu}^{\beta}\eta^{\nu\alpha}T_{\beta\nu,\alpha} = B_{\mu}^{\beta}T_{\beta\nu}{}^{,\nu} = 0, \qquad (6.48)$$

as was required, after sifting with the delta function, using an index-raising operation (6.7), and using conservation property (6.27). The second right-hand side term of (6.47) may be treated similarly.

6.3.15 Finding exponent b

The left- and right-hand sides of Eq. (6.47) must balance in their q- and T-dependences. Because the left-hand side is a separated function $q_{\mu\nu}^b$ of $q_{\mu\nu}$ times a function of the $T_{\alpha\beta}$, the right-hand functions B_μ^β must likewise be so separated. For the case $\beta \neq \nu$, $\gamma = \nu$, Eq. (6.47) gives

$$B_\mu^\beta \equiv B_{1\mu}^\beta(q_{\alpha\beta})B_{2\mu}^\beta(T_{\alpha\beta}),\tag{6.49a}$$

$$B_{1\mu}^\beta = Cq_{\mu\nu}^b, \ B_{2\mu}^\beta = \frac{\partial F_{\mu\nu}(T_{\alpha\beta})}{\partial T_{\beta\gamma}}.\tag{6.49b}$$

Quantity C is a constant. The first Eq. (6.49b) says that

$$B_{1\mu}^\beta = Cq_{\mu 0}^b = Cq_{\mu 1}^b \ldots = Cq_{\mu 3}^b.\tag{6.50}$$

Since the $q_{\mu\nu}$ are independent degrees of freedom of the theory they cannot in general be equal. The only alternative possibility is that $B_{1\mu}^\beta$ be independent of all q, $B_{1\mu}^\beta = Const.$, so that

$$b = 0.\tag{6.51}$$

Also, by Eq. (6.49a), $B_\mu^\beta = B_\mu^\beta(T_{\alpha\beta})$ alone.

6.3.16 Finding functions $F_{\mu\nu}(T_{\alpha\beta})$

Evaluating Eq. (6.47) at $\beta = \nu$, $\gamma = \mu \neq \nu$ and using (6.51) gives

$$\frac{\partial F_{\mu\nu}}{\partial T_{\mu\nu}} = \tfrac{1}{2}B_\mu^\mu(T_{\alpha\beta}).\tag{6.52}$$

We used the symmetry (6.25) of T. Doing the same operations with, now, $\nu \to \nu'$, gives the same right-hand side as (6.52). Therefore,

$$\frac{\partial F_{\mu\nu}}{\partial T_{\mu\nu}} = \frac{\partial F_{\mu\nu'}}{\partial T_{\mu\nu'}} \ \text{for all } (\nu, \nu').\tag{6.53}$$

The general solution must be that each side is independent of ν (and ν'), hence $(\mu\nu)$, and hence $T_{\mu\nu}$. Therefore Eq. (6.53) may be integrated to give a general linearity between $F_{\mu\nu}$ and $T_{\mu\nu}$. The solution is

$$F_{\mu\nu} = \tfrac{1}{2}(B_\mu^\beta T_{\beta\nu} + B_\nu^\beta T_{\beta\mu}) + C_{\mu\nu}, \ B_\mu^\beta = Const., C_{\mu\nu} = Const.\tag{6.54}$$

This may be verified by back substitution into Eq. (6.47) with $b = 0$.

6.3.17 Exercise

Carry through the substitution, verifying that (6.54) is indeed a solution of Eq. (6.47).

6.3.18 Efficiency parameter κ

By Eqs. (6.44) and (6.51), the efficiency parameter is here

$$\kappa = 1/2. \tag{6.55}$$

This is the same value as for the electromagnetic case (Eq. (5.39)). Hence, the weak-field gravitational theory is inefficient, only allowing $I/J = 1/2$ of the bound information to be utilized in the intrinsic information I. This indicates that the theory is inefficient, in the sense that it is an approximation. The use of EPI in a generally strong-field scenario, and taking into account quantum gravitational effects, would yield a value of $\kappa = 1$. This is verified in Chap. 11.

6.3.19 Constants B_μ^β, $C_{\mu\nu}$ from form invariance

Using Eqs. (6.44) and (6.54) gives a wave equation in the amplitudes,

$$\Box q_{\mu\nu} = \tfrac{1}{2}(B_\mu^\beta T_{\beta\nu} + B_\nu^\beta T_{\beta\mu}) + C_{\mu\nu}. \tag{6.56}$$

We demand that this be form-invariant under Lorentz- and gauge transformations. Noticing that B_μ^β and $C_{\mu\nu}$ are constant tensors, by Eqs. (6.29) it must be that

$$B_\mu^\beta = B\delta_\mu^\beta \text{ and } C_{\mu\nu} = C\eta_{\mu\nu}, \ B, \ C \text{ constants.} \tag{6.57}$$

6.3.20 Identification of amplitudes with weak-field metric

Using this in Eq. (6.56), along with the symmetry $T_{\mu\nu} = T_{\nu\mu}$, gives

$$\Box q_{\mu\nu} = BT_{\mu\nu} + C\eta_{\mu\nu}. \tag{6.58}$$

This has the form of a wave equation in the amplitudes $q_{\mu\nu}$ due to a *gravitational* source $T_{\mu\nu}$. Since the weak-field gravitational equations have precisely such a form, this implies that the probability amplitude $q_{\mu\nu}$ and weak-field metric $\overline{h}_{\mu\nu}$ are proportional,

$$\overline{h}_{\mu\nu}(x^\alpha) = \sqrt{\Delta V} q_{\mu\nu}(x^\alpha), \tag{6.59}$$

where ΔV is a constant *four-volume*. (Note: As with the analogous electromagnetic Eq. (5.48), for uniqueness of definition we ignore the possible presence of an added right-hand side function whose $\Box$ is zero.) The correspondence (6.59) states that the weak-field metric at a position $x_{\mu\nu}^\alpha$ is, alternatively, the square-root of the probability density for a fluctuation $x_{\mu\nu}^\alpha$ from an ideal position $\theta_{\mu\nu}^\alpha$. This correspondence is also suggested by remarks of Lawrie (1990) and by the fact that, in the source-free case, the resulting field equation will go over into the wave equation for a particle of the gravitational field, i.e., the *graviton*. See Sec. 6.3.24.

The proportionality factor $\sqrt{\Delta V}$ in Eq. (6.59) is obtained from the fact that $\overline{h}_{\mu\nu}$ is unitless whereas, by Eq. (6.23), $q_{\mu\nu}$ is the square-root of a PDF on four-position and, hence, must have units of reciprocal area.

The metric $\overline{h}_{\mu\nu}$ is a classical (macroscopic), continuous quantity. It is usual to assume that it suffers from quantum fluctuations that dominate within *x*-intervals that are on the scale of an elemental length L (Misner *et al.*, 1973, pp. 1190, 1191). We show in the next section that L is the Planck length. The approach will also predict that the cosmological constant Λ is zero.

The *four-volume* interval corresponding to L is L^4. Assume that within such volume intervals quantum fluctuations dominate. Hence, *fluctuations x^α can now be identified as quantum in origin* (cf. last paragraphs of Sec. 5.3). By correspondence (6.59), $q_{\mu\nu}$ must also randomly fluctuate within such volumes. However, a macroscopic observer cannot see these details, instead seeing an average of $q_{\mu\nu}$ over volume L^4. Effectively, then, L^4 is an uncertainty volume, and all of four-space is subdivided into contiguous four-cells of size L^4 within which the details of $q_{\mu\nu}$ and $\overline{h}_{\mu\nu}$ cannot be seen.

6.3.21 Weak-field equation

Let us identify ΔV with the volume L^4. Squaring Eq. (6.59) and using Eq. (6.23) gives

$$p_{\mu\nu}(x^\alpha)\Delta V = \overline{h}_{\mu\nu}^2 \qquad (6.60)$$

for the probability of a fluctuation x^α lying within the elemental four-interval. (Note that because the left-hand side of (6.60) is essentially a differential, (6.60) tends to agree with the weak-field assumption $\overline{h}_{\mu\nu}^2 \ll 1$.)

Now use Eq. (6.59) on the left-hand side of Eq. (6.58) and set $B = -16\pi G/c^4\sqrt{\Delta V}$, $C = 2\Lambda/\sqrt{\Delta V}$, Λ the cosmological constant. Then Eq. (6.58) becomes the weak-field equation in the Lorentz gauge,

$$\Box \overline{h}_{\mu\nu} = -\left(\frac{16\pi G}{c^4}\right)T_{\mu\nu} + 2\Lambda\eta_{\mu\nu}. \qquad (6.61)$$

The symmetry (6.7), (6.25) in $\eta_{\mu\nu}$ and $T_{\mu\nu}$ implies symmetry in an $\overline{h}_{\mu\nu}$ obeying (6.61) as well. Therefore there are 10, rather than 16, independent degrees of freedom $\overline{h}_{\mu\nu}$.

6.3.22 Field equations in $q_{\mu\nu}$; the Planck length

The wave equation (6.58) in the probability amplitudes $q_{\mu\nu}(x)$ may be further specified. (Note: these are to be distinguished from probability amplitudes on the metric $g_{\mu\nu}$ itself.) Since $q_{\mu\nu}$ is purely a determinant of *quantum* fluctuations (Sec. 6.3.20), its field equation should be independent of specifically gravitational parameters such as G or Λ. Use of Eq. (6.59) in (6.61) gives, after multiplying both sides by quantity $c\hbar/\sqrt{\Delta V}$,

$$c\hbar\Box q_{\mu\nu} = -a_1 T_{\mu\nu} + a_2 \eta_{\mu\nu} \qquad (6.62a)$$

$$a_1 \equiv \frac{16\pi\hbar G}{\sqrt{\Delta V}c^3}, \quad a_2 \equiv \frac{2c\hbar\Lambda}{\sqrt{\Delta V}}. \qquad (6.62b)$$

Constants a_1 and a_2 will be determined by the choice of ΔV.

Each side of Eq. (6.62a) has units of energy/volume. Hence, since $T_{\mu\nu}$ also has these units, a_1 is unitless. Also, since Eq. (6.62a) defines the quantum quantity $q_{\mu\nu}$, by our previous reasoning the right-hand side must be independent of gravitational parameters, notably G. Then, in (6.62b), a_1 must be both unitless and independent of gravitational parameter G. The choice

$$\sqrt{\Delta V} = \hbar G/c^3 \qquad (6.63)$$

accomplishes these aims, to an arbitrary numerical multiplier of the right-hand side. Next, since $\Delta V = L^4$, elemental length L is

$$L = \sqrt{\frac{\hbar G}{c^3}}. \qquad (6.64)$$

This is the Planck length. Thus, the Planck length follows naturally from the information approach.

6.3.23 Zero cosmological constant

Next, consider the constant a_2. Using the solution (6.63) in the second Eq. (6.62b) gives

$$a_2 = 2c^4\left(\frac{\Lambda}{G}\right). \qquad (6.65)$$

Again, this quantity should not depend upon gravitational parameters. Possible solutions are $\Lambda = 0$ or $\Lambda \propto G$. But if the latter were true, since Λ/G is not unitless the proportionality constant would be another gravitational parameter. Hence the only satisfactory solution is $\Lambda = 0$.

6.3.24 A d'Alembert's wave equation: gravitons

With use of $\Lambda = 0$ and Eqs. (6.62b) and (6.63), the field equations (6.62a) become

$$\Box q_{\mu\nu} = -\frac{16\pi}{c\hbar} T_{\mu\nu}. \qquad (6.66)$$

For a zero stress-energy source $T_{\mu\nu}$ this becomes a d'Alembert wave equation which, as we saw in Chaps. 4 and 5, describes particles with integer values of spin, i.e., Bose–Einstein particles. Because of the double subscripting the spin is here 2, defining in the usual way the graviton. Hence, in the source-free scenario $q_{\mu\nu}$ is a probability amplitude for the position of a graviton. This is a verification of the identification between $q_{\mu\nu}$ and $\bar{h}_{\mu\nu}$ that was made in Sec. 6.3.20.

6.3.25 Graviton creation and absorption

Since $q_{\mu\nu}^2 = q_{\mu\nu}^2(x, y, z, ct)$ is a four-dimensional PDF, its integral over all (x, y, z) is $q_{\mu\nu}^2(ct) \equiv p_{\mu\nu}(ct)$, a PDF on time t of detection. This can be zero at certain times, indicating no possible detection or, equivalently, the annihilation of a graviton. A subsequent detection event would then characterize the creation of a graviton. This implies that gravitons can be created and absorbed, in time, by sources and sinks of the gravitational field.

6.3.26 Probability on fluctuations

By the use of Eqs. (6.23) and (6.59) in Eq. (2.23), we get as a PDF on the fluctuations x^α

$$p(x^\alpha) = \frac{1}{10L^4} \sum_{\mu\nu} \bar{h}_{\mu\nu}^2(x^\alpha) \qquad (6.67)$$

where $(\mu\nu)$ range over their 10 non-symmetric pairs. In the source-free case (Sec. 6.3.24) we found that this represents the PDF on the position of a graviton. However, in cases where there are sources present the fluctuations x^α no longer represent simply the positions of gravitons since the equation (6.66) defining amplitudes $q_{\mu\nu}$ is no longer of the Helmholtz form. The x^α then generally represent uncertainties in field positions (Sec. 6.3.2).

6.3.27 Uniqueness and normalization properties

The weak-field Eq. (6.61) with zero cosmological constant (Sec. 6.3.23) is of the same form as the electromagnetic wave Eq. (5.51). The only difference between the two is the higher dimensionality of Eq. (6.61). Therefore, by the reasoning in Secs. 5.1.25–5.1.28, the weak-field equation likewise has unique, normalizeable solutions.

6.3.28 Reference

Most of the material in this chapter is from the paper by Cocke and Frieden (1997).

6.4 Einstein field equation and equations of motion

The transition from the weak-field Eq. (6.61) to the general field case may be made by the usual bootstrapping argument (Misner *et al.,* 1973, p. 417). That is, the only tensor whose linear approximation is $\Box \bar{h}_{\mu\nu}$ is $g_{\mu\nu}R - 2R_{\mu\nu}$, where $R_{\mu\nu}$ is the Ricci curvature tensor. Also, the source-free constant should now be proportional to the metric $g_{\mu\nu}$ instead of $\eta_{\mu\nu}$. Accordingly Eq. (6.61) becomes

$$ R_{\mu\nu} - \tfrac{1}{2}g_{\mu\nu}R = \left(\frac{8\pi G}{c^4}\right)T_{\mu\nu} - \Lambda g_{\mu\nu}, \qquad (6.68) $$

the Einstein field equations. From before, $\Lambda = 0$.

It is known (Misner *et al.,* 1973, p. 480) that the field equations can be used to derive the *equations of motion* for a test particle in the field. Hence, the equations of motion follow from EPI as well.

6.5 Overview

The Einstein field equation may be derived via EPI in almost 1:1 fashion with that of Maxwell's equations in Chap. 5. As in the latter derivation, there are two overall steps to accomplish: (1) derive the weak-field wave equation for the Lorentz condition; and (2) raise the rank of the weak-field equation to the tensor level of the Einstein equation by a simple argument.

The smart measurements are of the metric $g_{\mu\nu}$ at ideal four-positions $\theta^\alpha_{\mu\nu}$. The input space measurements are at positions $y^\alpha_{\mu\nu}$ with 'noise' values $x^\alpha_{\mu\nu} \equiv (x, y, z, ct)_{\mu\nu}$. The noise is quantum in origin, so that the probability amplitudes $q_{\mu\nu}(x^\alpha)$ are on fluctuations in measured four-position due to quantum effects on the scale of the Planck length. These arguments allow us to derive the Planck length in Sec. 6.3.22.

The information I in the data positions obeys Eq. (6.24). The invariance principles that are used to find information J are the Lorentz condition (6.28) and conservation of flow (6.27) for the stress-energy. The total number N of degrees of freedom $q_{\mu\nu}(x^\alpha)$ is fixed at 10 by the two-index nature of all quantities and symmetry under interchange of the indices.

With information J determined, the 10 amplitudes $q_{\mu\nu}(x^\alpha)$ are found to obey a wave equation (6.58) due, in particular, to a *gravitational source* $T_{\mu\nu}$. This has the form of the weak-field gravitational equation. Hence, it implies that the amplitudes $q_{\mu\nu}$ are linear (6.59) in the corresponding metric elements $\bar{h}_{\mu\nu}$. This results in the weak-field equation (6.61), where the cosmological constant Λ arises as a constant of integration $C_{\mu\nu}$ in Eq. (6.54). By a dimensional argument in Sec. 6.3.23 the cosmological constant is found to be zero. The same kind of argument fixes the Planck length L as obeying proportionality to Eq. (6.64).

Because of the correspondence between quantities $q_{\mu\nu}$ and $\bar{h}_{\mu\nu}$, there is a wave equation (6.66) for the probability amplitudes corresponding to the weak-field metric equation (6.61). The existence of this probability amplitude implies the existence of a PDF (6.67) on four-fluctuation x^α. Its marginal PDF on the time has properties that are consistent with the hypothesis that gravitons are created and absorbed.

The concept of a 'prior measurement' was defined in Sec. 2.1.1. The derived Eqs. (6.61) and (6.66) show that each 'prior measurement' $y_{\mu\nu}^\alpha$ defines a new degree of freedom $q_{\mu\nu}$ or $\bar{h}_{\mu\nu}$ of the theory. This is an intuitively correct result. It implies that, in the sense of acquired information Eq. (6.17), *ten* prior measurements of the metric tensor are sufficient to define classical gravitational theory.

EPI regards the Einstein field equation as having a *fundamentally statistical* nature. In analogy with the electromagnetic case in Chap. 5, the statistics are an outgrowth of admitting that any measurement of the gravitational metric must suffer from random errors in four-position. That is, the measurer does not know precisely where and when the metric is measured.

It is interesting to compare this viewpoint with that of conventional gravitation theory. The latter, of course, derives from a framework of *deterministic* Riemannian geometry. In this chapter, by contrast, we showed that gravitation grows out of the concept of Fisher information, a geometrical measure of distance in a statistical parameter space (Secs. 1.4.3, 1.5). By this approach, gravitation may be viewed as arising from a framework of *statistical* geometry. See also Amari (1985) on the subject of the mathematics of distance (information) measures in statistical parameter space; and Caianiello (1992) on some physical interpretations.

7

Classical statistical physics

7.1 Goals

Classical statistical physics is usually stated in the non-relativistic limit, and so we restrict ourselves to this limit in the analyses to follow. However, as usual, we initiate the analysis on a covariant basis.

The overall aim of this chapter is to show that many classical distributions of statistical physics, defining both equilibrium and non-equilibrium scenarios, follow from a covariant EPI approach. Such equilibrium PDFs as the Boltzmann law on energy and the Maxwell–Boltzmann law on velocity will be derived. Non-equilibrium Hermite–Gaussian PDFs on velocity will also be found. Finally, some recently discovered inequalities linking entropy and Fisher information will be derived.

7.2. Covariant EPI problem

7.2.1 Physical scenario

Let a gas be composed of a large number M of identical molecules of mass m within a container. The temperature of the gas is kept at a constant value T. The molecules are randomly moving and mutually interacting through forces due to potentials. The particles randomly collide with themselves and the container walls. All such collisions are assumed to be perfectly elastic. The mean velocity of any particle over many samples is zero.

Measurements E and μ are made of the energy and momentum, respectively, of a randomly selected particle. This defines an *(energy, momentum) joint event*, and is a Fourier-space counterpart to the (space, time) event characterizing the quantum mechanical analysis in Chap. 4. The measurements perturb the amplitude functions $\mathbf{q}(E, \mu)$ of the problem, initiating the EPI process.

179

Denote the true energy and momentum of the particle as $(\theta_E, \boldsymbol{\theta}_\mu)$, respectively, where the subscript μ signifies momentum (and is not a numerical index!), and $\boldsymbol{\theta}_\mu \equiv (\theta_{\mu 1}, \theta_{\mu 2}, \theta_{\mu 3})$ are the usual Cartesian components. By Eq. (2.1) the intrinsic date $\mathbf{y} \equiv (y_E, \mathbf{y}_\mu)$ obey

$$y_E \equiv E = \theta_E + x_E, \quad E_0 \leqslant y_E \leqslant \infty, \tag{7.1a}$$

$$\mathbf{y}_\mu = \boldsymbol{\theta}_\mu + \mathbf{x}_\mu, \quad \mathbf{y}_\mu = (y_{\mu 1}, y_{\mu 2}, y_{\mu 3}), \tag{7.1b}$$

$$\mathbf{x}_\mu \equiv c\boldsymbol{\mu} \tag{7.1c}$$

with fluctuations x_E and $\mathbf{x}_\mu$. Subscripts 1, 2, 3 denote Cartesian components. All momentum coordinates $(\mathbf{y}_\mu, \boldsymbol{\theta}_\mu, \mathbf{x}_\mu)$ are expressed in units of the speed of light c, as in Eq. (7.1c), so as to share a common unit with the energy coordinates (see next section).

The particle obeys a joint PDF $p(x_E, \mathbf{x}_\mu)$ on the four-fluctuation $(x_E, \mathbf{x}_\mu)$. We first want to know its marginal PDFs $p(x_E)$ and $p(\mathbf{x}_\mu)$ or, equivalently, the corresponding probability amplitudes $q(x_E)$ and $q(\mathbf{x}_\mu)$ (see the second Eq. (2.18)). After finding $p(x_E)$ we will use Eq. (7.1a) to find the required law $p(E)$ on the energy. The EPI procedure (Sec. 3.4) is used.

7.2.2 Choice of Fisher coordinates

Given the aims of the problem, and the requirement (Sec. 3.5.8) that the Fisher coordinates be a four-vector, the chosen Fisher coordinates are the four-momentum $(ix_E, c\boldsymbol{\mu})$ (see Eq. (4.17)). This means that we have an imaginary energy coordinate and real momentum coordinates,

$$x_0 \equiv ix_E, \quad \mathbf{x}_\mu \equiv c\boldsymbol{\mu} \equiv (c\mu_1, c\mu_2, c\mu_3) \tag{7.2}$$

in the usual Cartesian components. It will be seen that the imaginary energy coordinate, in particular, is necessary for preserving normalization in the predicted $p(E)$.

7.2.3 Fisher information quantities

The basic unknowns of the problem are the probability amplitudes

$$q_n(x_E), \ n = 1, \ldots, N_E \text{ and } q_n(\mathbf{x}_\mu), \ n = 1, \ldots, N. \tag{7.3}$$

By Eq. (2.18) the Fisher information quantities $I(E)$ and $I(\boldsymbol{\mu})$ for the energy and momentum obey, respectively,

$$I(E) = -4 \int dx_E \sum_n \left(\frac{dq_n(x_E)}{dx_E} \right)^2 \text{ and} \tag{7.4}$$

$$I(\boldsymbol{\mu}) = 4 \int d\mathbf{x}_\mu \sum_n \sum_{m=1}^{3} \left(\frac{\partial q_n(\mathbf{x}_\mu)}{\partial x_{\mu m}} \right)^2. \tag{7.5}$$

The minus sign in Eq. (7.4) is a consequence of the imaginary nature of coordinate x_0 in Eq. (7.2); see Appendix C.

7.2.4 Bound information quantities

Corresponding to each information functional $I(E)$ and $I(\boldsymbol{\mu})$ is a bound information functional $J(E)$ and $J(\boldsymbol{\mu})$ to be determined. By the EPI principle (3.16), (3.18) we have two EPI problems to solve:

$$I(E) - J(E) = extrem., \quad I(E) - \kappa J(E) = 0 \tag{7.6}$$

and

$$I(\boldsymbol{\mu}) - J(\boldsymbol{\mu}) = extrem., \quad I(\boldsymbol{\mu}) - \kappa J(\boldsymbol{\mu}) = 0. \tag{7.7}$$

We first solve problem (7.6) for $p(E)$.

7.3 Boltzmann probability law

Here our goal is to determine $p(E)$. This will be found by first determining $p(x_E)$ and then using Eq. (7.1a) as a Jacobian transformation (see Frieden, 1990) from coordinate x_E to E. For convenience, where all coordinates are understood to represent energies, we drop superfluous subscripts E and use notation

$$x \equiv x_E, \ E \equiv y_E, \ N \equiv N_E \text{ and}$$

$$J(E) \equiv J[\mathbf{q}] \text{ where } \mathbf{q} \equiv (q_1, \ldots, q_N). \tag{7.8}$$

The latter designates information J to be a functional of amplitudes $\mathbf{q}$; see Sec. 0.2.

7.3.1 Fisher information

As we saw in Eq. (7.4), in this imaginary-coordinate case Eq. (2.19) becomes negative,

$$I(E) \equiv I = -4 \int dx \sum_n \left(\frac{dq_n}{dx} \right)^2, \ q_n = q_n(x). \tag{7.9}$$

The $\mathbf{q}$ are the unknown probability amplitudes of the problem. Also, by Eq. (2.23) the PDF relates to the $\mathbf{q}$ as

$$p(x) = \frac{1}{N} \sum_n q_n^2(x). \tag{7.10}$$

7.3.2 Finding J

As in the preceding two chapters, we find the unknown information functional $J[\mathbf{q}]$ by demanding that both requirements (3.16) and (3.18) (requirements (7.6) here) give the same answer $\mathbf{q}$. Note that a more general form $J[\mathbf{q}, x]$ is not needed; see Sec. 7.4.10.

We first carry through requirement (3.16). Represent

$$J[\mathbf{q}] = 4\int dx \sum_n J_n(q_n). \tag{7.11}$$

Components J_n then need to be found. Using Eq. (7.9), extremum principle (3.16) becomes

$$K \equiv I - J = -4\int dx \sum_n (q_n'^2 + J_n(q_n)). \tag{7.12}$$

The Euler–Lagrange Eq. (0.34) for the problem is

$$\frac{d}{dx}\left(\frac{\partial \mathscr{L}}{\partial q_n'}\right) = \frac{\partial \mathscr{L}}{\partial q_n}, \quad q_n' \equiv \frac{dq_n}{dx}. \tag{7.13}$$

The Lagrangian $\mathscr{L}$ is the integrand of Eq. (7.12),

$$\mathscr{L} = -4\sum_n (q_n'^2 + J_n(q_n)). \tag{7.14}$$

The solution $\mathbf{q}$ therefore obeys a condition

$$q_n'' = \frac{1}{2}\frac{\partial J_n}{\partial q_n}, \quad n = 1, \ldots, N. \tag{7.15}$$

Now we turn to requirement (3.18). Using Eqs. (5.13) and (7.9) gives

$$I = 4\int dx \sum_n q_n q_n''. \tag{7.16}$$

Using this with Eq. (7.11) gives

$$I - \kappa J = 4\int dx \sum_n (q_n q_n'' - \kappa J_n) \equiv 0. \tag{7.17}$$

The solution to this problem is the microscale Eq. (3.20), which is here

$$q_n q_n'' = \kappa J_n. \tag{7.18}$$

Therefore the common solution $\mathbf{q}$ to conditions (3.16) and (3.18) may be found. By Eqs. (7.15) and (7.18), $\mathbf{q}$ obeys

$$\frac{1}{2}\frac{\partial J_n}{\partial q_n} = \frac{\kappa J_n}{q_n} = \frac{1}{2}\frac{dJ_n}{dq_n} \tag{7.19}$$

where the latter is because $J_n = J_n(q_n)$ alone. Eq. (7.19) may be integrated, giving

$$J_n(q_n) = A_n q_n^{2\kappa}, \; A_n = Const. \geqslant 0. \tag{7.20}$$

This may be back-substituted into either of Eqs. (7.15) or (7.18) to get the same solution (as we demanded in the preceding). Either substitution gives a requirement

$$q_n'' = \alpha_n^2 q_n^{2\kappa - 1}, \; \alpha_n^2 \equiv \kappa A_n \geqslant 0. \tag{7.21}$$

7.3.3 Fixing κ

The solution $\mathbf{q}$ to Eq. (7.21) depends decisively upon the value of parameter κ. At this point we do not know κ, nor have we used any prior information about the physics of the measurement scenario. By the EPI procedure, this should take the form of an invariance principle.

The integral of $p(x)$ must be unity regardless of the particular form of $p(x)$. This is true regardless of the observer's reference frame; and is the statement of invariance that we use. If J is a normalization integral, this represents particularly weak prior information in the sense that any PDF obeys normalization. A phenomenon that only obeys normalization is said to exhibit 'maximum ignorance' in its independent variable.

We will find that, in fact, the choice of κ that makes J a normalization integral leads to the Boltzmann law form of solution for $p(E)$. That is, the Boltzmann law represents a scenario of maximum ignorance about the energy. Let

$$\kappa = 1. \tag{7.22}$$

By Sec. (5.1.18), this represents a scenario of maximum data efficiency, i.e., where all of the bound information is relayed into the data. Hence, we have a net situation of maximum ignorance in the face of maximum data efficiency.

7.3.4 General solution

For the particular case Eq. (7.22), Eq. (7.21) becomes

$$q_n'' = \alpha_n^2 q_n. \tag{7.23}$$

The general solution is exponential,

$$q_n = B_n \exp(-\alpha_n x) + C_n \exp(+\alpha_n x), \; \alpha_n \geqslant 0, \; B_n, \; C_n = Const. \tag{7.24}$$

Since the energy x is bounded below but of unlimited size above (Eq. (7.1a)), Eq. (7.24) could not represent a normalizable PDF $p_n \equiv q_n^2$ unless all $C_n = 0$. Hence

$$q_n = B_n \exp(-\alpha_n x). \tag{7.25}$$

7.3.5 Non-oscillatory nature of solution

We can see, now, why the imaginary Fisher coordinate ix_E was used (Eq. (7.2)). Use of a real coordinate instead would result in the negative of the right-hand side of Eq. (7.16). Proceeding with the derivation on this basis gives, instead of the exponential solution (7.25), a sinusoidal solution (unless we take $\kappa = -1$, which seems unphysical). The resulting $p_n \equiv q_n^2$ is likewise sinusoidal. It is, therefore, non-normalizable since, by Eqs. (7.1a), x is of unlimited size.

7.3.6 Nature of information game

It should be remarked that, with I here negative, by Eq. (7.17) so is J. Regardless of this, however, the observer in the game of Sec. 3.4.12 is still taking independent data, i.e., attempting to (now) maximize $-I$. Likewise, although J is negative nature is still attempting to minimize $|J| = -I$ through increased blur. One can embrace both cases of real and imaginary coordinates, then, by a game for which the aim of each opponent is to maximize the *amplitude* of his/her information I regardless of its sign.

7.3.7 Value of N, resulting p(x)

There are no physical constraints in this problem that need, or can utilize, the distinct solutions q_n given by (7.25). Therefore, for this problem effectively

$$N = 1. \tag{7.26}$$

Then by Eqs. (7.10), (7.25) and (7.26)

$$p(x) = B^2 e^{-2\alpha x}. \tag{7.27}$$

7.3.8 PDF p(E)

By Eq. (7.1a), the random variable E is simply a shifted version of $x_E = x$. Since $dE/dx = 1$, the PDF on E is, from Eq. (7.27),

$$p(E) = Ce^{-2\alpha E}, \ C \equiv B^2 e^{2\alpha\theta}. \tag{7.28}$$

The evaluation of the parameters α, C of this PDF requires more specific knowledge of the physical scenario, in particular a number for the lower bound to E.

7.3.9 Lower bound to energy

By inequality (7.1a) the domain of energy E is $E_0 \leqslant E < \infty$. Suppose we also know the mean value $\langle E \rangle$. It is easily shown that the constants C and α in Eq. (7.28) that satisfy both normalization and $\langle E \rangle$ give a

$$p(E) = b^{-1} \exp\left[-(E - E_0)/b\right], \ b = (\langle E \rangle - E_0), \ \text{for} \ E \geqslant E_0 \qquad (7.29)$$

and $p(E) = 0$ for $E < E_0$. This is a simple exponential law. It is interesting that the exponent must always be negative, indicating monotonically decreasing probability for increasing measured energy.

The nonrelativistic limit of Eq. (4.17) with fields inserted is, of course, that energy $E = E_{kin} + V$ where E_{kin} is the kinetic energy of the particle and V is its potential energy at the given point of detection. We can always add a constant to V without affecting any physical observable. Hence, subtract the constant E_0 from it. Then Eq. (7.29) becomes a simple exponential form

$$p(E) = \langle E \rangle^{-1} e^{-E/\langle E \rangle}, \ E \geqslant 0. \qquad (7.30)$$

7.3.10 Exercise

Verify that the PDF (7.28) becomes (7.29) by imposing normalization and a known mean value $\langle E \rangle$ upon (7.28).

7.3.11 Boltzmann law

We now express $\langle E \rangle$ in terms of physical quantities. Designate by E_t the total energy in the gas. As there are M identical particles, it is elementary that

$$\langle E_t \rangle = M \langle E \rangle. \qquad (7.31)$$

Energy E of Sec. 1.8.7 is called E_t here. In this notation, by Eq. (1.41) and either (1.44) or (1.45) we have

$$\overline{p}V = -\frac{dE_t}{dV} V = -\left\langle \frac{E_t}{V} \right\rangle V = -\langle E_t \rangle \sim T \ \text{or} \ T_E, \qquad (7.32)$$

the latter the Fisher temperature. This assumes a well-mixed, perfect gas. More specifically, if the gas particles are constrained to move without rotation within the container (three degrees of freedom per molecule) it is well-known that

$$\langle E_t \rangle = 3MkT/2 \qquad (7.33)$$

where k is the Boltzmann constant. This is usually called the 'equipartition of energy' law. Then by (7.31) $\langle E \rangle = 3kT/2$, so that Eq. (7.30) becomes

$$p(E) = (3kT/2)^{-1}e^{-2E/3kT}, \ E \geqslant 0, \tag{7.34}$$

the Boltzmann law for a three-dimensional gas.

7.3.12 Transition to discrete states

But of course the energies of the gas particles are not indefinitely continuous. Depending upon the type of gas particle that is present, one of the quantum equations (4.28) or (4.42) or (D9) governs the energy levels. The result is that any particle energy value E is quantized as $E \to E_j$, $j = 0, 1, \ldots$ where integer j denotes a state of the particle. Thus, by the proportionality between a probability density and its corresponding absolute probability, Eq. (7.30) now describes an absolute probability

$$P(E_j) \equiv P_j = Ce^{-E_j/\langle E \rangle}, \ C = const., \ j = 0, 1, \ldots \tag{7.35}$$

If we now use result (4.17) with the usual replacement $E \to E - V$ (ignoring any vector potential $\mathbf{A}$), we get

$$P_j = Ce^{-[E_{kin}(\boldsymbol{\mu}_j)+V(\mathbf{r}_j)]/\langle E \rangle}, \tag{7.36a}$$

$$E_{kin} = (c^2\mu^2 + m^2c^4)^{1/2} - mc^2 \tag{7.36b}$$

$$\approx \mu^2/2m \tag{7.36c}$$

in the non-relativistic limit. The constant C has absorbed a constant factor $\exp(-mc^2/\langle E \rangle)$. Quantity $\mathbf{r}_j$ is a random position of the particle, and E_{kin} and V are its kinetic and potential energies, respectively. Both E_{kin} and V are random variables, since they are functions of the random variables $\boldsymbol{\mu}_j$ and $\mathbf{r}_j$. Hence, Eq. (7.36a) becomes a joint probability law

$$P(E_{kin\,j}, V_j) = P_E(E_{kin\,j})P_V(V_j), \quad \text{where} \tag{7.37a}$$

$$P_E(E_{kin\,j}) \sim e^{-E_{kin}(\boldsymbol{\mu}_j)/\langle E \rangle} \quad \text{and} \quad P_V(V_j) \sim e^{-V(\mathbf{r}_j)/\langle E \rangle}. \tag{7.37b}$$

This allows us to form discrete probabilities $P_r(\mathbf{r}_j)$ and $P_\mu(\boldsymbol{\mu}_j)$ on momentum $\boldsymbol{\mu}_j$ and position $\mathbf{r}_j$ as follows.

A given potential value V is associated with a fixed number of position values $\mathbf{r}_j$. As examples, if $V \sim 1/r$ then the association is unique; or if $V \sim r^2$ then there are two values $\mathbf{r}_j$ for each V. The upshot is that $P_r(\mathbf{r}_j) \sim P_V(V_j)$. In the same way, $P(\boldsymbol{\mu}_j) \sim P_E(E_{kin\,j})$. Then by Eqs. (7.37b)

$$P(\mathbf{r}_j) \sim e^{-V(\mathbf{r}_j)/\langle E \rangle} \quad \text{and} \tag{7.38a}$$

$$P(\boldsymbol{\mu}_j) \sim e^{-E_{kin}(\boldsymbol{\mu}_j)/\langle E \rangle}. \tag{7.38b}$$

The latter will act as a check on the answer we get for $p(\boldsymbol{\mu})$ below. The former is often used in a continuous limit, as follows.

7.3.13 Barometric formula

Suppose that the gas in question is in a gravitational field. We want to predict the number density of particles as a function of the altitude z. Assuming that the altitude is small compared to the earth's radius, the potential is $V(\boldsymbol{r}) \equiv V(z) = mgz$, where g is the acceleration due to gravity. Placing this in the continuous version of Eq. (7.38a) gives

$$p(z) = p(0)e^{-mgz/\langle E\rangle}. \qquad (7.39)$$

By the law of large numbers (Frieden, 1991) this defines as well the required number density of particles. Unfortunately, this formula does not hold experimentally. The problem traces from our assumption that the temperature T is constant; of course it varies with altitude z.

7.4 Maxwell–Boltzmann velocity law

This distribution law $p(v)$ on the magnitude of the velocity fluctuation is found as follows. First we find $p(\boldsymbol{x}_\mu)$, the equilibrium PDF on c times the momentum fluctuations $\boldsymbol{\mu}$. Once known, this readily gives $p(\boldsymbol{\mu})$, the PDF on momentum or, by $\boldsymbol{\mu} = m\mathbf{v}$, the desired PDF $p(v)$.

To find $p(\boldsymbol{x}_\mu)$, we proceed as in Sec. 7.3, solving for the unknown bound information J by simultaneously solving the two EPI problems (7.7). The essential differences are that, here, (a) the information efficiency parameter κ has already been fixed, by Eq. (7.22), at value $\kappa = 1$; (b) the Fisher coordinates are *real* values $c\boldsymbol{\mu}$ as compared with the imaginary coordinate ix_E used previously; and (c) the bound information is represented more generally, with an explicit dependence upon $\boldsymbol{\mu}$. As will be seen, the result for PDF $p(\boldsymbol{\mu})$ will confirm result (7.38b) in the non-relativistic limit (7.36c). Since most of the steps repeat those of Sec. 7.2, we can proceed at a slightly faster pace.

7.4.1 Bound information J

Here represent

$$J[\mathbf{q}] = 4\int d\mathbf{x} \sum_n J_n(q_n, \mathbf{x}), \quad \mathbf{x} \equiv \mathbf{x}_\mu \qquad (7.40)$$

as simpler notation. We allow for an explicit dependence upon the momenta $\mathbf{x}$ as well as upon the amplitudes $\mathbf{q}$. This more general representation than the corresponding one (7.11) for energy will permit a wider scope of solutions to the problem including, in fact, non-equilibrium solutions! An explicit x-dependence could have been used in Eq. (7.11) as well, but with no change in the Boltzmann result (7.34). See Sec. 7.4.10 below.

We now proceed to solve the two EPI problems (7.7).

7.4.2 EPI extremum problem

By Eqs. (7.5) and (7.40) the first problem (7.7) is

$$I(\boldsymbol{\mu}) - J(\boldsymbol{\mu}) = 4 \int d\mathbf{x} \sum_n \left[\sum_{m=1}^{3} \left(\frac{\partial q_n}{\partial x_m} \right)^2 - J_n(q_n, \mathbf{x}) \right] = extrem. \qquad (7.41)$$

The Euler–Lagrange Eq. (0.34) for the solution is

$$\sum_m \frac{\partial}{\partial x_m} \left(\frac{\partial \mathscr{L}}{\partial q_{nm}} \right) = \frac{\partial \mathscr{L}}{\partial q_n}, \quad n = 1, \ldots, N$$

$$\qquad (7.42)$$

$$q_{nm} \equiv \frac{\partial q_n}{\partial x_m}.$$

With the Lagrangian $\mathscr{L}$ the integrand in Eq. (7.41), the solution is the differential equation

$$\sum_m \frac{\partial^2 q_n}{\partial x_m^2} = -\frac{1}{2} \frac{\partial J_n(q_n, \mathbf{x})}{\partial q_n}. \qquad (7.43)$$

7.4.3 EPI zero-root problem

By Eqs. (7.5) and (7.40) the second problem (7.7) is

$$I - \kappa J = -4 \int d\mathbf{x} \sum_n \left[q_n(\mathbf{x}) \sum_m \frac{\partial^2 q_n}{\partial x_m^2} + J_n(q_n, \mathbf{x}) \right] = 0, \qquad (7.44)$$

where we used $\kappa = 1$ and the partial integration (5.13) of I. The solution is the microscale Eq. (3.20), which is here

$$q_n \sum_m \frac{\partial^2 q_n}{\partial x_m^2} = -J_n(q_n, \mathbf{x}). \qquad (7.45)$$

7.4.4 Simultaneous solution

The simultaneous solution $\mathbf{q}$ and J_n to conditions (7.43) and (7.45) obviously obeys

$$\frac{1}{2} \frac{\partial J_n(q_n, \mathbf{x})}{\partial q_n} = \frac{J_n(q_n, \mathbf{x})}{q_n}. \qquad (7.46)$$

This may be integrated in J_n and q_n to give

$$J_n(q_n, \mathbf{x}) = q_n^2 f_n(\mathbf{x}), \; f_n(\mathbf{x}) \geqslant 0, \tag{7.47}$$

for some positive functions $f_n(\mathbf{x})$. These arose as additive 'constants' (in $\mathbf{q}$ and J_n) during the integration above. The positivity requirement arises since e to a real number is $\geqslant 0$.

7.4.5 *Nature of functions $f_n(x)$*

Substituting the result (7.47) into either Eq. (7.43) or (7.45) produces the same solution (as we required),

$$\nabla^2 q_n = -q_n(\mathbf{x}) f_n(\mathbf{x}), \tag{7.48}$$

where the Laplacian ∇^2 is with respect to coordinates x_i, $i = 1, 2, 3$. This shows that the form of functions $f_n(\mathbf{x})$ directly affects that of the output amplitudes $\mathbf{q}$ and, hence, the PDF $p(\mathbf{x})$.

We have not yet input into the development anything 'momentum-like' about the x_i. Some plausible assumptions of this kind can be made. First, at equilibrium, the probability of a given momentum should not depend upon its direction. Hence, $p(\mathbf{x})$ should be even in each component x_i. Second, $p(\mathbf{x})$ should depend upon each x_i in the same way. Third, we take a non-relativistic approach by which all velocities are small, so that the x_i are small as well.

These considerations imply that $f_n(\mathbf{x})$ should be expandable as a power series in even powers of the modulus x of $\mathbf{x}$,

$$f_n(\mathbf{x}) = A_n + B_n x^2, \; A_n, \; B_n = const. \tag{7.49}$$

It is not necessary to include terms beyond the quadratic because x is small. The constants need to be defined.

7.4.6 *Hermite–Gauss solutions*

Substituting series (7.49) into Eq. (7.48) gives as a solution for $\mathbf{q}$ the differential equation

$$\nabla^2 q_n(\mathbf{x}) + (A_n + B_n x^2) q_n(\mathbf{x}) = 0. \tag{7.50}$$

A separation of variables

$$q_n(\mathbf{x}) = q_{n1}(x) q_{n2}(y) q_{n3}(z), \; \mathbf{x} \equiv (x, y, z) \tag{7.51}$$

gives three distinct differential equations

$$q''_{ni}(x_i) + (A_{ni} + Bx_i^2) q_{ni}(x_i) = 0, \; i = 1, 2, 3, \; B_n \equiv B, \; \sum_{i=1}^{3} A_{ni} \equiv A_n. \tag{7.52}$$

The coordinate $x_i \equiv x$, y or z, in turn. Each such equation becomes a parabolic cylinder differential equation (Abramowitz and Stegun, 1965) if the constants obey

$$A_{ni} = \frac{(n_i + 1/2)}{a_0^2}, \quad \sum_{i=1}^{3} n_i \equiv n, \tag{7.53a}$$

$$A_n = \frac{n + 3/2}{a_0^2}, \quad B = -\frac{1}{4a_0^4}, \quad a_0 = const. \tag{7.53b}$$

Then Eq. (7.52) has a Hermite–Gaussian solution

$$q_{ni}(x_i) = e^{-x_i^2/4a_0^2} 2^{-n_i/2} H_{n_i}(x_i/a_0\sqrt{2}), \quad i = 1, 2, 3. \tag{7.54}$$

Also, in order to satisfy the requirement (7.47) of positive $f_n(\mathbf{x})$ by Eqs. (7.49) and (7.53) the net momentum must obey

$$x^2 \leqslant 2(2n + 3)a_0^2. \tag{7.55}$$

This is indeed obeyed in the non-relativistic (small x) limit.

7.4.7 Superposition states

Eq. (7.53a) states that there is a degeneracy of product solutions for each index value n. Eqs. (7.51), (7.53a) and (7.54) give a solution

$$q_n(\mathbf{x}) = e^{-|\mathbf{x}|^2/4a_0^2} 2^{-n/2} \sum_{\substack{ijk \\ i+j+k=n}} a_{nijk} H_i(x/a_0\sqrt{2}) H_j(y/a_0\sqrt{2}) H_k(z/a_0\sqrt{2}),$$

$$a_{nijk} = const. \tag{7.56}$$

The Hermite polynomials are defined as (Abramowitz and Stegun, 1965)

$$H_n(x) = n! \sum_{m=0}^{[n/2]} (-1)^m \frac{(2x)^{n-2m}}{m!(n-2m)!} \tag{7.57}$$

where the notation $[b]$ means the largest integer not exceeding b. The lowest-order polynomials are

$$H_0(x) = 1, \quad H_1(x) = 2x, \quad H_2(x) = 4x^2 - 2. \tag{7.58}$$

Using Eqs. (2.23) and (7.56), the PDF on momentum fluctuations is
$$p(\mathbf{x}) =$$

$$Ae^{-|\mathbf{x}|^2/2a_0^2} \left\{ 1 + \left[\sum_{\substack{nijk \\ i+j+k=n \\ n>0}} a_{nijk} H_i(x/a_0\sqrt{2}) H_j(y/a_0\sqrt{2}) H_k(z/a_0\sqrt{2}) \right]^2 \right\},$$

$$A = a_{0000}^2/N. \tag{7.59}$$

The '1' is from the sum evaluated at $n = 0$. The constants a_{nijk} are proportional to those in (7.56).

7.4.8 Interpretation of general solution

Because of the free parameters a_{nijk} in Eq. (7.59), the PDF on momentum obeys a multiplicity of solutions. Any particular one is defined by a set of the a_{nijk}. What can this mean? Since EPI Eq. (3.16) seeks a stationary solution, the result (7.59) indicates that there is a multiplicity of such solutions to this problem. Of these, one must represent the equilibrium solution, approached as time $t \to \infty$, with the others representing stationary solutions *en route* to equilibrium. Which one is the equilibrium solution?

Since our Fisher coordinates $\mathbf{x} \equiv c\boldsymbol{\mu}$ are purely real here, the extrema that are attained in I by the solutions (7.56) must be true minima (Sec. 1.8.8). Then, by the I-theorem (1.30), the equilibrium solution must be the stationary solution that attains the *smallest* minimum among all possible minima. Which solution has the smallest minimum in I?

Observing Eq. (7.59), we see that in the absence of the terms in the sum (for $n > 0$) the overall $p(\mathbf{x})$ would be a smooth Gaussian. The presence of the $n > 0$ terms causes subsidiary maxima and minima, or ripples, in the curve. This represents an increased gradient content over the Gaussian. Hence, this must represent an increased level of information I, by Eq. (7.4). The result is that the Gaussian solution attains the absolute minimum in I and, therefore, represents the equilibrium solution we sought.

The remaining solutions therefore represent stationary solutions *en route* to the Gaussian equilibrium solution. That is, they are examples of non-equilibrium solutions. In fact, Rumer and Ryvkin (1980) previously found these Hermite–Gauss functions to be non-equilibrium solutions that follow from the Boltzmann transport equation. Our particular Hermite–Gauss solution (7.59) corresponds to a particular choice of these authors' expansion coefficients. We find it rather remarkable that the EPI approach, which avoids any use of the Boltzmann equation, attained essentially the same class of solution.

The fact that the non-equilibrium solutions of (7.59) are stationary appears to indicate that they are, in some sense, more probable or more frequent in occurrence than other types of non-equilibrium solutions. This conjecture ought to be testable by experimental observation or by Monte Carlo computer simulation.

7.4.9 Correspondence between derived $P(E_{kin\,j})$ and $p(\mu)$ functions

Eqs. (7.36c) and (7.38b) predict that, in the non-relativistic limit,

$$P(E_{kin\,j}) \to P(\boldsymbol{\mu}_j) \sim e^{-\mu^2/2m\langle E \rangle} \tag{7.60}$$

while the equilibrium case of solution (7.59) predicts that

$$p(\boldsymbol{\mu}) \sim A e^{-\mu^2/2a_0^2} \tag{7.61}$$

for some choice of a_0. Both expressions were derived for the non-relativistic case and, therefore, should agree. The fact that they do agree is a verification of the overall theory. This also confirms the argumentation of the previous section.

7.4.10 A retrospective on the p(E) derivation

We assumed, with some loss of generality, that each $J_n = J_n(q_n)$ alone in Eq. (7.11). A possible x-dependence, as $J_n(q_n, x)$ was left out. Now we can see why it was. As at Eq. (7.56), a superposition of stationary solutions would result, only one of which is the required equilibrium solution $p(E)$. Again, the smoothest one would represent the equilibrium solution, and this is the simple exponential form (7.25) as previously derived! The remaining solutions would represent non-equilibrium laws, as in Sec. 7.4.8.

7.4.11 Equivalent constrained I problem

We remarked in Sec. 1.8.8 that a constrained minimization (of I) superficially resembles the EPI approach. Using the quadratic form (7.49) in Eq. (7.47) leads to a minimization problem (7.41) with a 'constraint' term

$$-4 \int dx \sum_n q_n^2(\mathbf{x})(A_n + Bx^2) = -4 \left[\sum_n A_n \int dx\, q_n^2(\mathbf{x}) + B \int d\mathbf{x}\, x^2 \sum_n q_n^2(\mathbf{x}) \right]$$

$$= \lambda_1 \int d\mathbf{x}\, p(\mathbf{x}) + \lambda_2 \int d\mathbf{x}\, x^2 p(\mathbf{x}), \quad \lambda_1, \lambda_2 = const. \tag{7.62}$$

Normalization of $p(\mathbf{x})$ and of each PDF $q_n^2(\mathbf{x})$ was used, along with Eq. (2.23). Hence, in this case the EPI approach is *mathematically* equivalent to a 'constrained minimization' approach Eq. (0.39), where the constraints are those of normalization and a fixed second moment.

The correspondence is interesting but, in fact, largely coincidental. Two factors should be considered.

First, in contrast to constrained minimization, EPI has a definite mechanism for finding the effective constraint term. This is the bound information J which, by Eq. (3.18), *must be proportional to I.*

Second, to obtain the correct result by EPI a minimal number of effective constraints must be used. That is, nature imposes a minimal set of constraint conditions upon the Fisher information, and not a maximal set. As empirical evidence for this, in all of the derivations up to this point only one or, at most, two constraints have effectively been imposed via the information J. In fact, adding the most trivial of constraints – normalization – to, say, the electromagnetic derivation of Chap. 5 gives as the output wave equation the Proca equation (Jackson, 1975) and not the electromagnetic Eq. (5.51). This may be easily verified. Or, adding a constraint on mean energy to the single normalization constraint in Sec. 7.3 would lead, not to the Boltzmann law, but to the square of an Airy $Ai(E)$ function as the predicted $p(E)$ (Frieden, 1988).

Thus, EPI is not an *ad hoc* approach wherein all constraints known to affect $p(\mathbf{x})$ are tacked onto the minimization of I. It would give incorrect results if used in this way. Instead, principle (3.18) of proportionality between I and J *must be used* to form the effective constraints through the action of J. A constrained minimization approach Eq. (0.39), by contrast, offers no such systematic method of finding its constraints.

7.4.12 *Equilibrium law on magnitude of velocity*

Eq. (7.61) may be evaluated in the particular case of zero potential energy. Then Eq. (7.33) holds, and since $\langle E_t \rangle = M\langle E_{kin} \rangle = M\langle \boldsymbol{\mu}^2 \rangle/2m$, gives $\langle \boldsymbol{\mu}^2 \rangle = 3mkT$. The constants A, a_0 in (7.61) that satisfy normalization and the given moment $\langle \boldsymbol{\mu}^2 \rangle$ then define a

$$p(\boldsymbol{\mu}) = \frac{1}{(2\pi)^{3/2} a_0^3} e^{-\mu^2/2a_0^2}, \ a_0^2 = 3mkT \tag{7.63}$$

for our particles with three degrees of freedom. This equation is of the separated form $p(\boldsymbol{\mu}) = p(\mu_1)p(\mu_2)p(\mu_3)$ in the three components of $\boldsymbol{\mu}$. Here, by contrast, we want the PDF on the magnitude μ. This may readily (if tediously) be found by first transforming coordinates from the given rectangular coordinates (μ_1, μ_2, μ_3) to spherical coordinates (μ, θ, ϕ). The formal solution is

$$p(\mu, \theta, \phi) = |J(\mu_1, \mu_2, \mu_3/\mu, \theta, \phi)| p(\mu_1, \mu_2, \mu_3) \tag{7.64}$$

where J is the Jacobian of the transformation (Frieden, 1991). After evaluating the Jacobian and integrating out over θ and ϕ, the solution is

$$p(\mu) = A\mu^2 e^{-3\mu^2/2a_0^2}, \ A = \sqrt{2/\pi} 3^{3/2}/a_0^3. \tag{7.65}$$

This is the Maxwell–Boltzmann law on the magnitude of the momentum. Transformation of this law to $p(v)$ via relation $\mu = mv$ then gives the usual Maxwell–Boltzmann law on the velocity.

7.5 Fisher information as a bound to entropy increase

Like the I-theorem (1.30), the following is a new finding that arises out of the use of classical Fisher information. As with the I-theorem, the output takes the form of an inequality. This is in contrast with the preceding EPI derivations which produce equality outputs. Thus, the derivation will not be a direct use of the EPI principle, but (as with EPI) will use the concepts of Fisher information and an invariance principle to form a fundamental physical law.

7.5.1 Scenario

A system consists of one or more particles moving randomly within an enclosure. Denote the probability density for finding a particle at position $r = (x, y, z)$ within the enclosure at the known time t as $p(r|t)$. A particle measurement r is made.

The enclosure is isolated. Hence, no particles leave the box and no new particles enter at its boundary $r = B$. Then the measurement r must lie within the enclosure, or

$$\int d\mathbf{r}\, p(\mathbf{r}|t) = 1, \tag{7.66}$$

a condition of normalization.

7.5.2 Shannon entropy

Denote by $H(t)$ the Shannon entropy of the system as evaluated at time t. This has the form Eq. (1.13),

$$H(t) = -\int d\mathbf{r}\, p(\mathbf{r}|t) \ln p(\mathbf{r}|t). \tag{7.67}$$

Suppose that the Shannon entropy represents the Boltzmann entropy as well. Then, by the Second Law of thermodynamics,

$$H_t \equiv \frac{dH}{dt} \geqslant 0. \tag{7.68}$$

(In this section it is convenient to denote derivatives by subscripts without commas.) This establishes a definite lower bound to the change in entropy, but is there an upper bound as well? If so, what is it?

7.5.3 Invariance condition

Since no particles either enter or leave the enclosure, the particles obey an equation of conservation of flow,

$$p_t(r|t) + \nabla \cdot P(r, t) = 0, \tag{7.69}$$

where P is a measure of flow whose exact nature depends upon the system. Denote the components of P as (P_1, P_2, P_3). Thus, numbered subscripts denote vector components, whereas (from before) letter subscripts denote derivatives.

7.5.4 Dirichlet boundary conditions

Assume that there is no net flow of particles across the boundaries of the enclosure. That is,

$$P(r, t) \bigg|_B = 0. \tag{7.70}$$

Hence, P obeys Dirichlet boundary conditions (Eq. (5.55)). Also, assume that if the boundary is at infinity then

$$\lim_{r \to \infty} P(r, t) \to 0 \tag{7.71}$$

faster than $1/r^2$.

Since the enclosure is isolated, there must be vanishing probability that a particle is on the boundary,

$$p(r|t) \bigg|_B = 0. \tag{7.72}$$

Hence p also obeys Dirichlet boundary conditions. Also, for a boundary at infinity let p obey

$$\lim_{r \to \infty} p(r|t) \to 0 \tag{7.73}$$

faster than $1/r^2$. The latter is required by the condition (7.66) of normalization.

We will have need to evaluate the quantity $P \ln p$ at the boundary. By conditions (7.70) and (7.72) this product is of the indeterminate form $-0 \cdot \infty$. Assume that the logarithmic operation 'weakens' the ∞ so that condition (7.70) dominates the product,

$$P \ln p \bigg|_B = 0. \tag{7.74}$$

7.5.5 Derivation

The partial derivative $\partial/\partial t$ of Eq. (7.67) gives

$$H_t = -\frac{\partial}{\partial t}\int d\mathbf{r}\, p \ln p = -\int d\mathbf{r}\, p_t \ln p - \int d\mathbf{r}\, p(1/p)p_t \qquad (7.75)$$

after differentiating under the integral sign. The second right-hand integral gives

$$\int d\mathbf{r}\, p(1/p)p_t = \frac{\partial}{\partial t}\int d\mathbf{r}\, p = 0 \qquad (7.76)$$

by normalization (7.66).

Next, use the flow Eq. (7.69) in the first right-hand integral of Eq. (7.75). This gives

$$H_t = \int d\mathbf{r}\, \nabla\cdot\mathbf{P} \ln p \equiv \iiint dz\, dy\, dx \left[\frac{\partial}{\partial x}P_1 + \frac{\partial}{\partial y}P_2 + \frac{\partial}{\partial z}P_3\right]\ln p. \qquad (7.77)$$

Consider the first right-hand term. The innermost integral is

$$\int dx\, \frac{\partial P_1}{\partial x}\ln p = P_1 \ln p\bigg|_{\mathbf{B}} - \int dx\,(P_1/p)p_x = 0 - \int dx\,(P_1/p)p_x \qquad (7.78)$$

by Dirichlet condition (7.74).

Analogous results follow for the second and third right-hand terms in Eq. (7.77). The result is that

$$H_t = -\int d\mathbf{r}\, \mathbf{P}\cdot\nabla p/p. \qquad (7.79)$$

Squaring the latter and factoring the integrand gives

$$H_t^2 = \left[\int d\mathbf{r}\,(\mathbf{P}/\sqrt{p})\cdot(\sqrt{p}\nabla p/p)\right]^2. \qquad (7.80)$$

Temporarily replace the integral $d\mathbf{r}$ by a very fine sum over index m, and also execute the dot product as a sum over components n. This gives

$$H_t^2 = \left[\sum_{nm}\left(\frac{P_{nm}}{\sqrt{p_m}}\right)\left(\frac{\sqrt{p_m}}{p_m}\nabla_n p_m\right)\right]^2. \qquad (7.81)$$

Now, the *Schwarz inequality* states that for any two quantities A_{nm}, B_{nm}

$$\left[\sum_{nm}A_{nm}B_{nm}\right]^2 \leq \sum_{nm}A_{nm}^2 \sum_{nm}B_{nm}^2. \qquad (7.82)$$

Comparing Eqs. (7.81) and (7.82) suggests that we identify

$$A_{nm} = \frac{P_{nm}}{\sqrt{p_m}}, \quad B_{nm} = \frac{\sqrt{p_m}}{p_m} \nabla_n p_m. \tag{7.83}$$

Then the two equations show that

$$H_t^2 \leqslant \sum_{nm} \frac{P_{nm}^2}{p_m} \sum_{nm} \frac{(\nabla_n p_m)^2}{p_m}. \tag{7.84}$$

Going back from the fine sum over index m to the original integral dr gives

$$H_t^2 \leqslant \sum_n \int \frac{P_n^2(r, t)}{p(r|t)} \sum_n \int dr \frac{[\nabla_n p(r|t)]^2}{p(r|t)}. \tag{7.85}$$

Replacing the sums over components n by dot product notation gives

$$H_t^2 \leqslant \int dr \frac{P \cdot P}{p} \int dr \frac{\nabla p \cdot \nabla p}{p}. \tag{7.86}$$

As in Sec. 2.4.2, assume that the PDF $p(r|t)$ obeys shift invariance. Since one measurement r has been made (Sec. 7.5.1), the Fisher information obeys Eq. (2.17) with *its* index $n = 1$. This tempts us to associate the second integral in Eq. (7.86) with I. However, note the following potential complication: the coordinates in Eq. (2.17) are *fluctuations* from the ideal value, i.e., noise values, whereas our coordinates r in Eq. (7.86) are data component values. Can the second integral in Eq. (7.86) therefore still represent the information? The answer is yes, since there is only a constant shift between the two sets of coordinates. (The proof is left to the interested reader.) Hence, Eq. (7.86) becomes

$$H_t^2 \leqslant I \int dr \frac{P \cdot P}{p}, \quad I \equiv I(t) = \int dr \frac{\nabla p(r|t) \cdot \nabla p(r|t)}{p(r|t)}. \tag{7.87}$$

This shows that the entropy change during a small time interval is bounded from above. The bound is proportional to the square root of the Fisher information in a position measurement. This upper bound is what we set out to find.

7.5.6 *Entropy bound for classical particle flow*

Let the system consist, now, of many material particles. The particles interact under the influence of any potential, and they also collide with each other and with the boundary walls located at position $r = B$. In such a scenario, the flow vector is

$$P(r, t) = p(r|t)v(r, t) \tag{7.88}$$

where v is the particle velocity (Lindsay, 1951, p. 284). Inequality (7.87) will hold for such a system, provided conditions (7.69)–(7.74) hold. We show that this is the case.

Since the system is isolated, there is no net flow of particles in or out so that condition (7.69) holds by definition.

Condition (7.70) is now

$$p(r|t)\mathbf{v}(r, t)\Big|_B = 0. \tag{7.89}$$

In order to satisfy condition (7.70), one or the other of the two factors must be zero on the boundary. If $\mathbf{v} = 0$ but $p \neq 0$ on the boundary that implies that once a particle is on the boundary it cannot move away from it. And since $p \neq 0$ there, ultimately every particle will be on the boundary! The system degenerates into a collapsed state. We will not consider this kind of specialized solution. Hence the solution to (7.89) is taken to be

$$p(r|t)\Big|_B = 0. \tag{7.90}$$

By Eq. (7.88), this satisfies (7.70), stating that no particles cross the boundary.

Since no particles cross the boundary, normalization condition (7.66) must hold. But for normalization to hold, it must be that

$$\lim_{r\to\infty} p(r|t) \to 0 \tag{7.91}$$

faster than $1/r^2$. Hence condition (7.73) holds.

By Eqs. (7.88) and (7.91), condition (7.71) holds. Also, condition (7.72) is now satisfied by Eq. (7.90).

Finally, condition (7.74) holds because, by Eq. (7.88)

$$P \ln p = \mathbf{v}p \ln p \tag{7.92}$$

and of course

$$\lim_{p\to 0} p \ln p = 0. \tag{7.93}$$

The zero limit for p is taken because of Eq. (7.90).

Since all of the requirements (7.70) through (7.74) hold, the inequality (7.87) holds for this problem.

7.5.7 Exercise

Prove the assertion (7.93) using l'Hopital's rule.

7.5.8 *Reduction of problem*

Combining Eqs. (7.87) and (7.88) gives

$$\frac{dH\,(t)}{dt} \leqslant \sqrt{I \int dr\, v^2 p} \;\; \text{or}$$

$$\frac{dH\,(t)}{dt} \leqslant \sqrt{I \langle v^2 \rangle} \tag{7.94}$$

by the definition of the expectation $\langle \cdot \rangle$. The positive sign for the square root is chosen because, by the second law, the change in H must be positive (or zero). Eq. (7.94) says that the rate of change of H is bounded jointly by the Fisher information I in a measurement r and the root-mean square velocity. In some cases, the latter is a constant so that there is a direct proportionality $dH/dt \leqslant C\sqrt{I}$, $C = const$. This occurs, for example, for the random scenario of Sec. 7.2.1 where, in addition, there are no forces on the particles. Then Eq. (7.33) holds, and with no potentials present $\langle E_t \rangle = \langle E_{kin} \rangle = Mm\langle v^2 \rangle/2 = 3MkT/2$, so that

$$\langle v^2 \rangle \equiv C^2 = 3kT/m. \tag{7.95}$$

The result is the interesting expression

$$\frac{dH\,(t)}{dt} \leqslant I(t)^{1/2} \left(\frac{3kT}{m} \right)^{1/2}. \tag{7.96}$$

Classical particle flow describes the motion of an ideal fluid. Hence, for such a medium the rate of change of entropy is bounded above by the square root of the Fisher information.

7.5.9 *Entropy bounds for electromagnetic flow phenomena*

The derivation in Sec. 7.5.5 shows that any flow phenomenon will obey the basic inequality (7.87) if Dirichlet boundary conditions (7.70)–(7.74) are obeyed. The phenomenon does not have to describe classical fluid flow specifically, as in the preceding sections. Quite a wide range of other phenomena obey the required boundary conditions, or can be restricted to cases that *do* obey the boundary conditions.

For example, the charge density ρ and current j obey the conservation of flow Eq. (5.7). Then, if we assume Dirichlet boundary conditions for ρ and j we have a mathematical correspondence $\rho(r, t) \rightarrow p(r|t)$, $j(r, t) \rightarrow P(r, t)$ between electromagnetic quantities and quantities of the derivation Sec. 7.5.5. Thus, for an entropy of charge density

$$H_\rho(t) \equiv -\int d\boldsymbol{r}\,\rho(\boldsymbol{r},\,t)\ln\rho(\boldsymbol{r},\,t),\ \rho \geqslant 0 \tag{7.97}$$

and an information quantity

$$I_\rho(t) \equiv \int d\boldsymbol{r}\,\frac{\nabla\rho\cdot\nabla\rho}{\rho}, \tag{7.98}$$

result (7.87) becomes

$$\left(\frac{dH_\rho}{dt}\right)^2 \leqslant I_\rho \int d\boldsymbol{r}\left(\frac{\boldsymbol{j}\cdot\boldsymbol{j}}{\rho}\right). \tag{7.99}$$

For a single moving charge, where $\boldsymbol{j} = \rho\mathbf{v}$, this simplifies to

$$\left(\frac{dH_\rho}{dt}\right)^2 \leqslant I_\rho \int d\boldsymbol{r}\,\rho v^2. \tag{7.100}$$

Similar results grow out of flow Eq. (5.8), with now the correspondences $q_4 \equiv \phi \to p$, $q_n \equiv A_n \to P_n$. The electromagnetic potentials take on the roles of p and P. Result (7.87) becomes

$$\left(\frac{dH_\phi}{dt}\right)^2 \leqslant c^2 I_\phi \int d\boldsymbol{r}\,\frac{\boldsymbol{A}\cdot\boldsymbol{A}}{\phi},\ I_\phi = \int d\boldsymbol{r}\,\frac{\nabla\phi\cdot\nabla\phi}{\phi}. \tag{7.101}$$

The entropy is, here, that of the scalar potential ϕ,

$$H_\phi(t) \equiv -\int d\boldsymbol{r}\,\phi(\boldsymbol{r},\,t)\ln\phi(\boldsymbol{r},\,t),\ \phi \geqslant 0. \tag{7.102}$$

This is a new concept. Note that it is mathematically well defined if one adds enough of a constant to the potential to keep it positive during the 'ln' operation. As with the Fisher information I_ϕ, it measures the 'spread-out-edness' of the function ϕ over space $\boldsymbol{r}$.

7.5.10 Entropy bounds for gravitational flow phenomena

These comments apply as well to the gravitational flow phenomena (6.27) for the stress-energy tensor and (6.28) for the weak-field tensor. The entropy inequality (7.87) would hold for these tensors as well. However, the entropy of stress-energy and the entropy of the weak field would be new concepts that need interpretation.

7.5.11 Entropy bound for quantum electron phenomena

A flow equation for the quantum electron may be obtained as follows. Multiply the Dirac Eq. (4.42) on the left by the Hermitian adjoint $\psi^\dagger$ (transpose of ψ^*);

multiply the Hermitian adjoint of (4.42) on the right by ψ, and subtract the two results. This gives a flow equation

$$\frac{\partial}{\partial t} p(\boldsymbol{r},\, t) + \nabla \cdot \boldsymbol{P}(\boldsymbol{r},\, t) = 0, \text{ where} \tag{7.103a}$$

$$p(\boldsymbol{r},\, t) = \psi^{\dagger}\psi, \ \psi \equiv \psi(\boldsymbol{r},\, t), \text{ and} \tag{7.103b}$$

$$\boldsymbol{P}(\boldsymbol{r},\, t) = -c\psi^{\dagger}[\alpha]\psi. \tag{7.103c}$$

Quantity $[\alpha]$ is the Dirac vector of matrices $[\alpha_x,\, \alpha_y,\, \alpha_z]$ defined by Eqs. (4.43).

We show, next, that conditions (7.70) and (7.72) are obeyed. The wave function ψ must continuously approach zero as $\boldsymbol{r}$ approaches any boundary $\boldsymbol{B}$ to measurement space (Schiff, 1955, pp. 29, 30). Since, by Eqs. (7.103b,c), both p and $\boldsymbol{P}$ increase quadratically as ψ necessarily

$$p(\boldsymbol{r},\, t) \bigg|_{\boldsymbol{B}} = 0, \tag{7.104a}$$

$$\boldsymbol{P}(\boldsymbol{r},\, t) \bigg|_{\boldsymbol{B}} = 0. \tag{7.104b}$$

Since we had $p(t) = 1$, by Eq. (4.30b) ψ obeys normalization requirement (7.66). Then if the boundary is at infinity, $p(\boldsymbol{r}|t)$ must fall off with r faster than $1/r^2$. This satisfies requirement (7.73). Properties (7.71) and (7.74) remain to be verified.

We turn to property (7.71), which requires the boundary $\boldsymbol{B}$ be at infinity. Because property (7.70) holds, by Eq. (7.103b) ψ must fall off with r as $1/r$ or faster. Hence, by Eq. (7.103c), $\boldsymbol{P}$ must fall off with r as $1/r^2$ or faster. Hence, requirement (7.71) is obeyed.

Finally, we consider requirement (7.74). By Eq. (7.103c) the x-component P_1 of $\boldsymbol{P}$ obeys

$$P_1 = -c(\psi_1\psi_2\psi_3\psi_4)^* \begin{bmatrix} 0 & 0 & 0 & 1 \\ 0 & 0 & 1 & 0 \\ 0 & 1 & 0 & 0 \\ 1 & 0 & 0 & 0 \end{bmatrix} \begin{bmatrix} \psi_1 \\ \psi_2 \\ \psi_3 \\ \psi_4 \end{bmatrix} \tag{7.105}$$

where we used the matrices (4.43), (4.44) defining $[\alpha_x]$. After the matrix products in Eq. (7.105) are carried out, the result is

$$P_1 = -c(\psi_1^*\psi_4 + \psi_2^*\psi_3 + \psi_3^*\psi_2 + \psi_4^*\psi_1). \tag{7.106}$$

Then, by Eqs. (7.103b, c) and (7.106), we have

$$P_1 \ln p = -c(\psi_1^*\psi_4 + \psi_2^*\psi_3 + \psi_3^*\psi_2 + \psi_4^*\psi_1)$$

$$\times \ln(|\psi_1|^2 + |\psi_2|^2 + |\psi_3|^2 + |\psi_4|^2). \tag{7.107}$$

Requirement (7.74) addresses the limiting form of this expression as $r \to \mathbf{B}$ where, by Eqs. (7.103b) and (7.104a), all components $\psi_i = 0$, $i = 1\text{-}4$. There is arbitrariness in how we choose the components to approach the boundary. We first evaluate ψ_2 and ψ_3 on the boundary, letting $\psi_2 = \psi_3 = 0$ in (7.107). This gives

$$P_1 \ln p \,\Big|_{\mathbf{B}} = -2c \, Re \, (\psi_1^* \psi_4) \ln \left(|\psi_1|^2 + |\psi_4|^2\right). \qquad (7.108)$$

The right-hand side is of the form $0 \ln 0$, and so has to be evaluated in a limiting process. Now we will have ψ_1 and ψ_4 approach the boundary.

Denote a given boundary point by $\mathbf{R}$. Expand each of ψ_1 and ψ_4 in Taylor series about the point $\mathbf{R}$, dropping all quadratic and higher-power terms since the limit $r \to \mathbf{R}$ will be taken:

$$\psi_i(r, t) = \psi_i(\mathbf{R}, t) + dr \cdot \nabla \psi_i(\mathbf{R}, t), \quad \text{where } dr = \mathbf{R} - r, \, i = 1, 4 \quad (7.109)$$

$$\text{and } \lim dr \to 0 \qquad (7.110)$$

now defines the boundary.

The first right-hand term in (7.109) is zero, since all $\psi_i = 0$ on the boundary. Then, substituting Eq. (7.109) into Eq. (7.108) gives

$$P_1 \ln p \,\Big|_{\mathbf{B}} = -2c \, Re \, [(dr \cdot \nabla \psi_1^*)(dr \cdot \nabla \psi_4)] \ln \left[|dr \cdot \nabla \psi_1|^2 + |dr \cdot \nabla \psi_4|^2\right] \,\Big|_{dr=0}.$$

$$(7.111)$$

Taking $dy = dz = 0$ eliminates terms in these differentials,

$$P_1 \ln p \,\Big|_{\mathbf{B}} = -2c(dx)^2 \, Re \, [(\psi_{1x}^*)(\psi_{4x})] \ln \left[(dx)^2 |\psi_{1x}|^2 + (dx)^2 |\psi_{4x}|^2\right] \,\Big|_{dx=0}.$$

$$(7.112)$$

(Note that $\psi_{1x} \equiv \partial \psi_1 / \partial x$, etc., is our derivative notation.) This is of the form $(Au) \ln (Bu)$, $u = (dx)^2$. In the limit $dx \to 0$ it gives 0, as at Eq. (7.93).

Retracing the steps (7.105)–(7.112) for the other components P_2, P_3 gives the same result. Hence requirement (7.74) is satisfied.

Hence, we have shown that the equation of flow (7.103a) and all required boundary value conditions (7.70)–(7.74) hold. This means that the inequality (7.87) follows. It is fruitful to compute the inner product $\mathbf{P} \cdot \mathbf{P}$ in (7.87) for our particular problem. This is

$$\mathbf{P}\cdot\mathbf{P} \equiv \mathbf{P}^\dagger \mathbf{P} = (-c\psi^\dagger[\alpha]\psi)^\dagger(-c\psi^\dagger[\alpha]\psi)$$

$$= c^2(\psi^\dagger[\alpha]^\dagger\psi)(\psi^\dagger[\alpha]\psi)$$

$$= c^2\psi^\dagger[\alpha]^\dagger|\psi|^2[\alpha]\psi$$

$$= 3c^2|\psi|^2\psi^\dagger[1]\psi$$

$$= 3c^2|\psi|^4. \tag{7.113}$$

In the preceding, the first line is by definition (7.103c). The second line uses the well-known theorem that the Hermitian adjoint of a product is the product of Hermitian adjoints in reverse order. The third line defines $|\psi|^2 \equiv \psi^\dagger\psi$. The fourth line uses $[\alpha]^\dagger = [\alpha]^*$ since $[\alpha]$ is diagonal. It also uses the property Eq. (E9) for the squares of all three components of $[\alpha]$. The fifth line uses the defining property of the identity matrix [1].

Therefore, inequality (7.87) becomes

$$\left(\frac{\partial H}{\partial t}\right)^2 \leqslant I\int dr\,\frac{3c^2|\psi|^4}{|\psi|^2} = 3c^2 I\int dr\,|\psi|^2 = 3c^2 I, \quad \text{or} \quad \left(\frac{1}{c}\frac{\partial H}{\partial t}\right)^2 \leqslant 3I. \tag{7.114}$$

Normalization property (7.66) was used. This is a remarkably simple result. It states that, at each instant of time, the rate of change of the entropy of an electron is bounded above by the square root of three times the instantaneous Fisher information. The latter is the information in a smart measurement of the spatial position of the electron. The fundamental role played by the speed of light in connecting H and I is also of interest.

7.5.12 Historical note

The derivation Eqs. (7.66)–(7.87) was first given by Nikolov (1992) in a personal correspondence with this author. The derivation was later independently discovered by Plastino and Plastino (1995). The quantum mechanical application in Sec. 7.5.11 was published in Nikolov and Frieden (1994), except for the specific answer Eq. (7.114).

7.6 Overview

The EPI approach to statistical physics is, as usual, four-dimensional. The Fisher measurement coordinates are $(ix_E, c\boldsymbol{\mu})$ in energy-momentum space. By contrast, space-time coordinates (ir, ct) were used in derivation of quantum mechanics in Chap. 4. Hence, relativistic quantum mechanics arises out of EPI as applied to *space-time* coordinates, whereas statistical mechanics arises out of EPI as applied to *energy-momentum* coordinates. Of course, according to quantum mechanics (Eq. (4.4)) the two coordinate choices are *complementary*. Hence, EPI unifies quantum mechanics and statistical mechanics as resulting from the choice of one or the other of a pair of complementary coordinate spaces. It is interesting that in both derivations the use of mixed imaginary and real coordinates is essential.

Our chosen coordinates $(iE, c\boldsymbol{\mu})$ constitute a *four-dimensional* 'phase space', in the usual language of statistical mechanics. This is in contrast with the standard approach, which takes place in a higher, six-dimensional $(\boldsymbol{\mu}, \mathbf{x})$ momentum-position phase space (Rumer and Ryvkin, 1980). That the Boltzmann and Maxwell–Boltzmann distribution laws can arise out of a lower-dimensioned analysis is of interest.

Both the Boltzmann probability law on energy and the Maxwell–Boltzmann law on momentum were seen to arise out of a common value κ for the information efficiency. That value was $\kappa = 1$, indicating complete efficiency in the transfer of information from the physical phenomenon to the data (see also material below Eq. (3.18)).

We note that the Fisher coordinates $\mathbf{x}_\mu$ for the momentum problem are purely real. This means that the 'game' (Sec. 3.4.12) between the observer and the 'demon' is physical here. Both the general EPI solution (7.59) and the Maxwell–Boltzmann Eq. (7.65) follow as payoffs of the contest for maximum information.

A serendipitous result was the prediction (7.59) of multiple solutions for the PDF on momentum. These correspond to (a) the equilibrium Maxwell–Boltzmann PDF on momentum fluctuations, and to (b) other stationary PDFs *en route* to equilibrium. The latter are non-equilibrium solutions, and fall into a category of Hermite–Gauss solutions previously found by Rumer and Ryvkin (1980) as solutions to the Boltzmann transport equation.

These non-equilibrium PDFs on momentum should have as counterparts non-equilibrium PDFs on the energy. The latter should be obtainable by replacement of the dependence $J(\mathbf{q})$ for the energy in Eq. (7.8) by a series of the form $J(\mathbf{q})(A + Bx + \ldots)$ as in Eqs. (7.47), (7.49) for the momentum. See Sec. 7.4.10 for further discussion.

It was found that, if a PDF obeys an equation of continuity of flow (7.69), then its entropy increase must be limited by the Fisher information in a system measurement. The exact relation is the inequality (7.87). One might regard this result as an addendum to the second law of thermodynamics. That is, entropy shall increase, but (now) by not too much! This result has wide applicability.

As we showed, it applies to classical particle flow (7.96), electromagnetic flow (7.99) and (7.101), gravitational flow (Sec. 7.5.10), and Dirac electron flow (7.114). The latter is a particularly interesting expression, bringing in the speed of light in a new way. The general result (7.87) should have many other applications.

A further problem in statistical physics – turbulent flow – is analyzed by EPI in Chap. 11.

8

Power spectral $1/f$ noise

8.1 The persistence of $1/f$ noise

Consider a real, temporal signal $X(t)$ defined over a time interval $0 \leqslant t \leqslant T$. It has an associated (complex) Fourier spectrum

$$Z_T(\omega) \equiv T^{-1/2} \int_0^T dt X(t)\, e^{-i\omega t} \equiv (Z_r(\omega),\, Z_i(\omega)) \qquad (8.1)$$

and an associated 'periodogram'

$$I_T(\omega) = |Z_T(\omega)|^2 = Z_r^2(\omega) + Z_i^2(\omega). \qquad (8.2)$$

Functions $Z_r(\omega)$, $Z_i(\omega)$ are, respectively, the real and imaginary parts of $Z_T(\omega)$. Define a power spectrum

$$S(\omega) = \lim_{T \to \infty} \langle I_T(\omega) \rangle. \qquad (8.3)$$

The brackets $\langle \cdot \rangle$ denote an average over an ensemble of signals.

If $S(\omega) \approx Const.$ then the signal is said to be 'white noise'. Of course most physical phenomena exhibit a varying power spectrum. The most common of these is of the form

$$S(\omega) = A\omega^{-\alpha}, \ A = const., \ \alpha \approx 1. \qquad (8.4)$$

This is usually called a '$1/f$ noise power spectrum' or, simply, '$1/f$ noise'. Typical white-noise and $1/f$-noise signal traces are shown in Fig. 8.1.

A tremendously diverse range of phenomena obey $1/f$ noise. Just a partial list includes: voltage fluctuations in resistors, semiconductors, vacuum tubes and biological cell membranes; traffic density on a highway, economic time series, musical pitch and volume, sunspot activity, flood levels on the river Nile, and the rate of insulin uptake by diabetics. See respective references: Handel (1971), Hooge (1976), Bell (1980), Johnson (1925), Holden (1976), Musha and Higuchi (1976), Granger (1966), Voss and Clarke (1978), Mandelbrot and Wallis (1969), and Campbell and Jones (1972).

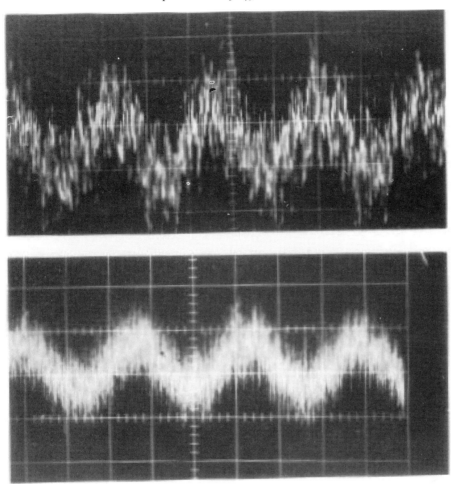

Fig. 8.1. A sinusoidal signal with superimposed noise: (top) $1/f$ noise; (bottom) white noise. (Reprinted from Motchenbacher and Connelly, copyright 1993, by permission of John Wiley & Sons, Inc.)

Numerous mathematical models have been advanced for achieving $1/f$ noise under differing conditions. Examples are models of fractal shot noise (Lowen and Teich, 1989), filtered white Gaussian noise (Takayasu, 1987), fractionally integrated white noise (Barnes and Allan, 1966), fractal Brownian motion (Mandelbrot, 1983), and a diffusion process driven by a white noise boundary condition (Jensen, 1991). Most of these models are based upon a situation of underlying white noise, which undergoes a modification to become $1/f$ noise.

However, the last mentioned model (Jensen, 1991) is of particular interest since a diffusion process obeys increasing entropy (Wyss, 1986), which implies increasing disorder. This reminds us of the *H*-theorem Eq. (1.29) and, more to the point, the *I*-theorem Eq. (1.30). This suggests that we attack the problem of deriving $1/f$ noise for $S(\omega)$ by the use of EPI. Some further justification for the use of this approach follows. (Of course the ultimate justification is that it works.)

8.2 Temporal evolution of tone amplitude

It is instructive to follow the evolution of a typical time signal $X(t)$ in terms of the Fisher information in a measurement. It will be shown that, as $T \to \infty$, the disorder of $X(t)$ increases and consequently $I \to$ a minimum value. Then, as we reasoned in Chap. 1 and Sec. 3.1.1, this suggests the use of EPI.

Consider the gedanken measurement experiment of Fig. 8.2. Time signal $X(t)$ is a musical composition, say, a randomly selected violin sonata. As time progresses the signal $X(t)$ is, of course, produced over increasing time intervals $(0, T_0), (0, T_1), (0, T_2), \ldots$, where $T_0 < T_1 < T_2 \ldots$. The 'ideal' parameter θ of the EPI approach is as follows.

Suppose that a note ω occurs in the zeroth interval $(0, T_0)$ with the complex amplitude

$$Z_0(\omega) \equiv \theta(\omega) \equiv (\theta_r(\omega), \theta_i(\omega)) \tag{8.5}$$

in terms of its real and imaginary parts. This is the ideal complex parameter value.

However, the observer is not listening during the zeroth interval. Instead, he observes during the *n*th interval the complex spectral amplitude

$$Z_n(\omega) = y_n \tag{8.6}$$

in our generic data notation (2.1). From such observation he is to best estimate $\theta(\omega)$. How should the mean-square error e^2 in such an estimate vary with the chosen interval number n or (equivalently) time duration T_n?

For the interval number $n = 0$ the acquired data would be $Z_n(\omega) \equiv \theta(\omega)$, so of course e^2 would be zero. Suppose that the next interval, $(0, T_1)$ includes the ideal interval $(0, T_0)$ plus a small interval. See Fig. 8.2. Then, by Eq. (8.1), its Fourier transform $Z_1(\omega)$ 'sees' the ideal interval plus a small tail region. Hence, $Z_1(\omega)$ will depart from $\theta(\omega)$ by only a small amount. Likewise, an optimum estimate of $\theta(\omega)$ made on this basis should incur small error e^2.

Next, a measurement $Z_2(\omega)$ based upon observation of $X(t)$ over time interval $(0, T_2)$ is made. It should incur a slightly larger error, since interval $(0, T_2)$ incurs more 'tail' of $X(t)$ than its predecessor. Hence, the error after an estimate is made will likewise go up.

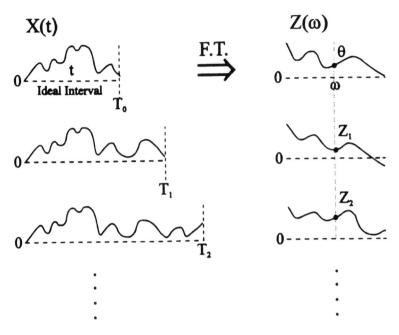

Fig. 8.2. Gedanken measurement–estimation experiment. The unknown tone amplitude $\theta(\omega)$ is caused by signal $X(t)$ over ideal interval $(0, T_0)$. Subsequent tone
amplitudes $Z_1(\omega)$, $Z_2(\omega)$, ... are due to listening over ever-longer time intervals.
(Reprinted from Frieden and Hughes, 1994.)

It is obvious that this trend continues indefinitely. The error $e^2 \to \infty$ with
the time.

It will turn out that the underlying PDF is Gaussian, so that the optimum
estimate achieves *efficiency* (Exercise 1.6.1). Then the equality is achieved in
Eq. (1.1), and the Fisher information I obeys

$$I = \frac{1}{e^2}, \qquad (8.7)$$

in terms of the error e^2. Then from the preceding paragraph $I \to 0$ with time,
its absolute minimum value. By the reasoning in Sec. 3.1.1 this is its
equilibrium value as well, so that the EPI principle should resultingly give the
PDF attained at equilibrium. The power spectrum $S(\omega)$ attained at equilibrium
will also be so determined since, for this problem, I relates uniquely to the
power spectrum, as found next.

8.3 Use of EPI principle

Since our measurement is of the (complex) spectral amplitude $Z = (Z_r, Z_i)$, the Fisher information (2.15) obeys (with $N = 1$)

$$I = \int dZ_r \, dZ_i \frac{(\partial p/\partial \theta_r)^2 + (\partial p/\partial \theta_i)^2}{p} \tag{8.8}$$

in terms of the joint likelihood $p(Z_r, Z_i|\theta_r, \theta_i)$ of the data. As usual, we assume that an observation (Z_r, Z_i) using a real instrument causes perturbation of the system PDF $p(Z_r, Z_i|\theta_r, \theta_i)$. This initiates the EPI process.

8.3.1 Information I in terms of power spectrum S(ω)

For a certain class of time signals $X(t)$ the information expression (8.8) may be directly related to the power spectrum.

By the form of Eq. (8.4), $1/f$ noise exhibits long term memory. Recalling that a frequency $\omega \sim 1/t$, where t is a time lapse, by (8.4) a $1/f$ noise process has very large power at very long time lapses (very small ω). This means that the joint statistics of $(Z(t), Z(t'))$ at two different times t, t' do not simply depend upon the relative time difference $(t - t')$. The joint statistics must depend upon the absolute time values t, t'. A process $X(t)$ with this property is called nonstationary (Frieden, 1991).

A wide class of nonstationary signals $X(t)$ was recently defined and analyzed (Solo, 1992). This is called the class of intrinsic random functions (IRF$_0$) of order zero (Matheron, 1973). Such a class of signals is not stationary in its values $X(t)$ but *is stationary* in its changes $X(t + \tau) - X(t)$, $\tau \geqslant 0$. There is evidence that $1/f$ noise processes obey just such a property of stationarity (Keshner, 1982). Hence, we assume the process $X(t)$ to be an IRF$_0$ process.

It is known that such a process obeys a central limit theorem. Thus, $p(Z_r, Z_i|\theta_r, \theta_i)$ turns out to be separable normal. The Fisher information for a *single* normal random variable is found next.

By Eq. (1.2)

$$I = \int dx \frac{p'^2}{p} = \int dx p \left(\frac{p'}{p}\right)^2 = \left\langle \left(\frac{\partial \ln p}{\partial x}\right)^2 \right\rangle. \tag{8.9}$$

Then, by $p(x) = A \exp(-(x - \theta)^2/2\sigma^2)$, A constant, we get $\ln p = B - (x - \theta)^2/2\sigma^2$, so that $\partial \ln p/\partial x = -(x - \theta)/\sigma^2$. Squaring this and taking the mean gives, by Eq. (8.9),

$$I = \frac{1}{\sigma^2} \tag{8.10}$$

after using $\langle(x - \theta)^2\rangle \equiv \sigma^2$. Solo (1992) showed further that, for the IRF_0 process, $\sigma^2 = S(\omega)/2$ and each of the two random variables (Z_r, Z_i) has the same variance. Then by the added nature of Eq. (8.8) the resulting I is simply twice the value in Eq. (8.10),

$$I = \frac{2}{\sigma^2} = \frac{4}{S(\omega)}. \qquad (8.11)$$

This is the information in observing Z at a single frequency ω. Let us now generalize the data scenario of Sec. 8.2 to include observations over any band of frequencies, denoted as Ω, which *excludes* the pure d.c. 'tone' $\omega = 0$; the latter has no physical reality. Band Ω can be narrow or wide. The IRF_0 process is also independent over all frequencies. Then by the additivity of information (Sec. 2.4.1) the total information is the integral

$$I = 4 \int_\Omega d\omega \frac{1}{S(\omega)}. \qquad (8.12)$$

The information I is, then, a known functional of the power spectrum $S(\omega)$. We needed such an expression in order to start the EPI calculation of the equilibrium $S(\omega)$. However, note that this EPI problem differs *conceptually* from all others in preceding chapters. Here, the unknowns of the problem are not probability amplitudes $\mathbf{q}$. Indeed, the form of the PDF $p(Z_r, Z_i|\theta_r, \theta_i)$ (and hence, amplitude) over all frequencies *is known* to be independent normal. Instead, the unknowns are the power spectrum values $S(\omega)$ over the bandpass interval $(0, \Omega)$. These are unknown *parameters* of the known PDF – essentially the variances (see below Eq. (8.10)).

Hence, we are using EPI here to find, not a continuous PDF, but the identifying parameters $S(\omega)$ of a PDF of known form. This might be considered a parameterized version of the EPI procedure.

8.3.2 Finding information J

The use of EPI principle (3.16), (3.18) requires knowledge of informations I and J. The former obeys Eq. (8.12). The bound information J must be found.

Let the functional J be represented by the general form

$$J = 4 \int_\Omega d\omega F[S(\omega), \omega] \qquad (8.13)$$

where F is a general function of its indicated arguments. Then by Eq. (8.12) and (8.13) the physical information (3.16) is

$$K = 4 \int d\omega \left[\frac{1}{S(\omega)} - F[S(\omega), \omega] \right] \equiv extrem. \qquad (8.14)$$

As in Chaps. 5–7 we find J by simultaneously solving EPI Eqs. (3.16) and (3.18) for a common solution, here $S(\omega)$.

The Euler–Lagrange Eq. (0.34) for the extremum in Eq. (8.14) is

$$\frac{d}{d\omega}\left(\frac{\partial \mathscr{L}}{\partial S'}\right) = \frac{\partial \mathscr{L}}{\partial S}, \quad S' \equiv \partial S/\partial \omega, \quad \text{where } \mathscr{L} = 1/S - F. \tag{8.15}$$

The result is that F satisfies

$$\frac{1}{S^2} + \frac{\partial F}{\partial S} = 0. \tag{8.16}$$

On the other hand, the EPI solution (3.18) requires that I, J and κ obey

$$I - \kappa J = 4\int d\omega \left[\frac{1}{S} - \kappa F\right] \equiv 0. \tag{8.17}$$

The microlevel solution Eq. (3.20) requires this to be true at each value of the integrand,

$$\frac{1}{S} = \kappa F. \tag{8.18}$$

Simultaneously solving Eqs. (8.16) and (8.18) requires that

$$\kappa F/S = -\partial F/\partial S. \tag{8.19}$$

This simple differential equation has a general solution

$$F[S(\omega), \omega] = G(\omega)S(\omega)^{-\kappa}, \tag{8.20}$$

where $G(\omega) \geqslant 0$ is some unknown function. Substituting form (8.20) for F into Eq. (8.17) gives

$$I - \kappa J \equiv 0 = 4\int d\omega[S(\omega)^{-1} - \kappa G(\omega)S(\omega)^{-\kappa}]. \tag{8.21}$$

Function G is found as follows.

8.3.3 *Invariance principle*

As usual we require an invariance principle (see Sec. 3.4.5). Let the unknowns κ, G and S of the problem obey invariance to *an arbitrary change of scale in* ω. Then the output physics should look the same regardless of scale. This is a 'self-similarity' requirement for the overall EPI approach to the problem. It also closely resembles the invariances Eq. (3.36) to Lorentz transformation – a linear rotation of coordinates – that were required of the informations. By comparison, here we have a stretch of the coordinate ω, another linear operation.

First we need to find how a given $S(\omega)$ transforms under a change of scale. Define a new variable

$$\omega_1 = a\omega, \ a > 0. \tag{8.22}$$

We suppose that $\int d\omega S(\omega)$ represents the total power in the process. Then this should be a constant independent of any change of coordinates. Let $S_1(\omega_1)$ represent the power spectrum in the new coordinate system. Then (Frieden, 1991, p. 100)

$$S_1(\omega_1) \, d\omega_1 = S(\omega) \, d\omega. \tag{8.23}$$

Combining this with Eq. (8.22) yields

$$S_1(\omega_1) = (1/a)S(\omega_1/a). \tag{8.24}$$

Eq. (8.21) is the EPI principle (3.18) as applied to this problem. Hence, suppose that $G(\omega)$ satisfies Eq. (8.21). In the new coordinate system, the information (8.21) becomes

$$I_1 - \kappa J_1 = 4 \int d\omega_1 \left[\frac{1}{S_1(\omega_1)} - \kappa G(\omega_1) S_1(\omega_1)^{-\kappa} \right]. \tag{8.25}$$

We demand that $G(\omega)$ make this zero as well, in view of the invariance for EPI principle (3.18) that was postulated. Now using identity (8.24) gives

$$I_1 - \kappa J_1 = 4 \int d\omega_1 \left[\frac{a}{S(\omega_1/a)} - \kappa G(\omega_1) a^\kappa S(\omega_1/a)^{-\kappa} \right] \equiv 0. \tag{8.26}$$

Change variable back to ω via transformation (8.22). Eq. (8.26) becomes

$$I_1 - \kappa J_1 = 4a^2 \int d\omega \left[\frac{1}{S(\omega)} - \kappa G(a\omega) a^{\kappa-1} S(\omega)^{-\kappa} \right] \equiv 0. \tag{8.27}$$

Now compare this expression with the like one (8.21) in the original coordinate system. For Eq. (8.27) to attain the invariant value $I_1 - \kappa J_1 = 0$ function $G(\omega)$ must be such that the integrand of Eq. (8.27) is proportional to that of Eq. (8.21). This gives a requirement

$$\kappa G(\omega) S(\omega)^{-\kappa} = \kappa G(a\omega) a^{\kappa-1} S(\omega)^{-\kappa}. \tag{8.28}$$

After cancellations, this becomes

$$G(a\omega) = a^{1-\kappa} G(\omega). \tag{8.29}$$

The general solution to this is

$$G(\omega) = B\omega^{1-\kappa}, \ B = const. \tag{8.30}$$

This is the function $G(\omega)$ we sought.

8.3.4 Power spectrum

From Eq. (8.20), we therefore have $F = B\omega^{1-\kappa} S^{-\kappa}$. Substituting this into Eq. (8.18) gives the solution

$$S(\omega) = A\omega^{-1}, \ A = const. \tag{8.31}$$

The EPI approach derives the $1/f$ noise power spectrum. This result follows for any value of κ, except as noted next.

A minor point: If $\kappa = 1$ exactly then one finds that the preceding steps do not lead to a solution for $S(\omega)$. However, they do if the *limit* $\kappa \to 1$ is instead taken, and with the constant $B = 1$ (or else A blows up in (8.31)).

8.3.5 Exercise – the question of bandwidth Ω

In all the preceding we assumed the data Z values to be observed over a continuous bandwidth Ω of frequencies. This amounts to an infinite number N of data values! But in fact, the analysis (8.21)–(8.31) could have been carried through, with the same result (8.31), at but $N = 1$ frequency. Show this. Hence, any number N of data suffice to derive the $1/f$ law by the EPI approach.

8.3.6 Value of efficiency parameter κ

The result (8.31) is seen to be independent of the choice of parameter κ. From Eq. (3.18), κ represents the ratio of the total acquired Fisher information I in the data to the information J that is intrinsic to the phenomenon. Hence, κ represents an efficiency parameter for the relay of information from phenomenon to data.

Until now, a given physical phenomenon has been characterized by a unique value for κ. As examples, we found value $\kappa = 1$ for quantum mechanics (Chap. 4) and for statistical physics (Chap. 7), and value $\kappa = 1/2$ for electromagnetic theory (Chap. 5) and gravitational theory (Chap. 6) Here, for the first time, we have a phenomenon that is characterized by *any level* of efficiency κ.

8.3.7 Assumption of an efficient estimator

At Eq. (8.7) we assumed the existence of an efficient estimator. (*Note*: this sense of the word 'efficiency' is independent of the concept of 'information efficiency' κ used in Sec. 8.3.6). In fact, the separable normal law $p(Z_r, Z_i|\theta_r, \theta_i)$ assumed in Sec. (8.3.1) for the data values gives rise to an efficient estimator; see Sec. 1.6.1.

8.3.8 Reference

The idea of using EPI to derive the $1/f$ noise law was published in Frieden and Hughes (1994). However, many aspects of the derivation in this chapter differ from those in the reference. The EPI approach has evolved since then.

8.4 Overview

The EPI principle regards the $1/f$ power spectral law (8.31) as an equilibrium state of the signal and its spectrum $Z(\omega)$. The assumption is that, as time $t \to \infty$, the PDF $p(Z_r, Z_i|t)$ on the data Z at one (or more) frequencies approaches an equilibrium form $p(Z_r, Z_i)$ whose power spectrum has the required $1/f$ form. (*Note*: for simplicity, in this section we suppress the fixed parameters θ_r, θ_i from the notation.) The equilibrium state is to be achieved by extremizing the physical information K growing out of the PDF $p(Z_r, Z_i)$.

The requisite invariance principle (see Sec. 3.4.5) is invariance to linear change of scale (Sec. 8.3.3). A physical assumption is that the time signal $X(t)$ is an IRF$_0$ random process. This assumption does not, by itself, imply $1/f$ noise (Solo, 1992).

We note that the Fisher coordinates for the problem, Z_r and Z_i, are purely real. Then it follows (Sec. 3.4.12) that the information transfer game is played here. The payoff of the contest is the $1/f$ law.

The $1/f$ noise solution (8.31) follows under uniquely general circumstances. There is nearly complete freedom in (a) the values κ, and (b) the number N of degrees of freedom, where $N \geqslant 1$. No other phenomenon in this book is derived under such general conditions.

Perhaps these factors account for the ubiquitous nature of the $1/f$ law. In every other application of EPI, a given phenomenon is associated with a given pair (κ, N). For example, the Dirac equation (Chap. 4) is described by pair values (1, 8). By contrast, $1/f$ noise encompasses many different phenomena, as the list in Sec. 8.1 attests. The implication, then, is that $1/f$ noise should be associated with, and arise out of, a *range* of pair values (κ, N), exactly as EPI predicts.

The result (8.31) is useful in predicting 'pure' $1/f$ noise, i.e., noise with an exponent $\alpha = 1$ in definition (8.4). However, this leaves out physical phenomena for which $\alpha \neq 1$. The assumption that $X(t)$ is an IRF$_0$ random process may have been decisive in this regard, only permitting derivation of the $\alpha = 1$ result. A more general class of random functions might exist for deriving Eq. (8.4) with a general value of α.

9

Physical constants and the $1/x$ probability law

9.1 Introduction

The universal physical constants c, e, $\hbar$, etc., are the cornerstones of physical theory. They are, by definition, the fundamental numbers that define the different phenomena that make up physics. The size of each sets the scale for a given field of physics. They are by their nature as unrelated as any physical numbers can be.

Naturally, their magnitudes depend upon the system of units chosen, i.e., whether c.g.s, m.k.s, f.p.s. or whatever. However, regardless of units, they cover a tremendous range of magnitudes (see Table 9.1 below, from Allen (1973)). There is great mystery surrounding these constants, particularly their values. By their definition as fundamentally unrelated numbers, they must be independent. Hence, no analytical combination of them should equal rational fractions, for example. To get around the fact that their values change with choice of units, many physicists advocate using unitless combinations of the constants as new, more 'fundamental' constants. Examples are

$$\alpha_e \equiv \frac{e^2}{\hbar c} = \frac{1}{137.0360 \ldots}, \quad \frac{m_n}{m_p} = 1.001\,38 \ldots, \quad \text{and} \quad \frac{m_e}{m_p} = \frac{1}{1836.12 \ldots} \quad (9.1)$$

where α_e is the electromagnetic fine structure constant and m_n, m_p and m_e are, respectively, the masses of the neutron, proton and electron.

We have found in preceding chapters that the fixed nature of the constants is, in fact, demanded by EPI theory. However, their magnitudes have not been fixed by the theory (nor by any accepted theory to date). No one knows why they have their particular values. It has been conjectured that they have these values because, if they were different, we would not be here to observe them. That is, physical conditions such as the 'moderate' range of temperatures we experience on Earth would not necessarily be in effect, ruling out our existence here. This is an example of use of the 'anthropic principle' (Dicke, 1961;

Table 9.1. *The fundamental constants*

Quantity	Magnitude	Log_{10}	$(-30, -18)$	$(-18, -6)$	$(-6, 6)$	$(6, 18)$	$(18, 30)$
Velocity of light c	2.99×10^{10}	$+10$				X	
Gravitational G	6.67×10^{-8}	-7		X			
Planck constant h	6.63×10^{-27}	-26	X				
Electronic charge e	4.80×10^{-10}	-10		X			
Mass of electron m	9.11×10^{-28}	-27	X				
Mass of 1 amu	1.66×10^{-24}	-24	X				
Bolzmann constant k	1.38×10^{-16}	-16		X			
Gas constant R	$8.31 \times 10^{+7}$	$+8$				X	
Joule equivalent J	4.19×10^{0}	0			X		
Avogadro number N_A	6.02×10^{23}	$+24$					X
Loschmidt number n_0	2.69×10^{19}	$+19$					X
Volume gram-molecule	$2.24 \times 10^{+4}$	$+4$			X		
Standard atmosphere	$1.01 \times 10^{+6}$	$+6$				X	
Ice point	$2.73 \times 10^{+2}$	$+2$			X		
Faraday $N_A e/c$	$9.65 \times 10^{+3}$	$+4$			X		
		Totals	$\longrightarrow 3$	$\longrightarrow 3$	4	3	$\longrightarrow 2$

Carter, 1974). Thus, the values of the constants are, somehow, scaled to accommodate the presence of human beings. This is flattering, if true: it means that our presence has a profound – if indirect – effect upon physical laws and, hence, the Universe!

9.2 Can the constants be viewed as random numbers?

There are no analytical relationships that allow us to know one fundamental physical constant uniquely in terms of the others. If there were, the constants would not be fundamental. The absence of *analytical* relationships suggests *statistical* relationships: i.e., that the constants are a random sample of independent numbers from some 'master' probability law $p(y)$, where y is the magnitude of any physical constant.

Another argument for randomness (Barrow, 1991) is as follows. The constants that we deem to be 'universal' arise out of a limited, four-dimensional worldview. But if, as proponents of 'string theory' suggest, there are additional dimensions present as well, then our observed universal constants may be *derived* from more fundamental constants that exist in other dimensions. And the process of projection into our limited dimensionality might have a random component, e.g., due to unknown quantum gravitational fluctuations.

Admitting, then, that the universal constants might have been randomly generated, what form of probability law should they obey?

The EPI principle has been formulated to derive such laws. The EPI view is that all physical phenomena – even apparently deterministic ones – arise as random effects. This includes, then, the physical constants. Let us see what probability law EPI can reasonably come up with for the physical constants.

9.3 Use of EPI to find the PDF on the constants

Recall that, to use EPI, a physical principle of invariance must be incorporated as an input. This leads to a dilemma. What single invariance principle can be used for the present scenario – one that includes *all* the physical constants and, therefore, all physical effects and all possible sets of units!

9.3.1 Invariance principle

In fact there is such an invariance principle and, remarkably, it grows out of the *problem* of units previously mentioned. Recall the arbitrary nature of the system of units for the numbers y. Now, a general PDF $p(y)$ changes its functional form as the units are changed. For example, if each y is merely rescaled to a value $z = ay$, $a =$ const., then the PDF $p_z(z)$ on z relates to the PDF $p_y(y)$ as $p_z(z) = a^{-1} p_y(z/a)$ (see derivation Eqs. (9.18)–(9.20)). Thus $p_z(z)$ is not of the original functional form $p_y(z)$ unless there is *no* change of units, i.e., $a = 1$.

The Fisher information that is acquired in a measurement will generally depend upon the units as well. For example, if $p_y(y)$ is Gaussian then $I(y) = 1/\sigma_y^2$ (as found at Eq. (8.10)) whereas $I(z) = 1/a^2\sigma_y^2$. Likewise, the bound information J will generally change its value as the units change. The result is that the value of the physical information $K = I - J$ will generally depend upon the units. If it is extremized in one set of units it might no longer be so in another, violating Eq. (3.16) of EPI. Likewise, EPI Eq. (3.18) does not necessarily remain zero under a change of units. The upshot is that all of the physical laws of preceding chapters were derived to hold within one (the given) system of units. It is well known, for example, that Maxwell's equations change slightly (in proportionality constants) according to the choice of units.

Now, one of the tenets of EPI theory (Secs. 3.4.5, 3.4.14) is that the physical information for each phenomenon should be an invariant in some sense (depending upon the phenomenon). Here we are confronted, fundamentally, with a problem of the arbitrariness of units. We therefore confront the problem directly, and postulate the following invariance principle.

The solution $p(y)$ must be such that EPI principle Eqs. (3.16) and (3.18) are obeyed independent of the choice of units.

Of course, as in other chapters, it is assumed that the phenomenon is, initially, in a definite set of units. That is, the population y is the list of ordinary physical constants directly as they occur in physical tables such as Table 9.1, in some definite set of units (c.g.s., m.k.s., etc.). An important ramification is that the conventional *unitless ratios* of constants, as in Eqs. (9.1), will *not* be utilized.

9.3.2 A cosmological effect

What is the ideal value θ of a physical constant? θ, of course, plays a key role in EPI theory. First, should each constant have a different value of θ?

If we were seeking to analyze the fluctuations in the measurements of but one physical constant (say, the speed of light), then logically θ would be the ideal value of that constant, e.g., $2.99 \ldots \times 10^8$ m/s for the speed of light. However, such is not our case. Our population consists of all the physical constants as sampled from a single PDF law. Therefore we need to know *the single* number θ that can represent the common ideal value of all of them taken together. Such a number is suggested by recent developments in cosmology.

Vilenkin (1982, 1983) and others suggest that the universe arose out of 'nothingness'. That is, everything we sense is a fluctuation from some unknown vacuum state. We posit that the physical constants, as well, were fixed at the vacuum state fluctuation. Then the physical constants are fluctuations from pure nothingness, i.e., a value zero. Hence, we take the 'ideal' value of every physical constant to be

$$\theta = 0. \tag{9.2}$$

It is interesting that this particular number is independent of any (multiplicative) change of units. It is an absolute, as a theory of the constants would require. It also agrees with the spirit of the invariance principle of Sec. 9.3.1.

An immediate consequence is that an observed value y of a physical constant is pure fluctuation, i.e.

$$y = \theta + x = x. \tag{9.3}$$

This allows us to regard x *per se* as the observed value of a physical constant and, consequently, the PDF $p(x)$ as the unknown law to be found.

9.3.3 Fisher coordinates for the problem

The coordinates must be relativistic invariants (Sec. 3.5), i.e., four-vectors. On the other hand, each observed constant x is a relativistic invariant, i.e., does not change its apparent value when viewed from a system that is in uniform motion with respect to the laboratory. It is also, of course, a scalar quantity. Preserving four-vector notation, and as usual taking one of the coordinates to be imaginary, we therefore use as Fisher coordinates the vector

$$\mathbf{x} = (ix, 0, 0, 0). \tag{9.4}$$

The net effect is merely that of a scalar, imaginary coordinate ix. (See Appendix C for the meaning of such a coordinate from the standpoint of estimation theory.) Imaginary coordinates were previously used in Chaps. 4, 5, 6 (hidden in tensor notation) and 7.

9.3.4 Fisher information

Assume, for simplicity, that $N = 1$ amplitude function $q(x)$ suffices to describe the phenomenon. Then the Fisher information due to the scalar coordinate $x_0 = ix$ in Eq. (9.4) is, by Eq. (2.19),

$$I = -4 \int dx q'^2(x), \ p(x) \equiv q^2(x), \ q' = dq/dx. \tag{9.5}$$

(Note similar negative Fisher I components in Eqs. (4.9), (5.6), (6.17) and (7.9); again, due to imaginary coordinates.)

9.3.5 Type of EPI solution

The EPI principle Eqs. (3.16), (3.18) require knowledge of the functionals I and J, and the coefficient κ. We know I from Eq. (9.5). That leaves κ and J as unknowns. As discussed in Sec. 3.4.6, there are two basically different approaches, designated as (a) and (b), for determining these quantities. In type (b) the quantities are known directly from the form of the invariance principle. Such a solution was found in Chap. 4. By comparison our invariance principle as given in Sec. 9.3.1 is of type (a). Here, as discussed in Sec. 3.4.6, the solution κ, J, $p(x)$ to the problem must be found as the *simultaneous* solution to EPI conditions Eqs. (3.16) and (3.18). We carry this through in sections to follow.

9.3.6 Bound information J

The bound information J is supposed to bring into the theory any physical sources of the measured phenomenon. For example, it brought the electromagnetic four-current $((j, \rho)$ in Eq. (5.9)) and the gravitational stress tensor ($T_{\mu\nu}$ in Eq. (6.30)) into those, respective, scenarios. However, in the present problem, there is no well-defined 'source' of the physical constants. Therefore, the most general representation for the unknown functional J is

$$J = 4 \int dx\, j(q, x). \tag{9.6}$$

Functional j must be found.

According to plan we seek the common solution $q(x)$ to the EPI extremum principle (3.16) and the 'zero' principle (3.18). We next implement these, in turn.

9.3.7 EPI extremum solution

Here we want to solve

$$K \equiv I - J = -4 \int dx(q'^2 + j) = extrem., \tag{9.7}$$

where we used Eqs. (9.5) and (9.6). Hence we use the Euler–Lagrange Eq. (0.13), rewritten as

$$\frac{d}{dx}\left(\frac{\partial\mathcal{L}}{\partial q'}\right) = \frac{\partial\mathcal{L}}{\partial q}. \tag{9.8}$$

The Lagrangian is, by Eq. (9.7),

$$\mathcal{L} = -4(q'^2 + j), \tag{9.9}$$

so that

$$\frac{\partial\mathcal{L}}{\partial q'} = -8q' \text{ and } \frac{\partial\mathcal{L}}{\partial q} = -4\frac{\partial j}{\partial q}. \tag{9.10}$$

Hence, Eq. (9.8) gives as a solution

$$q'' = \frac{1}{2}\frac{\partial j}{\partial q}. \tag{9.11}$$

9.3.8 EPI zero solution

By Eq. (3.18), here we want to solve

$$I - \kappa J \equiv 0 = -4\int dx(q'^2 + \kappa j). \tag{9.12}$$

Eqs. (9.5) and (9.6) were used.

By use of the integration-by-parts result Eq. (5.13), Eq. (9.12) becomes

$$4\int dx(qq'' - \kappa j) = 0. \tag{9.13}$$

As usual we use the microscale solution Eq. (3.20), by which the integrand is zero at each x,

$$q'' = \frac{\kappa j}{q}. \tag{9.14}$$

This permits us to easily implement the next step.

9.3.9 Common solution q

The common solution to Eqs. (9.11) and (9.14) obeys

$$\frac{1}{2}\frac{\partial j}{\partial q} = \frac{\kappa j}{q} \text{ or } \frac{\partial j}{j} = 2\kappa\frac{\partial q}{q}. \tag{9.15}$$

This may be partially integrated, to give

$$\ln j = 2\kappa \ln q + C(x),$$

$$\text{or } j = q^{2\kappa}B(x), \ B(x) \geqslant 0. \tag{9.16}$$

Function $B(x)$ must be found.

9.3.10 Use of invariance principle

The invariance principle of Sec. 9.3.1 will enable function $B(x)$ to be found. The solution Eq. (9.16) may be used in Eq. (9.12) to give

$$I - \kappa J \equiv 0 = -4 \int dx [q'^2(x) + \kappa q^{2\kappa} B(x)]. \tag{9.17}$$

According to the invariance principle of Sec. 9.3.1, this should remain zero under an arbitrary change of units. Then it also should remain zero under a linear change of coordinates, from x to

$$x_1 = ax, \ a > 0. \tag{9.18}$$

Let subscript 1 designate quantities as computed in the new coordinate system. We need to compute quantity $I_1 - \kappa J_1$ and demand that this be zero as well. This means that we have to compute quantities I_1 and J_1.

Quantity I_1 is found as follows. Since an event x_1 happens as often as a corresponding event x via Eq. (9.18), their probabilities are equal, or,

$$p_1(x_1)|dx_1| = p(x)|dx|. \tag{9.19}$$

Then by Eq. (9.18) this gives

$$p_1(x_1) = \frac{1}{a} p\left(\frac{x_1}{a}\right). \tag{9.20}$$

Operating d/dx_1 and squaring gives

$$p_1'^2(x) = \frac{1}{a^4} p'^2\left(\frac{x_1}{a}\right). \tag{9.21}$$

By Eq. (9.5), the Fisher information I_1 in the new coordinate system is

$$I_1 = -4 \int dx_1 q_1'^2(x_1) = -\int dx_1 \frac{p_1'^2(x_1)}{p_1(x_1)}, \ p_1 \equiv q_1^2. \tag{9.22}$$

Using Eqs. (9.20) and (9.21) in Eq. (9.22), and changing variable x_1 back to x via Eq. (9.18), gives

$$I_1 = -\frac{4}{a^2} \int dx q'^2(x), \ p = q^2. \tag{9.23}$$

Also, for later use,

$$q_1^2(x_1) \equiv p_1(x_1) = \frac{1}{a} p\left(\frac{x_1}{a}\right) = \frac{1}{a} q^2\left(\frac{x_1}{a}\right). \tag{9.24}$$

The middle equality is by use of Eq. (9.20).

We turn next to J_1. From Eqs. (9.6) and (9.16)

$$\kappa J_1 = 4\kappa \int dx_1 j(q_1, x_1) = 4\kappa \int dx_1 q_1(x_1)^{2\kappa} B(x_1). \tag{9.25}$$

Using Eq. (9.24) in Eq. (9.25) and changing variable back to x via Eq. (9.18) gives

$$\kappa J_1 = 4\kappa a^{1-\kappa} \int dx q^{2\kappa}(x) B(ax). \tag{9.26}$$

Therefore combining Eqs. (9.23) and (9.26) gives

$$I_1 - \kappa J_1 = -\frac{4}{a^2} \int dx [q'^2(x) + \kappa a^{3-\kappa} q^{2\kappa}(x) B(ax)] \equiv 0. \tag{9.27}$$

We now use the invariance principle of Sec. 9.3.1. Assume that the solution $q(x)$ satisfies Eq. (9.17). Then a linear change of coordinate (9.18) is performed. In the new coordinate system, the zero condition $I_1 - \kappa J_1 = 0$ is to hold as well. Hence, Eq. (9.27) is to be zero. Comparing this in form with Eq. (9.17), which already *is* zero, demands that the two be proportional. This requires that the far-right terms be equal,

$$\kappa q^{2\kappa} B(x) = \kappa a^{3-\kappa} q^{2\kappa}(x) B(ax), \text{ or } B(x) = a^{3-\kappa} B(ax). \tag{9.28}$$

What function $B(x)$ satisfies this? The following trial solution works,

$$B(x) = Cx^{\kappa-3}, \ C = const. \tag{9.29}$$

9.3.11 PDF solution

Substituting the latter into Eq. (9.16) gives the solution for j,

$$j = Cq^{2\kappa} x^{\kappa-3}. \tag{9.30}$$

Therefore by Eq. (9.11) the probability amplitude q obeys

$$q'' = \kappa C q^{2\kappa-1} x^{\kappa-3}. \tag{9.31}$$

This is a nonlinear, second-order differential equation. Remarkably, its general solution is a simple power law,

$$q(x) = Ax^{-1/2}, \ A = const. \tag{9.32}$$

(Kamke, 1948, p. 544). Of further interest is that this answer is independent of the value of κ. Any value of κ works here.

Exercise

Verify this answer, by direct substitution of a trial solution $q = Ax^{\alpha}$, α initially unknown, into Eq. (9.31).

The solution (9.32) also requires that the two constants A and C be related as

$$C = \frac{3}{4\kappa} A^{2(1-\kappa)}.$$ (9.33)

Substituting this into Eq. (9.29) gives

$$B(x) = \frac{3}{4\kappa} A^{2(1-\kappa)} x^{\kappa-3}.$$ (9.34)

Since $x > 0$ and $\kappa \geqslant 0$ this verifies that $B(x) \geqslant 0$, as was required by Eq. (9.16).
    From Eq. (9.32), the PDF is

$$p(x) = \frac{A^2}{x}.$$ (9.35)

Thus, the unknown PDF obeys a $1/x$ law. This has the form of a 'Zipf's law' (Woodroofe and Hill, 1975) as well. However, Zipf's law holds only for discrete random variables whereas our variable x is continuous. Therefore we prefer to simply call $p(x)$ a $1/x$, or 'reciprocal', PDF law.

9.3.12 Use of second EPI invariance

The preceding has used as an invariance principle the invariance of the *zero-solution* of EPI, Eq. (9.12), to a change of scale. But the invariance principle in Sec. 9.3.1 requires invariance in the extremum of Eq. (9.7) as well. That is, *both* aspects of the EPI principle should be invariant. Does the solution (9.35) also satisfy this second invariance?

Exercise

Following analogous steps to those preceding, show that exactly the same solution (9.35) results from demanding invariance in the extremum of Eq. (9.7) to a linear change of scale. *Hint*: The proof consists essentially of replacing the multiplier κ (but not exponent κ) by 1.

9.3.13 Normalization

Eq. (9.35) for $p(x)$ has a pole at the point $x = 0$. Hence, this PDF could not obey normalization if the origin were included in the range of x. Likewise, $p(x)$ does not approach zero fast enough as $x \to \infty$ to avoid a logarithmic infinity there. Hence, the range of x must be finite. From Table 9.1, we see that the presently known constants range in value from approximately value 10^{-28} to 10^{+24}. This suggests that, were enough constants known, the range would be numbers between some fixed number b and its reciprocal $1/b$, where $b = 10^{+30}$ (say). Then the normalization requirement

$$\int_{1/b}^{b} dx \, p(x) = 1, \text{ where } p(x) = A^2/x \tag{9.36}$$

fixes the value of A^2. The result is that

$$p(x) = \frac{1}{(2 \ln b)x} \text{ for } 1/b \leqslant x \leqslant b \tag{9.37}$$

and 0 for all other x. This is the required PDF on the constants as derived by EPI. It will be shown, below, to agree reasonably well with the actual occurrences of the numbers in Table 9.1.

9.3.14 Peculiar significance of the number 1

A direct calculation shows that the probability that a constant is less than or equal to 1 obeys

$$P(x \leqslant 1) = \int_{1/b}^{1} \frac{dx}{(2 \ln b)x} = \frac{1}{2}. \tag{9.38}$$

Remarkably, this result is independent of the size of b. Hence, if the PDF law (9.37) is correct, no matter what the range $1/b \leqslant x \leqslant b$ of the constants actually is, *their median value ought to be precisely 1*. Notice that this holds even if b is a huge number, say, much greater than the value 10^{30} imputed from the table.

The prediction (9.38) is actually obeyed pretty well by the constants in Table 9.1. Fraction 6/15 of them have a value $\leqslant 1$.

Why the value 1 should have this significance is a mystery. As is borne out below, our PDF (9.37) is invariant to a change of units. Therefore, the median of the constants is 1 *independent of the choice of units*. This gives added strength to the result and to the mystery.

9.4 Statistical properties of the 1/x law

The reciprocal law (9.37) has further statistical properties that give insight into the nature of the physical constants. As will be seen, there is something to be gained by viewing them as statistical entities.

9.4.1 Invariance to change of units

We show next that the reciprocal law maintains its form under a change of units. Subdivide the constants into classes i according to units. For example, constants in class $i = 1$ might all have the units of mass, those in class $i = 2$

might be lengths, etc. Let a fraction r_i of the constants be in class i, $i = 1, \ldots, n$. Of course

$$\sum_{i=1}^{n} r_i = 1, \tag{9.39}$$

a constant must be in one of the classes. Denote the PDF for the x of a specific class i as $p_i(x)$. Then by the law of total probability (Frieden, 1991) the net probability density for a value x, regardless of class, obeys

$$p(x) = \sum_{i=1}^{n} r_i p_i(x). \tag{9.40}$$

Suppose that the units are now changed. This means that, for class i, all numbers x are multiplied by a fixed factor b_i (the change of units factor). Denote the resulting PDF for class i as $P_i(x)$. Now, by Eq. (9.20)

$$P_i(x) = \frac{1}{b_i} p_i\left(\frac{x}{b_i}\right). \tag{9.41}$$

As with Eq. (9.40), the *net* PDF after the change of units obeys

$$P(x) = \sum_{i=1}^{n} r_i P_i(x). \tag{9.42}$$

Then combining Eqs. (9.41) and (9.42) gives

$$P(x) = \sum_{i=1}^{n} \frac{r_i}{b_i} p_i\left(\frac{x}{b_i}\right). \tag{9.43}$$

It is reasonable to assume that the x are identically distributed (independent of class i), i.e.,

$$p_i(x) = p(x). \tag{9.44}$$

Combining the last two equations gives

$$P(x) = \sum_{i=1}^{n} \frac{r_i}{b_i} p\left(\frac{x}{b_i}\right). \tag{9.45}$$

This gives the net new law $P(x)$ in terms of the old net law $p(x)$ (before the change of units). In the particular case (9.35), where the old law is the reciprocal law, Eq. (9.43) gives

$$P(x) = A^2 \sum_{i=1}^{n} \frac{r_i b_i}{b_i x} = \frac{A^2}{x} \sum_{i=1}^{n} r_i = \frac{A^2}{x} \tag{9.46}$$

by Eq. (9.39), or,

$$P(x) = p(x) \tag{9.47}$$

by Eq. (9.35). That is, the PDF for the constants in the new units is the same as in the old units. The reciprocal law is invariant to choice of units.

This result is actually indifferent to the assumption (9.44). Even if the individual classes had different PDF laws, when the net PDF $p(x)$ is a reciprocal law the net output law after change of units remains a reciprocal law (Frieden, 1986).

9.4.2 Invariance to inversion

Most of the constants are defined in equations that are of a multiplicative form. For example, the usual definition of Planck's constant h is in the relation

$$E = h\nu \tag{9.48}$$

where E is the energy of a photon and ν is its frequency. This relation could instead have been written

$$E = \nu/h \tag{9.49}$$

in which case h would have the reciprocal value to that defined in Eq. (9.48). Of course neither definition is more 'correct' than the other.

This illustrates an arbitrary aspect of the definition of the physical constants. But then, in the face of such arbitrariness of inversion, *does the concept of a PDF p(x) for the constants make sense?* Clearly it would only if the *same* law $p(x)$ resulted independent of any choice of the number r of inverted constants in a population of constants.

Let us, then, test our candidate law (9.37) for this property. Suppose that the reciprocated constants obey a PDF $p_r(x)$. Then by the law of total probability (Frieden, 1991) the PDF for the entire population obeys

$$p(x) = r p_r(x) + (1 - r)p(x), \ 0 \leqslant r \leqslant 1. \tag{9.50}$$

We can also relate the PDF $p_r(x_r)$ to the PDF $p(x)$. A number x_r is formed from a number x by the simple transformation

$$x_r = 1/x. \tag{9.51}$$

By the transformation identity Eq. (9.19) it must be that

$$p_r(x_r)|dx_r| = p(x)|dx|, \text{ so that } p_r(x_r) = p\left(\frac{1}{x_r}\right)\frac{1}{x_r^2} \tag{9.52}$$

by Eq. (9.51). Substituting this into Eq. (9.50) gives

$$p(x) = r p(1/x)x^{-2} + (1 - r)p(x) \tag{9.53}$$

as the relationship to be obeyed by a candidate law $p(x)$.

Trying our law $p(x) = A^2/x$, the right-hand side of Eq. (9.53) becomes

$$r A^2 x x^{-2} + (1 - r)A^2 x^{-1} = A^2 x^{-1}[r + (1 - r)] = A^2/x, \tag{9.54}$$

the left-hand side. Therefore, the $1/x$ law obeys invariance to reciprocation. Moreover it obeys it independent of any particular value of r, as was required.

9.4.3 Invariance to combination

Combinations of constants, such as the quantity e/m for the electron, are often regarded as fundamental constants in their own right. But if they are fundamental, they should obey the same law $p(x)$ that is obeyed by the fundamental constants x. Now, a general PDF does not have this property. For example, if numbers x and y are independently sampled from a Gaussian PDF, then the quotient $x/y = z$ is no longer Gaussian but, rather, Cauchy, with a PDF of the form $(1 + z^2)^{-1}$.

Nevertheless, our law $p(x) = A^2/x$ obeys the property of invariance to combination, provided the range parameter b large (which it certainly is − see Sec. 9.3.13). The proof is in Frieden (1986).

9.4.4 Needed corrections to the estimated physical constants

The $1/x$ law is a prior law of statistics, i.e., it governs *formation* of the constants. Obviously the law shows that small constants x are more probable to occur than are large constants. The result is that, if an unknown constant is observed through some measurement technique, its maximum probable value is *less than* the observed value. Furthermore, if many repetitions of the experiment are performed, the maximum probable value of the constant is *less than* the arithmetic mean of the experimental outcomes. See Frieden (1986). An example follows.

Suppose that a constant is measured N times, and that each measurement x_n is corrupted by added Gaussian noise of standard deviation σ. The conventional answer for a 'good' estimate of the constant is the sample mean

$$\bar{x} = N^{-1} \sum_{n=1}^{N} x_n \tag{9.55}$$

of the measurements. By contrast, the *maximum probable* value x_{MP} of the constant is (Frieden, 1986)

$$x_{MP} = \tfrac{1}{2}(\bar{x} + \sqrt{\bar{x}^2 - \sigma^2/4N}). \tag{9.56}$$

Hence, if N is of the order of 1, and/or the noise σ is appreciable, the maximum probable value of the constant is considerably less than $\bar{x}$. The sample mean *would not* be a good estimate.

9.4.5 The log-uniform law

Observing just the *exponents* of the numbers in Table 9.1, it appears that they are about uniformly distributed over a range of $(-28, +24)$. Now, the exponents are approximately the logarithms of the numbers. Hence, do the physical constants obey a uniform PDF in the logarithms of their values – a 'log-uniform' distribution?

There are statistical tests for checking out such a possibility, and we defer this question to a following section. For now, we take up the implication to the acquired $1/x$ law (Sec. 9.3). If a random variable x obeys a $1/x$ PDF, does the new random variable $y = \ln x$ obey a uniform PDF?

Once again using the transformation identity Eq. (9.19), we have

$$p_y(y)|dy| = p(x)|dx|, \text{ where } y = \ln x. \tag{9.57}$$

Since $dy/dx = 1/x = \exp(-y)$, we get

$$p_y(y) = p(e^y)e^y = (A^2/e^y)e^y = A^2 = const.,$$

$$\text{or} \tag{9.58}$$

$$p_y(y) = \frac{1}{2\ln b} \text{ for } -\ln b \leqslant y \leqslant +\ln b$$

after normalization. Hence, the $1/x$ PDF law is equivalent to a log-uniform law. This means that the exponents of the numbers x are evenly distributed over a finite, symmetric range of numbers. The numbers in the Table 9.1 should show this effect. We now check this out quantitatively.

9.5 What histogram of numbers do the constants actually obey?

9.5.1 Log-uniform hypothesis

As mentioned above, we suspect that the physical constants listed in Table 9.1 are log-uniform distributed. Let us now test out this hypothesis. Taking the logarithm of each item x in the table, we have essentially the powers of 10 as data: as listed, these are $+10$, -7, -26, -10, etc. These numbers range from -28 to $+24$. In view of the symmetric limits to the range of $y = \ln x$ in the theoretical model Eq. (9.58), we symmetrize the estimated limits to a range of -30 to $+30$. Thus the hypothesis is that

$$p_{\log x}(x) = \tfrac{1}{60} \text{ for } -30 \leqslant x \leqslant +30 \tag{9.59}$$

and 0 otherwise.

To test this hypothesis, divide the total range $(-30, +30)$ into 5 equal subranges of length $60/5 = 12$. These are the subranges $(-30, -18)$,

$(-18, -6)$, etc. as indicated in the table. The uniform law (9.59) predicts, then, that *the same number of events* $(15/5 = 3)$ should lie in each of the subranges. The actual numbers (see Table 9.1) are 3, 3, 4, 3 and 2, not bad approximations to the 3! We can quantify this agreement as follows.

9.5.2 Chi-square test – background

Given a die that is rolled N times, the number of occurrences of roll outcome 1 is denoted as m_1, etc., for roll outcomes 2, ..., 6. As an example, for $N = 10$ rolls, a typical set of roll occurrences is $(m_i, \ldots, m_6) = (2, 1, 3, 0, 1, 3)$. As a check, notice that the sum of the occurrences is $N = 10$.

The chi-square test is used to measure the degree to which the *observed* occurrence numbers m_i for a set of events i agree with the *expected* occurrence numbers on the basis of a hypothesis $P_1, \ldots, P_M$ for the probabilities of the events. Let the probability of an event i be P_i and suppose that there are N trials. The number m_i of events i in the N trials will tend to be exactly NP_i. However, the *actual* number of times event i occurs will randomly differ from this value. A measure of the disparity between an observation m_i and a corresponding theoretical probability P_i is, then, $(m_i - NP_i)^2$. If the numbers used for the P_i are correct then, from the preceding, this disparity will tend to be small.

However, if the P_i are actually *unknown*, and instead an estimated set of them (called a 'hypothesis' set) is used, then the above disparity will tend to be larger as the hypothesis becomes more erroneous. Hence, a total disparity, weighted appropriately, is used to test the probable correctness of hypothetical P_i values:

$$\sum_{i=1}^{M} \frac{(m_i - NP_i)^2}{NP_i} \equiv \chi^2. \tag{9.60}$$

This is called the 'chi-square' statistic.

Since the m_i are random, so must be the value of χ^2. The PDF for χ^2 is known. It obeys the (what else?) chi-square distribution. The number of event types M is a free parameter of the chi-square distribution. Quantity $(M - 1)$ is called the 'number of degrees of freedom' of the distribution. The chi-square distribution is tabulated for different values of the number of degrees of freedom.

On this basis, an observed value of χ^2, call it χ_0^2, has a known probability. For example, if χ_0^2 is very large (by Eq. (9.60), indicating strong disagreement between the observations m_i and the hypothetical P_i) the computed probability is very low. On the other hand, the maximum likelihood hypothesis is that events m_i that occur were *a priori* highly probable (that's why they occurred). Or, on the 'flip side', *unlikely events don't happen*. Hence, the low probability associated with a high value of χ_0^2 indicates a conflict: on the basis of the hypothetical probabilities P_i the value χ_0^2 was just too high. It shouldn't have occurred. The conclusion is, then, that either the observations m_i are wrong or the hypothetical P_i are wrong. Since the observations are presumed known to be correct, the onus is placed upon the hypothesis P_i. It is rejected.

The degree of rejection is quantified as follows. Denote the known PDF for χ^2 as $p_{\chi^2}(x)$. The probability α that a value of χ^2 at least as large as the observed value χ_0^2 will occur is

$$\int_{\chi_0^2}^{\infty} dx \, p_{\chi^2}(x) \equiv \alpha. \tag{9.61}$$

This can be computed. It is also tabulated (see, e.g., Frieden, 1991, Appendix B). Obviously, if the lower limit χ_0^2 in Eq. (9.61) is large, then α is small. And, from the preceding, a large χ_0^2 connotes rejection of the hypothesis P_i. If α is so small as to be less than (say) value 0.05, then the hypothesis P_i is rejected 'on the level of significance' $\alpha = 0.05$.

Or, by contrast, if α is found to be fairly close to 1, the hypothesis P_i may be accepted to that level of significance.

In this manner, the chi-square test may be used to accept or reject a hypothesis for an underlying probability law.

9.5.3 Chi-square test – performed

In Table 9.1 there are $M = 5$ possible occurrence intervals for the logarithm of a given constant to occur within. These are the intervals $(-30, -18)$, $(-18, -6)$, etc., indicated along the top row. Our hypothesis is the uniform one $P_i = 1/5$, $i = 1, \ldots, 5$ for the logged constants. Since there are $N = 15$ of them, quantities $NP_i = 3$ for all i. The observed values of m_i are (3, 3, 4, 3, 2), respectively (see bottom row in Table 9.1). These numbers as used in Eq. (9.60) give a computed

$$\chi_0^2 = \frac{(3-3)^2 + (3-3)^2 + (4-3)^2 + (3-3)^2 + (2-3)^2}{3} = 0.667. \tag{9.62}$$

This value of χ_0^2 coupled with the $M - 1 = 4$ degrees of freedom gives, using any table of the χ^2 distribution, a value $\alpha = 0.95$.

Hence, the disparity between the data m_i and the log-uniform hypothesis is so small that, if many sets of 15 data were randomly generated from the law, *for 95% of such data sets* the disparity χ_0^2 would exceed the observed disparity of 0.667. The hypothesis is well-confirmed by the data.

It should be noted, however, that because of the small population $N = 15$ of constants, the test-interval size was necessarily quite coarse – 12 orders of magnitude. Hence, the hypothesis of a log-uniform law was only confirmed on a coarse subdivision of points. It would require more constants to test on a finer scale. However, it is doubtful that significant departures from the flat curve would result, since 15 widely separated points already fall on that curve.

9.6 Overview

The preceding has shown that *the empirical data at hand* – the actual values of the physical constants – obey the hypothesis that their logarithms are uniformly distributed. As we showed in Sec. 9.4.5, this is equivalent to a $1/x$ probability law for the numbers themselves. Also, as was shown in Sec. 9.4.1, this result would hold *regardless of the set of units* assumed for the numbers. Hence, the result is not a 'fluke' of the particular (c.g.s.) units employed in the table.

The $1/x$ law is also invariant to arbitrary inversion of any number of the constants (Sec. 9.4.2); and to forming new constants by taking quotients of arbitrary pairs of the constants (Sec. 9.4.3).

The fact that the $1/x$ law holds independent of the choices of units, inversion and quotients means that it would have been observed regardless of human history. It is a truly *physical* effect.

Knowledge of such a law allows us to achieve improved estimates of physical constants. See Sec. 9.4.4. Given multiple measurements of a constant, its maximum-probable estimate is *not* the sample mean of the measurements. It is *less than* the sample mean, and obeys Eq. (9.56). This follows from the bias of a $1/x$ PDF toward small numbers: it predicts that, *a priori*, a number x is more probable to be small than to be large.

A fascinating property of the $1/x$ law Eq. (9.37) is that its median value is 1 (Sec. 9.3.14). Essentially, the value 1 lies halfway on the range of physical phenomena. It is noteworthy that this result holds independent of units, inversion and combination, since the $1/x$ law itself is invariant under these choices. Therefore, the median value of 1 is a *physical effect*. On the other hand, we reasoned (Sec. 9.3.2) that the ideal value of any physical constant is 0. Hence, the values 0 and 1 are cardinal points in the distribution of the physical constants.

The fact that every range of exponents given in Table 9.1 should contain, theoretically, the same number of constants, can be used to crudely predict the values of new constants. For example, there should be additional constants in the range $(18, 30)$ of exponents (in the c.g.s. units of the table, of course).

As we saw, the law that is predicted by EPI is validated by the data. This, in turn, tends to validate the model assumptions that went into the EPI derivation. Among these are: (a) the argument of Barrow (1991) in favor of randomness among the physical constants (Sec. 9.2); and (b) the argument by Vilenkin (1982, 1983) that the Universe – and its constants – arose out of nothingness (Sec. 9.3.2.).

As with the $1/f$ power spectral law derivation (Chap. 8), the principle of symmetry utilized by the $1/x$ PDF derivation is one of invariance to scale change (compare Secs. 8.3.3 and 9.3.10). A further similarity is that both the $1/f$ law and the $1/x$ law derive via EPI for any choice of the efficiency constant κ. Hence, as with the $1/f$ law (Sec. 8.4), the $1/x$ law applies over a vast range of differing phenomena. This makes sense, since its population consists of the numbers that define all known physical phenomena!

What we have basically found is that the PDF for a population of numbers from *mixed phenomena* (apples, oranges, meters, grams, etc.) should be of the $1/x$ form. The fundamental physical constants are of this type. But, so also are other populations. An example is the totality of numbers that an individual 'sees' during a typical year. Perhaps a more testable example is the population of numbers that are output from the central processor of a mainframe computer at a general-purpose facility such as a university. It would be interesting to check if the histogram of these numbers follows a $1/x$ form.

10

Constrained-likelihood quantum measurement theory

10.1 Introduction

In preceding chapters, EPI has been used as a *computational procedure* for establishing the physical laws governing various measurement scenarios. We showed, by means of the optical measurement model of Sec. 3.8, that EPI is, as well, a *physical process* that is initiated by a measurement. Specifically, it arises out of the interaction of the measuring instrument's probe particle with the object under measurement. This perturbs the system probability amplitudes, which perturbs the informations I and J, etc., as indicated in Figs. 3.3 and 3.4. The result is that EPI derives the phenomenon's physics as it exists at the input space to the measuring device.

We also found, in Sec. 3.8, the form of the phenomenon's physics at the output to the measuring instrument. This was given by Eq. (3.51) for the output probability amplitude function.

The analysis in Sec. 3.8 was, however, severely limited in dimensionality. A one-dimensional analysis of the measurement phenomenon was given. A full, covariant treatment would be preferable, i.e., where the space-time behavior of all probability amplitudes were determined.

Such an analysis will be given next. It constitutes a covariant *quantum theory of measurement*. This covariant theory will be developed from an entirely different viewpoint than that in Sec. 3.8. The latter was an analysis that focussed attention upon the probe–particle interaction and the resulting wave propagation through the instrument. The covariant approach will concentrate, instead, on *the meaning* of the acquired data to the observer as it reflects upon the quantum state of the measured particle. Hence, whereas the previous analysis was purely physical in nature, this one will be 'knowledge based' as well as physical.

Such a knowledge based orientation is consistent with our thesis (and

Wheeler's) that the measurer is part of the measured phenomenon. For such an observer, *acquired knowledge reflects physical state as well.* We return to this thesis in the Overview section.

This may be quantified as follows. Suppose that the observer measures the three-position $\bar{r}$ of a particle. This single piece of information affects the observer's state of knowledge of the particle's position and, hence, the physical state of the wave function (as above). This is a kind of physical manifestation of the principle of maximum likelihood. If $\bar{r} = 0$ then $\psi(\bar{r})$ for the particle should, by the principle of maximum likelihood, *tend to be* high near $r = 0$. Such a result is ordinarily taken to provide a mere prediction, or estimate, of an unknown quantity. Now it graduates into a statement of physical fact. Thus, the 'prediction' takes the physical form of the 'reduction of the wave function', an effect that is derived in the analysis below (this was also derived in one dimension in Sec. 3.8).

The theory shares many similarities of form with earlier work of Feynman (1948); Mensky (1979, 1993); Caves (1986); and Caves and Milburn (1987). These investigators used a path integral-based theory of continuous quantum measurements. For brevity, we will call this overall approach 'Feynman–Mensky theory'. Valuable background material on alternative approaches to measurement may also be found in books edited by Pike and Sarkar (1987), and by Cvitanovic, Percival and Wirzba (1992).

Our development parallels Feynman–Mensky theory, but with important *caveats*. It is both (i) a natural extension (literally an 'add-on') to the preceding EPI theory, and (ii) a further use of the simple notions of classical statistics and information. In particular, the classical notion of the *log-likelihood* (Sec. 1.6.2) of a *fixed* set of measurements is used. We develop this theory next. Key results that parallel those of Mensky and other workers in the field will be pointed out as they arise in the development.

10.2 Measured coordinates

The measurements that were used in Chap. 4 to derive relativistic quantum mechanics were the space-time coordinates of a particle. As we found, this gives rise to the physical information $K \equiv K_0$ obeying Eq. (4.27). In that treatment, as here, time was assumed to be as unknown, and random, as the space coordinates of the particle. We now modify K_0 with a constraint to the effect that three-position measurements have been made at a sequence of *imperfectly measured* time values. Each has a random component obeying the marginal law Eq. (4.30b),

$$p_T(t) = \int d\boldsymbol{r}\psi^*(\boldsymbol{r}, t)\psi(\boldsymbol{r}, t), \tag{10.1}$$

where $\psi(\boldsymbol{r}, t)$ is the usual four-dimensional wave function. PDF $p_T(t)$ defines the experimental fluctuations t in time from the 'true' laboratory clock value θ_0. (*Caveat*: By contrast, Feynman–Mensky theory presumes the time values to be perfectly known.)

Consistent with the knowledge-based approach, each data value will contribute a constraint term to supplement (in fact, to be added to) the information K_0. That is, knowledge of the measurements will be used to mathematicallly *constrain* the information. This is a standard, and convenient, step of the Lagrange approach (see Sec. 0.3.8).

Hence, assume some measured time values τ_m and intervals,

$$\tau_m = \theta_{0m} + t_m \text{ and } (t_m, t_m + dt) \equiv t_m, m = 1, \dots, M \tag{10.2}$$

respectively, during which particle *three-position* measurements

$$\bar{\mathbf{y}}(t_1), \dots, \bar{\mathbf{y}}(t_M) \equiv \bar{\mathbf{y}}_1, \dots, \bar{\mathbf{y}}_M \equiv \bar{\mathbf{y}} \tag{10.3}$$

are collected. For simplicity of notation, we have suppressed the deterministic numbers θ_{0m} from the arguments of the $\bar{\mathbf{y}}$ values. The data $\bar{\mathbf{y}}$ are taken to be fixed, non-random numbers, which is the usual stance of classical likelihood theory (Sec. 1.6; Van Trees, 1968).

Caveat: Many versions of Feynman–Mensky theory instead treat the data $\bar{\mathbf{y}}$ as *random*, to be integrated over during normalization. As in the preceding, we treat them as *fixed* data – the usual orientation of classical likelihood theory.

For simplicity of language, we sometimes say that the measurements take place *at* a time t_m which really means *during* the infinitesimal time interval $(t_m, t_m + dt)$.

The data $\bar{\mathbf{y}}$ are acquired with an instrument that realistically adds random noise to each reading. If the noise-free measurements are $\mathbf{y}_1, \dots, \mathbf{y}_M \equiv \mathbf{y}$ then the data obey

$$\bar{\mathbf{y}} = \mathbf{y} + \boldsymbol{\gamma} \tag{10.4}$$

where $\gamma_1, \dots, \gamma_M \equiv \boldsymbol{\gamma}$ denotes the random noise values.

As with Eq. (10.2), the space coordinates are themselves stochastic, obeying

$$\mathbf{y} = \boldsymbol{\theta} + \boldsymbol{r}, \text{ where } \boldsymbol{\theta} \equiv \boldsymbol{\theta}_1, \dots, \boldsymbol{\theta}_M \tag{10.5}$$

denotes the ideal (classical mechanics) space positions of the particle. Measurements $\bar{\mathbf{y}}$ are taken so as to know the values $\mathbf{y}$.

Substituting Eq. (10.5) into Eq. (10.4) shows that the measurements $\bar{\mathbf{y}}$ are degraded from the ideal space position values $\boldsymbol{\theta}$ by *two* sources of 'noise', the intrinsic quantum mechanical fluctuations $\boldsymbol{r}$ and the instrument noise values $\boldsymbol{\gamma}$.

The aim of making the measurements is to observe the quantum fluctuations γ. Toward this end, if we define a quantity

$$\bar{r} \equiv \bar{y} - \boldsymbol{\theta} \tag{10.6}$$

as *associated* data, then these relate to the quantum fluctuations r as

$$\bar{r} = r + \gamma. \tag{10.7}$$

From Eq. (10.6), the associated data are fixed; they are the departures of the (fixed) data from the (fixed) ideal three-positions $\boldsymbol{\theta}$. Eq. (10.7) states that the associated data are degraded from the ideal fluctuations r by the added noise γ due to the instrument. For simplicity, in the following we will often simply call the 'associated data' the 'data'.

Our aim is to see how the acquisition of fixed data affect the output wave function $\psi(r, t)$ as predicted by EPI. It is intuitive that such an effect must exist. If, for example, the data value $\bar{r}_m$ is zero then, by the classical *principle of maximum likelihood*, there must have been *a forteriori* a relatively high probability for the true position r_m to be zero. Hence, the probability *amplitude* $\psi(r_m, t_m)$ should be relatively high at $r_m = 0$. The corresponding output law $\psi(r, t)$ from EPI should, therefore, show this effect as well. We now proceed to quantify the situation.

10.3 Likelihood law

Any model of measurement has to address the question of how the measuring instrument affects the ideal parameter values under measurement. According to classical statistics, that specifier is the *likelihood law* (Sec. 1.2.3). This is the probability of a data value $\bar{r}_m$ conditional upon its ideal value r_m, denoted as $p(\bar{r}|r_m)$. This probability is a single, scalar number that is obviously a quality measure of the instrument. Consequently, it is also a quality measure of the data. Hence, if this quality measure is large the data should be 'taken seriously' or, if small, should be ignored.

In fact, rather than the likelihood law *per se*, in classical statistics it is the *logarithm* of the likelihood law that has real import. For example, the maximum likelihood principle is conventionally expressed as $\ln p = Max$. (Van Trees, 1968). Also see Sec. 1.6.2. In statistical mechanics it is the *average of the log-likelihood*, i.e., the *entropy*, which has fundamental significance; likewise with standard communication theory (Shannon, 1948). For these reasons, and because 'it works', we adopt the log-likelihood as well.

In the preceding examples, the average arose out of the stochastic nature of each likelihood term defining the system. Likewise, in our case there are many data values present, and these are taken over a sequence of time values, some of which are *a priori* more probable than others; see Eq. (10.1). It is logical to give less weight to log-likelihood terms that are less probable to occur, and conversely.

The considerations of the two previous paragraphs suggest that we use, as our scalar measure of data quality,

$$\langle \ln [C^2 \, p(\bar{r}|r)] \rangle_{time} \equiv \sum_{m=1}^{M} p_T(t_m) \ln [C^2 p(\bar{r}_m|r_m)] \equiv L. \qquad (10.8)$$

This is the time-averaged log-likelihood, aside from the presence of an arbitrary constant parameter C which merely adds in a constant value $\ln(C^2)$. The negative of the remaining sum can be considered the 'entropy' of the measurements. The arbitrary additive constant value $\ln(C^2)$ is permitted since (like optical phase) only *changes* in the entropy are physically observable.

10.4 Instrument noise properties

Eq. (10.8) simplifies in the following way. Assume that the noise γ is 'additive', i.e., independent of the ideal values y in Eq. (10.4) and, consequently, independent of the ideal values r in Eq. (10.7). Then from (10.7) fluctuations $\bar{r}_m$ simply follow those of the noise γ_m. If $p_\Gamma(\gamma)$ denotes the PDF on the noise,

$$p(\bar{r}_m|r_m) = p_\Gamma(\bar{r}_m - r_m). \qquad (10.9)$$

The noise PDF p_Γ arises as the squared modulus of a generally complex probability amplitude w called the 'instrument function',

$$p_\Gamma(\gamma_m) \equiv w^*(\gamma_m)w(\gamma_m), \quad m = 1, \ldots, M. \qquad (10.10)$$

10.5 Final log-likelihood form

Combining Eqs. (10.8)–(10.10) gives

$$L = \sum_{m=1}^{M} p_T(t_m)[\ln w(\bar{r}_m - r_m) + c.c.] \qquad (10.11)$$

where the notation *c.c.* denotes the complex conjugate of the preceding term. *Note*: For purposes of simplicity, we ignore the constant C that actually multiplies amplitude function w in this equation and in equations through Eq. (10.21).

Combining Eqs. (10.1) and (10.11) gives our final likelihood form

$$L = L_0[\psi, \psi^*, w] + L_0[\psi, \psi^*, w^*], \quad \text{where functional}$$

$$L_0[\psi, \psi^*, w] \equiv \int d\mathbf{r}\, dt \psi^*(\mathbf{r}, t)\psi(\mathbf{r}, t)\sum_{m=1}^{M}\delta(t - t_m)\ln w(\bar{\mathbf{r}} - \mathbf{r}). \quad (10.12)$$

Thus, the likelihood L has two distinct contributions, one due to the instrument amplitude w and the other due to its complex conjugate w^*. The sifting property of the Dirac delta function δ was central to this result, allowing quantities $\bar{\mathbf{r}}$ and $\mathbf{r}$ in the arguments of w and w^* to once again be expressed continuously in the time. It also permitted an interchange in the orders of summation and integration.

10.6 EPI variational principle with measurements

We are now ready to find the effects of the measurements upon the EPI output wave functions. Previously, we found that the log-likelihood function measures the quality of the data. But, intuitively, *the higher the quality of the data, the stronger should be its effect upon the EPI principle*; and conversely. The question is, how do we implement this connection?

Taking our cue from Eq. (0.39), we weight and add the log-likelihood terms $L_0[\psi, \psi^*, w]$ and $L_0[\psi, \psi^*, w^*]$ to the information K_0. The resulting EPI variational principle will show a tradeoff between extremizing the original information K_0 and the new, add-on log-likelihood terms. The larger (stronger) are the log-likelihood values the more will they be extremized, *at the expense of K_0*; and vice versa.

Accordingly, the EPI variational principle Eq. (3.16) becomes

$$K_0[\psi, \psi^*] + \alpha_1 L_0[\psi, \psi^*, w] + \alpha_2 L_0[\psi, \psi^*, w^*] = Extrem. \quad (10.13)$$

The square brackets identify the amplitude functions that functionals K_0 and L_0 depend upon. Parameters α_1 and α_2 are Lagrange multipliers that are to be fixed so as to agree with the form of Feynman–Mensky theory, wherever possible.

10.7 Klein–Gordon equation with measurements

The Lagrangian (or information density) for the problem is the total integrand of Eq. (10.13), obtained by the use of Eqs. (4.27) and (10.12). Substituting this Lagrangian into the Euler–Lagrange Eq. (4.24) gives

$$-c^2\hbar^2\left(\nabla - \frac{ie\mathbf{A}}{c\hbar}\right)\cdot\left(\nabla - \frac{ie\mathbf{A}}{c\hbar}\right)\psi + \hbar^2\left(\frac{\partial}{\partial t} + \frac{ie\phi}{\hbar}\right)^2\psi + m^2c^4\psi$$

$$+ \psi\sum_{m=1}^{M}\delta(t - t_m)[\beta_1 \ln w(\bar{\mathbf{r}} - \mathbf{r}) + \beta_2 \ln w^*(\bar{\mathbf{r}} - \mathbf{r})] = 0. \quad (10.14)$$

This is the Klein–Gordon Eq. (4.28) plus two measurement terms. The latter terms have new constants β_1, β_2, linearly related to α_1, α_2. These need to be determined. Subscript n, present in Eq. (4.28), is dropped here since there is only one complex component $\psi_1 \equiv \psi$ (Sec. 4.1.22) in this scalar amplitude case.

Eq. (10.14) is extremely nonlinear in the time. It shows that *between* measurements, i.e. for times $t \neq (t_m, t_m + dt)$, the ordinary Klein–Gordon Eq. (4.28) re-emerges. Hence, EPI output equations describe the phenomenon between, and not during, measurements. This is consistent with the analysis in Sec. 3.8, where it was found that EPI determines the phenomenon at the input space to the instrument, i.e., just as the measurement is initiated (but before the data value is acquired). The generalization that Eq. (10.14) gives is that the EPI output must define the phenomenon, as well, *after* the measurement; that is, after the output convolution Eq. (3.51) is attained.

Eq. (10.14) also shows that *at* the measurements, i.e. for values of $t = (t_m, t_m + dt)$ the constraint terms infinitely dominate the expression. This is very nonlinear behaviour, and true of other theories of measurement as well (see the edited books cited as references at the beginning of the chapter). We defer evaluating this case in favor of its nonrelativistic limit in Sec. 10.9.

The foregoing results are by use of the first EPI principle, Eq. (3.16). We now proceed to the second half of EPI, Eq. (3.18). Unfortunately, this development has not yet been carried through, and can only be sketched. In analogy with the development in the preceding section, this should lead to properties of the *Dirac equation with measurements*.

10.8 On the Dirac equation with measurements

In Sec. 4.2.5, the total Klein–Gordon Lagrangian was factored to produce the 'pure' Dirac equation (the equation between measurements). Here, the Lagrangian is the total Lagrangian of Eq. (10.13); *this, then, requires factoring*. As in Chap. 4, replace scalar functions ψ and w by vector functions $\boldsymbol{\psi}$ and $\mathbf{w}$. Factorization approaches are given in Matveev and Salle (1991, pp. 26–7). We expect one factor to contain amplitudes $\boldsymbol{\psi}$ and $\mathbf{w}$, and the other $\boldsymbol{\psi}^*$ and $\mathbf{w}^*$. Equate, in turn, each factor to zero (by EPI principle Eq. (3.18)). The first factor will yield a first-order differential wave equation for the particle (involving wave function $\boldsymbol{\psi}$ and instrument function $\mathbf{w}$), while the second will yield a first-order differential equation for the anti-particle (in terms of wave function $\boldsymbol{\psi}^*$ and instrument function $\mathbf{w}^*$). The latter effects are analogous to those of Sec. 4.2.6.

Also, in analogy with Eq. (4.38), the wave function for the anti-particle will be that for the direct particle as evaluated at $-t$, and the anti-particle will be measured by the instrument function as evaluated at $-t$. The anti-particle both travels backward in time *and is detected* backward in time. The latter effect may be clarified as follows.

The physical behavior of the anti-particle completely mirrors that of the particle (Feynman and Weinberg, 1993), so that the anti-particle is often described as a 'virtual' image of the particle. This means that, when the particle is measured at time t, *automatically* the anti-particle is measured at time $-t$; no new measurement is actually made. The anti-particle measurement is as virtual as the particle it measures.

However, the 'virtual' particle and its measurement are as 'real' physically as are the 'actual' particle and its measurement. One man's particle is another's anti-particle. This gives us a way of exactly reconstructing the past, albeit relative to an arbitrary laboratory time origin and with no real gain in information: the past *measured* history of an anti-particle is redundant with the future measured history of its particle mate.

10.9 Schroedinger wave equation with measurements

For simplicity, consider the case of a general scalar potential $\phi(r, t)$ but a vector potential $A(r, t) = 0$. The non-relativistic limit of the Klein–Gordon part of Eq. (10.14) can be taken as before (see Appendix G). The result is a SWE (G8) (or (10.30)) plus measurement terms,

$$i\hbar \frac{\partial \psi}{\partial t} = -\frac{\hbar^2}{2m} \nabla^2 \psi + e\phi\psi + \psi \sum_{m=1}^{M} \delta(t - t_m)[\varepsilon_1 \ln w(\bar{r} - r) + \varepsilon_2 \ln w^*(\bar{r} - r)]$$

(10.15)

The constants ε_1, ε_2 are the new, adjustable Lagrange multipliers.

10.9.1 Transition to Feynman–Mensky theory

The result (10.15) is analogous to that of Mensky (1993), provided the constants ε_1, ε_2 take the values

$$\varepsilon_1 = i\hbar, \quad \varepsilon_2 = 0.$$

(10.16)

Setting $\varepsilon_2 = 0$, in particular, eliminates from Eq. (10.15) the only complex-conjugated term in the equation, which seems correct on the basis of simplicity alone. However, this term should remain present in the measurement-inclusive Dirac equation of Sec. (10.8) since there both ψ and ψ^* are present. The values Eq. (10.16) bring Eq. (10.15) into the Feynman–Mensky form

$$ i\hbar \frac{\partial \psi}{\partial t} = -\frac{\hbar^2}{2m} \nabla^2 \psi + e\phi\psi + i\hbar\psi \sum_{m=1}^{M} \delta(t - t_m) \ln w(\overline{r} - r). \qquad (10.17) $$

We analyze it next.

10.9.2 Analysis

As with the Klein–Gordon measurement Eq. (10.14), Eq. (10.17) is highly nonlinear in its time-dependence. At times $t \neq t_m$ between measurements, because $\delta(t - t_m) = 0$, (10.17) becomes the ordinary Schroedinger wave Eq. (10.30). However, *at* a time t within a measurement interval $(t_m, t_m + dt)$ the sum in Eq. (10.17) collapses to its mth term, and dominates the right-hand side, so that the equation becomes

$$ i\hbar \frac{\partial \psi}{\partial t} = i\hbar\psi\delta(t - t_m) \ln w(\overline{r}(t) - r(t)). \qquad (10.18) $$

We inserted back the time-dependences within the argument of w for clarity. Now multiply both sides of Eq. (10.18) by $dt/i\hbar\psi$ and integrate dt over the measurement interval $(t_m, t_m + dt)$, $dt \geqslant 0$.

The left-hand side becomes

$$ \int_{t_m}^{t_m+dt} dt \frac{1}{\psi}\frac{\partial \psi}{\partial t} = \int_{\psi_0}^{\psi_+} \frac{d\psi}{\psi} = \ln\left(\frac{\psi_+(r_m)}{\psi_0(r_m)}\right), \qquad (10.19a) $$

$$ \psi_+(r_m) \equiv \psi(r_m, \overline{r}_m, t_m + dt) \equiv \psi(r_m, t_m + dt), \qquad (10.19b) $$

$$ \psi_0(r_m) \equiv \psi(r_m, t_m). \qquad (10.19c) $$

The notation $\psi_+(r_m)$ denotes the wave function as evaluated within the measurement interval. The joint event $(r_m, \overline{r}_m)$ shown in the argument of ψ comes about because, by hypothesis, the event r_m occurs within the measurement interval, i.e., co-jointly with the occurrence of the measurement $\overline{r}_m$. The single event r_m is equivalent to the joint event $(r_m, \overline{r}_m)$.

The right-hand side of Eq. (10.18) becomes, because of the sifting property of the delta function,

$$ \ln w(\overline{r}_m - r_m). \qquad (10.20) $$

10.9.3 Bohm–Von Neumann measurement theory

The net result is that

$$ \ln\left(\frac{\psi_+(r_m)}{\psi_0(r_m)}\right) = \ln w(\overline{r}_m - r_m) \text{ or } \psi_+(r_m) = \psi_0(r_m)w(\overline{r}_m - r_m). \qquad (10.21) $$

With the understanding that all coordinates are evaluated at time $t = t_m$, the latter simplifies in notation to

$$\psi_+(r) \equiv \psi(r, \bar{r}) = C\psi_0(r)w(\bar{r} - r) \tag{10.22}$$

(cf. Eq. (3.53)). Here we have regained the joint-event notation $(r, \bar{r})$ of Eq. (10.19b), as well as the constant C multiplier of w that was suppressed after Eq. (10.10).

Eq. (10.22) agrees with the well-known measurement theory of Bohm (1952), Von Neumann (1955) and others. It shows that the wave function $\psi(r, \bar{r})$ at the measurement is essentially the wave function $\psi_0(r)$ just before the measurement times the instrument amplitude function. Or, the effect of a measurement is a *separation* of the system wave function from the instrument function. The effect is commonly called the 'reduction of the wave function' by the measurement, since the product should ordinarily be narrower than the input amplitude function ψ_0 (see Secs. 10.9.6, 10.9.7).

A similar wave function separation, Eq. (3.53), was found in the analysis of the optical measuring device. In fact Eq. (10.22) agrees with Eq. (3.53). The right-hand sides agree identically, after the use of Eq. (10.6). The left-hand sides also agree, when one recalls that r and $\bar{r}$ are, here, three-dimensional generalizations of x and $\bar{y}$.

10.9.4 Value of constant C

We assume that the measurement procedure is passive: the inferred PDF $p_+(r) \equiv |\psi_+(r)|^2$ at a measurement cannot exceed the PDF $p_0(r) \equiv |\psi_0(r)|^2$ prior to the measurement,

$$|\psi_+(r)|^2 \leq |\psi_0(r)|^2. \tag{10.23}$$

Then by Eq. (10.22), the constant C must have a value such that

$$|Cw(\gamma)| \leq 1, \text{ all } \gamma, \text{ or } C|w_{max}| = 1, \tag{10.24}$$

where w_{max} is the maximum value of w over its argument.

10.9.5 Initialization effect

In Sec. 10.9.2 we found that between measurements the wave function (10.17) is the simple Schroedinger equation, whereas *at* each measurement the wave function is 'instantaneously' perturbed into a form obeying Eq. (10.22). The upshot is that each measurement *merely re-initializes* the wave function, after which it freely 'evolves' according to the ordinary Schroedinger wave equation. Hence, with the understanding that the notation ψ means the output ψ_+ of the *previous* or $(m-1)$th measurement, we may rewrite Eq. (10.17) at the *m*th measurement as

$$i\hbar\frac{\partial\psi}{\partial t} = -\frac{\hbar^2}{2m}\nabla^2\psi + e\phi\psi + i\hbar\psi\delta(t-t_m)\ln w(\bar{r}-r). \qquad (10.25)$$

10.9.6 A one-dimensional example

In Fig. 10.1 we sketch a one-dimensional example $r = x$. Fig. 10.1a shows the system probability amplitude curve $\psi_0(x)$ just prior to the measurement, and the instrument amplitude curve $w(x)$ centered upon the measurement $\bar{x}$. According to the effect Eq. (10.22), the result of the measurement is a new probability amplitude curve $\psi_+(x)$ obeying

$$\psi_+(x) \equiv \psi(x, \bar{x}) = C\psi_0(x)w(\bar{x}-x) \qquad (10.26a)$$

(cf. Eq. (3.53)). This is shown in Fig. 10.1b. The new curve $\psi_+(x)$ is narrower than the input $\psi_0(x)$, showing that the measurement has narrowed the realm of possibilities for coordinate x of the particle. An increased amount of position information has been acquired by the measurer. (Note that this is in the specific sense of *Fisher information* about the unknown particle position; Shannon information does not measure the information about an unknown *parameter*.) The *narrowed* product form (10.26a) and the *gain* in Fisher information are fundamental effects of the measurement theory.

The output probability amplitude obeys

$$\psi(\bar{x}) \equiv \int dx\psi(x, \bar{x}) = C\int dx\psi_0(x)w(\bar{x}-x) \qquad (10.26b)$$

by Eq. (10.26a). This has the form of a convolution, and checks with the optical instrument result Eq. (3.51).

10.9.7 Exercises

One observes that the output curve has not generally been reduced to a single point of complete 'collapse'. That would have required a narrower measurement function $w(x)$, in fact to the limit of a delta function. Even the maximum probable position value after the measurement, indicated as coordinate $\hat{x}$ in the figure, is not quite the same as the measurement value $\bar{x}$. Why?

In general, the output curve $\psi_+(x)$ in Fig. 10.1b is narrower than *either* curve $\psi_0(x)$ or $w(x)$ of Fig. 10.1a. Prove that this is the case if both of the latter curves have the normal form (Gaussian with finite mean). (*Hint:* Form the product and complete the square in the exponent.)

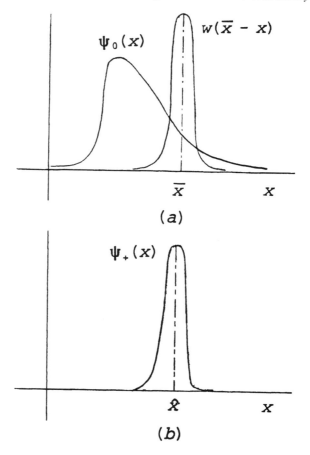

Fig. 10.1 (a) Wave function prior to measurement, and instrument measurement function; (b) wave function at measurement.

10.9.8 Passive Gaussian case

We continue next with the specifically Gaussian case

$$w(x) = \frac{1}{(2\pi\sigma^2)^{1/4}} e^{-x^2/4\sigma^2}, \quad C = (2\pi\sigma^2)^{1/4} \tag{10.27}$$

where we used Eq. (10.24) to find C. Gaussian noise is, of course, pervasive in nature because of the central limit theorem. The one-dimensional version of Eq. (10.25), including the multiplier C of w given by Eq. (10.27), becomes

$$i\hbar \frac{\partial \psi}{\partial t} = -\frac{\hbar^2}{2m} \psi'' + e\phi\psi - i\hbar\psi\delta(t - t_m)\frac{(\bar{x} - x)^2}{4\sigma^2}. \tag{10.28}$$

This shows that, in the case of a very poor, passive measuring instrument where the error variance obeys $\sigma^2 \to \infty$, the between-measurements Schroedinger wave equation essentially results once again. Making a measurement at $t = (t_m, t_m + dt)$ with such an instrument is equivalent to making almost no measurement at all. This is a reasonable result.

This result is confirmed, as well, by Eq. (10.26a) which becomes

$$\psi_+(x) = \psi_0(x)e^{-(\bar{x}-x)^2/4\sigma^2} \approx \psi_0(x)\left[1 - \frac{(\bar{x}-x)^2}{4\sigma^2}\right] \qquad (10.29)$$

in the limit $\sigma^2 \to \infty$ of a very poor instrument. Here ψ_+ is barely changed from ψ_0 by knowledge of the measurement $\bar{x}$. It suffers an *infinitesimal change* from the input state.

10.9.9 Time-reversal effects

(This section may be skipped in a first reading.) Between measurement times t_m, Eq. (10.17) becomes the ordinary Schroedinger wave equation

$$i\hbar\frac{\partial\psi}{\partial t} = -\frac{\hbar^2}{2m}\nabla^2\psi + e\phi\psi, \; \psi = \psi(\boldsymbol{r}, \, t). \qquad (10.30)$$

Under reversal of the time coordinate, i.e.,

$$t \to -t, \, dt \to -dt, \qquad (10.31)$$

and with complex conjugation, Eq. (10.30) becomes

$$i\hbar\frac{\partial\psi^*}{\partial t} = -\frac{\hbar^2}{2m}\nabla^2\psi^* + e\phi\psi^*, \; \psi^* = \psi^*(\boldsymbol{r}, \, -t). \qquad (10.32)$$

This assumes a real potential function ϕ as well. Comparing Eqs. (10.30) and (10.20) shows that Eq. (10.32) is again the Schroedinger wave equation, now in the wave function $\psi^*(\boldsymbol{r}, \, -t)$. Hence, if $\psi(\boldsymbol{r}, \, t)$ is a solution to the Schroedinger wave equation, so is $\psi^*(\boldsymbol{r}, \, -t)$, the 'time-reversed' solution.

One might think that, since measured lab time is usually regarded as moving forward, the time-reversed solution has no physical reality. However, it does indeed have some observed effects, one of which is the so-called 'Kramers degeneracy' of energy eigenfunctions (Merzbacher, 1970, p. 408). Another, as we will find next, is that of the *measured* time-reversed wave.

The Feynman–Mensky Eq. (10.17) has been derived as the nonrelativistic limit of Eq. (10.14), which was derived by EPI after varying the *complex conjugate* ψ^* of the wave function: see Eqs. (4.24) and (4.25). But the Euler–Lagrange approach for a complex function ψ should give valid results whether ψ or ψ^* is varied. Accordingly, let us now vary ψ. Carrying through the analogous steps that gave rise to Eq. (10.17) now gives, after a complex conjugation,

$$ih\frac{\partial\psi}{\partial t} = -\frac{\hbar^2}{2m}\nabla^2\psi + e\phi\psi - ih\psi\sum_{m=1}^{M}\delta(t - t_m)\ln w^*(\bar{r} - r). \qquad (10.33)$$

For times $t \neq (t_m, t_m + dt)$ between measurements, this is the same as Eq. (10.17). However, at a measurement time things are slightly different. Repeating the analogous steps to Eqs. (10.18)–(10.22) now gives

$$\psi(r, t + dt) = C\psi(r, t)/w^*(\bar{r} - r). \qquad (10.34)$$

Cross-multiplying Eq. (10.34) shows that, by the reasoning of Secs. 10.9.3, 10.9.6 and 10.9.7, $\psi(r, t)$ is now narrower than $\psi(r, t + dt)$. Or, the unmeasured wave is *narrower than* the measured wave, where we assumed that $dt > 0$. Consequently there is more Fisher information about particle position in the *unmeasured* state than in the *measured* state. By the reasoning in Sec. 10.9.6, this is unacceptable physically.

We noticed that the preceding result assumed a positive time differential dt. Other time values were also assumed to be positive. To try to obtain reasonable results, we now evaluate Eq. (10.34) for *negative* times and time intervals, i.e.

$$t \rightarrow -t,\ dt \rightarrow -dt,\ t_m \rightarrow -t_m,\ (t_m, t_m + dt) \rightarrow (-t_m - dt, -t_m). \quad (10.35)$$

(See, by comparison, Eqs. (10.31) for the time-reversed Schroedinger wave Eq. (10.32).) Since $-t_m - dt \leqslant -t_m$, $\psi(r_m, -t_m - dt)$ and $\psi(r_m, -t_m)$ now represent the wave functions just prior to and at the measurement, respectively. Eq. (10.34) becomes, with the replacements (10.35) and after a cross-multiplication and complex conjugation,

$$\psi^*(r, -t) = B\psi^*(r, -t - dt)w(\bar{r} - r) \qquad (10.36)$$

(subscripts m suppressed as before). B is a constant. This, now, shows the desired effect: the wave function $\psi^*(r, -t)$ at the measurement is *narrower than* the wave function $\psi^*(r, -t - dt)$ before measurement. Once again the Fisher information on position has been increased.

Also, Eq. (10.36) shows that, for negative times, the measurement function $w(\bar{r} - r)$ acts as a 'transfer function' of measurement for the system wave function ψ^*. That is, when the transfer function operates upon the system wave function in its unmeasured state the result is the system wave function in its measured state. By symmetry one would have expected the complex conjugated quantity $w^*(\bar{r} - r)$ to have filled that role here.

These results are for negative times. It is interesting that, even for positive times, Eq. (10.22) exhibits the same transfer function $w(\bar{r} - r)$. This seems to be consistent with the Feynman–Mensky Eq. (10.17), the basis for these effects, which likewise contains $w(\bar{r} - r)$ and not its complex conjugate.

Eq. (10.36) may be considered to be a time-reversed measurement effect, for the wave function ψ^*, that supplements the time-reversed Schroedinger wave Eq. (10.32), also for the wave function ψ^*.

10.10 Overview

The primary aim of this chapter was to develop a covariant, four-dimensional theory of measurement. In contrast with other theories of measurement, and consistent with the covariant approach of EPI, we allowed the measurement times to themselves suffer from random noise. A secondary aim was to confirm the one-dimensional measurement effects that were found in Sec. 3.8 from a different viewpoint.

The culmination of this measurement theory, Eq. (10.14) or Eq. (10.17), confirms that in the input space (see Fig. 3.3) of a measuring instrument (i.e., for time $t < t_m$) the laws of physics are exactly those that EPI predicts. In this case it is the Klein–Gordon law, as derived in Chap. 4, or the Schroedinger wave equation, as derived in Appendix D. The physical law *that contains* the measurement information is obtained *at* the measurement time $t = t_m$, with the result Eq. (10.22).

The result Eq. (10.22) shows *how* mathematically the EPI amplitude law is changed by the measurement, namely through simple multiplication by an instrument amplitude function. As we found, this confirms Eq. (3.53), the corresponding result for the optical measuring device of Sec. 3.8. Eq. (10.22) has been known for over 40 years, of course, independent of the measurement theory of this book (Bohm, 1952; Von Neumann 1955). The multiplication amounts to a transfer of the system statistics from an unmeasured to a measured state, where the instrument amplitude function w acts as the transfer function.

This view of measurement, supplemented by that in Sec. 3.8, defines EPI as a physical process. Its progression is as follows.

(1) The measuring instrument's probe particle perturbs the input, or object, amplitude function ψ_0 (see Sec. 3.8.1).
(2) Perturbing the complex wave function ψ_0 means, by Eq. (2.24), perturbing the real amplitude functions q_n, $n = 1, 2$ by amounts $\delta q_n(x)$.
(3) These cause a perturbation δI in the intrinsic information, through the defining Eq. (2.19) for I.
(4) This, in turn, causes an equal perturbation δJ in the bound information through the conservation law Eq. (3.13). As we saw in Sec. 3.8, *this law is identically obeyed in the presence of a unitary transformation* between conjugate spaces such as (r, ct) and $(\mu, E/c)$.
(5) The invariance principle (Sec. 3.4.5) causes the information densities i_n, j_n to obey proportionality, so that their sumtotal (integrated) values obey

$$I = \kappa J. \tag{10.37}$$

Alternatively, this equation follows identically, with $\kappa = 1$, when the invariance principle is the unitary transformation mentioned in item (4).

(6) Since the perturbations δI and δJ are equal (point (4) above), they must obey the EPI extremum principle Eq. (3.16),

$$I - J \equiv K = extrem. \tag{10.38}$$

(see Sec. 3.4.1). Together, Eqs. (10.37) and (10.38) comprise the EPI principle.

(7) The EPI principle is physically enacted. The 'knowledge game' of Secs. 3.4.11–3.4.13 is played if the coordinates are real, or if the time coordinate separates out due to energy conservation (see Exercise in Sec. 3.4.12).

(8) The outcome of the game is the physical occurrence of the law that defines the $q_n(\mathbf{x})$ or (by Eq. (2.24)) the complex $\psi_n(\mathbf{x})$.

(9) This probability law is randomly sampled by the measuring instrument, giving the data value, the acquisition of which was the aim of the whole procedure.

   This means that the act of seeking data is physically self-realizing; it elicits from nature the very probability law that randomly forms the data. One might say that EPI is an 'epistem*active*' process, in contrast with the more passive connotations of the word 'epistem*ology*'. That is, a question is asked, and the response is not only the requested answer, but the mechanism for determining the answer as well. A kind of local reality is so formed.

(10) The solution $\psi_n(\mathbf{x})$ and resulting information densities i_n, j_n define the future evolution and *dynamics* (Sec. 3.7) of the system until a new measurement is made.

These results imply two general modes of operation of the EPI principle: (a) a 'passive' mode, whereby EPI is used as a computational tool (as in other chapters); and (b) an 'active' mode, whereby real data are sought and EPI executes as a *physical process*. Mode (a) is utilized in a quest for *knowledge* of the physical law governing a measurement scenario. Mode (b) occurs during a quest for real data. As we saw, the quest elicits the law that generates the requested data.

We saw that an act of measurement elicits a physical law. This was a well-defined measurement of a particle in the presence of a well-defined potential function. Does EPI yield a physical law under less definite measurement conditions? For simplicity, specify the particle position by a single coordinate x. We found before (Secs. 3.8.5, 10.1) that the physical law will follow from any perturbation of the intrinsic wave function of the particle as, for example, occurs in the input port of the measuring instrument. However, some 'measurements' are of a less precise kind than as previously treated. Rather than measuring the position of a particle in the presence of a definite potential function $\phi(x, t)$, the measurement might be so poor, or indirect, as to effectively occur in the presence of any one randomly chosen potential over a class of possible potentials $\phi_i(x, t)$, $i = 1, \ldots, n$. In other words, any one of a class of possible particle interactions randomly took place. This occurs, e.g., if the 'measurement' determines that the particle lies *somewhere* within a large container, different parts of which are subjected to different potentials $\phi_i(x, t)$. This 'measurement' implies the knowledge that there is a range of possible *ideal* positions θ_i, $i = 1, \ldots, n$ present for the particle. This is a mode of operation of EPI that is, experimentally, less well-defined than mode (b), which has a single ideal position θ, but certainly more well-defined than mode (a) above. What will be the effective output amplitude law $\psi_{eff}(x, t)$?

We work in the non-relativistic limit. In the case of a definite measurement of position θ_i in the presence of a definite potential $\phi_i(x, t)$, the output law will be the ordinary SWE (10.30) for the wave function $\psi(x|i)$.

But here the potential $\phi_i(x, t)$ arises randomly. Suppose that it occurs according to a probability amplitude c_i. The c_i, $i = 1, \ldots, n$, define the degree of vagueness of the measurement. If all c_i are equal, then the measurement is maximally vague; a case of maximum ignorance. At the opposite extreme, if all $c_i = 0$ except for one $c_j = 1$, then we are back to definite measurement, mode (b).

Since the SWE is linear in $\psi(x|i)$, multiplying the equation by c_i and summing over i gives a net SWE in an effective probability amplitude $\psi_{eff}(x, t)$ obeying

$$\psi_{eff}(x) = \sum_{i=1}^{n} c_i \psi(x|i). \tag{10.39}$$

This is due to an effective potential $\phi_{eff}(x, t)$ obeying

$$\phi_{eff}(x, t) = \frac{\sum_i \phi_i(x, t) c_i \psi(x|i)}{\sum_i c_i \psi(x|i)}. \tag{10.40}$$

Eq. (10.39) has the form of a 'coherent', weighted superposition of amplitudes. Notice that in the case of all $c_i \geq 0$ it represents a broader PDF on x than in the case of a single definite measurement. Also, the effective potential (10.40) is widened, or blurred, by the superposition of individual potentials. These effects reflect an increased state of ignorance of the measurer due to the vagueness of the measurement scenario. Hence, under vague measurement conditions the output amplitude law is vague as well, although the output physical law is still the well-defined SWE.

Interestingly, there are cases where the coherence effect (10.39) is lost, and recourse must be made to a 'density function' type of average (Louisell, 1970),

$$\rho(x, x') = \sum_{i=1}^{n} P_i \psi^*(x|i)\psi(x'|i), \ P_i = |c_i|^2. \tag{10.41}$$

We see that, in the limit $x \to x'$, Eq. (10.41) becomes

$$p_{eff}(x) = \sum_i P_i p(x|i), \ p(x|i) \equiv |\psi(x|i)|^2. \tag{10.42}$$

This is an 'incoherent' or classical-statistics superposition of states. Comparison with Eq. (10.39) is germane.

If we let the difference $\Delta x \equiv x' - x \to 0$, as previously, from Sec. 1.4.3 we suspect that yet another transition to Fisher information is at hand; in this case, the average over all interaction states i. The specific relation is

$$\int dx \lim_{x' \to x} \frac{\partial^2 \rho(x, x')}{\partial x \partial x'} = \sum_{i=1}^{n} P_i I_i \equiv \langle I \rangle. \tag{10.43}$$

This may easily be verified by operating $\partial^2/\partial x \partial x'$ on Eq. (10.41); multiplying by the Dirac $\delta(x - x')$, integrating both sides dx' and using the sifting property Eq. (0.63b) of the delta function; then integrating both sides dx. The representation Eq. (3.38) for information I must also be used. We conclude that knowledge of the density matrix for a measurement scenario implies knowledge of the average Fisher information over the possible measurement states. This is another intriguing connection between quantum mechanics and classical measurement theory.

An underlying premise of EPI is that *the observer is part of the observed phenomenon* (see Sec. 10.1). Thus, a full quantum treatment of the data collection process should include the observer's internal interactions with the observed data values. This would require at least a five-dimensional analysis of the measuring apparatus and the observer as they jointly interact: four-dimensional space-time, plus one 'reading' internal to the observer. It is interesting that we did not have to take such an approach in the preceding – the use of Lagrange constraint terms served to avoid it. In effect, inputs of knowledge replaced detailed biological analysis.

However, a one-dimensional analysis that directly *includes* the observer may be performed using the coherent optical model of measurement given in Sec. 3.8. Simply regard the coherent optical device of Fig. 3.3 as a model for the visual workings of the eye. That is, the particle is in the object space of the eye, the lens L is the lens of the eye, the measurement space M is the retina, the data value $\bar{y}$ corresponds to a retinal excitation at that coordinate, etc. This provides a model for the eye–particle measurement interaction. Then, as found in Sec. 3.8, visual observation of the particle at $\bar{y}$ elicits its physics (the appropriate wave equation), both in the input space to the eye and at the retina (the latter by Eq. (3.51)). These effects serve to confirm the hypothesis made by Wheeler at the beginning of the book.

The development of this chapter did not achieve all our aims. We analyzed the EPI measurement procedure in the particular scenario of quantum mechanics, the subject of Chap. 4. However, other chapters used EPI to derive physical laws arising out of 'non-quantum' phenomena, such as classical electromagnetic theory, gravitational theory, etc. Would these be amenable to the same kind of measurement theory as developed here?

Also, there is an unresolved issue of dimensionality. Neither active nor passive observers can know what ultimate dimension to use in characterizing each measurement $\mathbf{x}$ of a particle or field. Coordinate $\mathbf{x}$ in definition (2.19) of I was made to be four-dimensional, by the dictates of covariance (Sec. 3.5.8). However, it might prove more productive to use a generally M-dimensional version of (2.19). In fact, *EPI operates on any level of dimensionality, giving the projection of the 'true' law into the M dimensions of the measurement space.* For example, use of $M = 1$ for a particle gives rise to the one-dimensional Schroedinger wave equation (see Appendix D), while use of $M = 4$ gives rise to either the Dirac equation or Klein–Gordon equation (Chap. 4). Use of higher values of M, such as values 10 or more, may give rise to string theory or a variant of it. Thus, EPI gives rise to a projection of the true law into the measurement space assumed. As with the famous shadows on the walls of Plato's cave, the ultimate, 'true' dimensionality of the phenomenon remains an unknown.

11

Research topics

11.1 Scope

Preceding chapters have dealt with relatively elementary applications of the EPI principle. These are problems whose answers are well-established within known experimental limits. Historically, such problems constituted 'first tests' of the principle. As we found, EPI agreed with the established answers, albeit with several extensions and refinements. The next step in the evolution of the theory is to apply it 'in the large', i.e., to more difficult problem areas, those of active, current research by the physics and engineering communities.

Two such are the phenomena of (i) quantum gravity and (ii) turbulence at low Mach number. We show, next, the current status of EPI work on these problems. Aspects of the approaches that are, as yet, uncertain are pointed out in the developments. It is hoped that, with time, rigorously correct EPI approaches to the problems will ensue. It is expected that final versions of the theory will closely resemble the first 'tries' given below.

11.2 Quantum gravity

11.2.1 Introduction

The central concept of gravitational theory is the metric tensor $g_{\mu\nu}(\mathbf{x})$. (For a review, see Exercise 6.2.5 and the material preceding it.) This defines the local distortion of space at a four-coordinate $\mathbf{x}$. The Einstein field Eq. (6.68) permits the metric tensor to be computed from knowledge of the gravitational source – the stress energy tensor $T_{\mu\nu}$ (Sec. 6.3.4). This is a deterministic view of gravity. That is, for a given stress energy tensor and initial conditions, a given metric tensor results. This view holds over macroscopic distances $\mathbf{x}$.

However, at very small distances the determinism is lost. It is believed that the metric tensor fluctuates *randomly* on the scale of the Planck length, Eq. (6.22), the order of 10^{-33} cm. This corresponds to a local curvature of 10^{33} cm^{-1}.

If the metric tensor fluctuates, what amplitude functions $\mathbf{q}[g_{\mu\nu}(\mathbf{x})]$ govern the joint fluctuations of all $g_{\mu\nu}$ at all $\mathbf{x}$? This is a probability amplitude with massively joint statistics. The determination of this amplitude function is a major aim of the field of study called quantum gravity. Many astronomers believe the wave function to obey a differential equation called the 'Wheeler–DeWitt' equation. We will attempt to show how this equation follows from the EPI procedure. As mentioned above, the derivation is not yet complete but should strongly indicate the direction to take.

11.2.2 Analogies with quantum mechanics

Assume, at first, a scalar quantum theory. It seems reasonable to model it after the one taken in Chap. 4 for deriving the scalar Klein–Gordon equation of relativistic quantum mechanics. Thus, there is a measurement space and its Fourier transform, a 'momentum' space. Also, Parseval's theorem is again used to compute the bound information J in momentum space as the equivalent amount of Fisher information I in measurement space. To keep the derivation brief, we freely use results from seminal papers by DeWitt (1967) and by Hartle and Hawking (1983).

11.2.3 Measurements and perturbations

For simplicity we restrict attention to spatially closed universes. This permits use of a three-metric g_{ij} in place of the general four-metric $g_{\mu\nu}$. (Recall from Sec. 6.2.1 that Latin indices go from 1 to 3 whereas Greek indices go from 0 to 3.)

As in the quantum mechanical development (Sec. 4.1.4), we pack the real probability amplitudes $\mathbf{q}$ as the real and imaginary parts of a *complex* scalar amplitude ψ. See Eq. (2.24). Presuming a scalar case, we use $N = 2$ functions $\mathbf{q}$ so that $\psi \equiv q_1 + iq_2$ (an irrelevant factor $1/\sqrt{2}$ is ignored). Thus we want to determine the wave function $\psi[g_{ij}(\mathbf{x})]$. It should be noted that each component g_{ij} is defined at every possible space-time coordinate $\mathbf{x}$ throughout the Universe. Since coordinate $\mathbf{x}$ is continuous, $\psi[g_{ij}(\mathbf{x})]$ is actually a *functional* (Sec. 0.2). It represents the wave function for every possible universe of metric tensor values; indeed, an all-inclusive wave function!

The observer measures the g_{ij} at a sequence of closely spaced points $\mathbf{x}$ throughout all of four-space. Thus, essentially an infinity of real measurements are made. Each measurement locally perturbs the overall wave function $\psi[g_{ij}(\mathbf{x})]$ with the result that it is perturbed everywhere, as $\delta\psi[g_{ij}(\mathbf{x})]$. As we saw (Chaps. 3, 10) this excites the EPI process. Since a unitary transformation (Eq. (11.18)) will be found to connect the space of values $g_{ij}(\mathbf{x})$ with a conjugate momentum space, EPI again holds rigorously (See Secs. 3.8.5, 3.8.7).

It is noted that the positions $\mathbf{x}$ are regarded as deterministic, in this problem.

11.2.4 Discrete aspect of the problem

Since the EPI approach is based upon a series of *discrete* measurements, we must replace the continuous position coordinate $\mathbf{x}$ by a fine subdivision $\mathbf{x}_1, \ldots, \mathbf{x}_N$ of such values. (A return to the continuous limit is, however, taken at the end of the derivation.) Thus, denote the value of the metric tensor g_{ij} at the four-position $\mathbf{x}_n$ as the discrete quantity g_{ijn}. Also, then

$$\psi \equiv \psi[g_{ij}(\mathbf{x})] = \psi(g_{ijn}). \tag{11.1}$$

The wave function is now no longer a functional, simplifying matters. It is an ordinary function of the random variables g_{ijn}, as EPI requires.

11.2.5 Fisher coordinates for the problem

The first step of the EPI approach is to identify the physical coordinates of the problem. These correspond to coordinates $\mathbf{y}$, $\boldsymbol{\theta}$, and $\mathbf{x}$ in the basic data Eq. (2.1). In particular, what is the 'ideal' value θ_{ijn} of g_{ijn}? Taking the cue from the quantum development in Sec. 4.1.2, we regard the ideal to be the classical, deterministic value $\overline{g}_{ijn}$ of the metric. This is the one that obeys the Einstein field Eq. (6.68) at a position $\mathbf{x}_n$. Then, since we want to find $\psi(g_{ijn})$, we must regard the g_{ijn} as the fluctuation values $\mathbf{x}$ in Eq. (2.1). Hence, our intrinsic data are the $y_{ij}(\mathbf{x}_n)$, where

$$y_{ij}(\mathbf{x}_n) = \overline{g}_{ij}(\mathbf{x}_n) + g_{ij}(\mathbf{x}_n) \text{ or } y_{ijn} = \overline{g}_{ijn} + g_{ijn} \tag{11.2}$$

for all i, j, n, and where the $\overline{g}_{ij}(\mathbf{x})$ obey the deterministic field Eq. (6.68).

11.2.6 The Fisher information in 'Riem' space

Eq. (2.19) expresses I in flat, rectangular coordinate space $\mathbf{x}$. The implied metric in (2.19) is (1, 1, 1, 1). We want to now express I in g_{ijn} space.

Temporarily revert to the case of a *continuous* four-position $\mathbf{x}$, at which the Fisher coordinates are g_{ij}. This is a generally curved space of coordinates.

Curved space has a Riemannian metric and is called 'superspace' or, more briefly, 'Riem' space. Coefficients G_{ijkl} obeying

$$G_{ijkl}(\mathbf{x}) \equiv \frac{1}{2\sqrt{g}}(g_{ik}g_{jl} + g_{il}g_{jk} - g_{ij}g_{kl}), \; g \equiv \det[g_{ij}] \quad (11.3)$$

may be regarded (DeWitt, 1967, Appendix A) as the metric of our Riem space. We adopt this metric as well.

What is a length in the Riem space defined by this metric? Consider a tensor function V^{ij} of generally all metric components g_{kl} evaluated at a continuous position $\mathbf{x}$. The squared magnitude of this tensor is then given as (Lawrie, 1990, p. 38)

$$|V(\mathbf{x})|^2 \equiv \sum_{ijkl} G_{ijkl}V^{ij}V^{*kl}, \; G_{ijkl} \equiv G_{ijkl}(\mathbf{x}), \; V^{ij} \equiv V^{ij}(\mathbf{x}). \quad (11.4)$$

Next, return to the *discrete* position case, where coordinates g_{ijn} are metric components g_{ij} evaluated at discrete coordinates $\mathbf{x}_n$. A function V in this space has the discrete representation $V = V^{ij}(\mathbf{x}_n) \equiv V^{ijn}$, so that the squared length $|V_n|^2$ is now

$$|V_n|^2 = \sum_{ijkl} G_{ijkln}V^{ijn}V^{*kln}, \; G_{ijkln} \equiv G_{ijkl}(\mathbf{x}_n). \quad (11.5)$$

Take the particular case of

$$V^{ijn} \equiv \frac{\partial \psi^*}{\partial g_{ijn}}. \quad (11.6)$$

This denotes the gradient with respect to Riem space coordinates g_{ij} of ψ^* as evaluated at the four-position $\mathbf{x}_n$. Then by Eq. (11.5)

$$|V_n|^2 \equiv |\text{grad}\,\psi(g_{ijn})|^2 = \sum_{ijkl} G_{ijkln}\frac{\partial \psi^*}{\partial g_{ijn}}\frac{\partial \psi}{\partial g_{kln}}. \quad (11.7)$$

Now, our random Fisher variables are the g_{ijn} (see Eq. (11.2)). Then by Eqs. (2.19) and (2.24), we get

$$I = 4\int d\mathbf{g}\sum_{n=1}^{N}|\nabla\psi(g_{ijn})|^2, \; d\mathbf{g} \equiv dg_{111}\ldots dg_{33N}. \quad (11.8)$$

The summation over n is required by the added dimensionality of each new discrete position. This result is analogous to the quantum mechanical result Eq. (4.2), with coordinates $\mathbf{g}$ replacing $(\mathbf{r}, t)$ of the quantum mechanical problem.

Combining Eqs. (11.7) and (11.8) gives the final information expression,

$$I = 4\int d\mathbf{g}\sum_{ijkln} G_{ijkln}\frac{\partial \psi^*}{\partial g_{ijn}}\frac{\partial \psi}{\partial g_{kln}}. \quad (11.9)$$

11.2.7 Free wave EPI solution

An interesting special case is that of nearly flat space and the absence of sources. As will be seen, this is analogous to the case of a Klein–Gordon particle with zero mass and charge.

We first seek the EPI extremum solution to Eq. (3.16),

$$K \equiv I - J = extrem. \tag{11.10}$$

The bound information J is a function of the sources that are present (Sec. 3.4.5). Hence, in the absence of sources,

$$J = 0. \tag{11.11}$$

The Euler–Lagrange Eq. (0.34) is here

$$\sum_{ijn} \frac{d}{dg_{ijn}} \left(\frac{\partial \mathscr{L}}{\partial \psi^{*}_{,ijn}} \right) = \frac{\partial \mathscr{L}}{\partial \psi^{*}}, \; \psi_{,ijn} \equiv \frac{\partial \psi}{\partial g_{ijn}}. \tag{11.12}$$

Because of Eq. (11.11), the Lagrangian $\mathscr{L}$ is simply the integrand of Eq. (11.9). From the latter we see that

$$\frac{\partial \mathscr{L}}{\partial \psi^{*}_{,ijn}} = 4 \sum_{kl} G_{ijkln} \frac{\partial \psi}{\partial g_{kln}}, \; \text{and} \; \frac{\partial \mathscr{L}}{\partial \psi^{*}} = 0. \tag{11.13}$$

The zero follows because information I of course does not depend explicitly upon ψ^{*} while J, which would have, is zero here.

Substituting results (11.13) into Eq. (11.12) gives a solution

$$\sum_{ijkl} \sum_{\mathbf{x}_{n}} G_{ijkl}(\mathbf{x}_{n}) \frac{\partial^{2}\psi}{\partial g_{ij}(\mathbf{x}_{n})\partial g_{kl}(\mathbf{x}_{n})} = 0 \tag{11.14}$$

where we show the explicit $\mathbf{x}$-dependence of all quantities. This is our free-wave solution according to EPI theory.

11.2.8 Free-wave Wheeler–DeWitt solution

In Eq. (11.14), the sum on n represents the contribution from all discrete four-positions. As was noted (Sec. 11.2.4), measurements can only be made at discrete positions. Nevertheless, let us take a quasi-continuous limit as such positions become ever more finely spaced. Denote the constant spacing between $\mathbf{x}$ values as $\Delta\mathbf{x}$. Multiply and divide Eq. (11.14) by $\Delta\mathbf{x}^{2}$. In the limit of small $\Delta\mathbf{x}$ it becomes

$$\Delta\mathbf{x} \int d\mathbf{x} \sum_{ijkl} G_{ijkl}(\mathbf{x}) \frac{\delta^{2}\psi}{\delta g_{ij}(\mathbf{x})\delta g_{kl}(\mathbf{x})} = 0. \tag{11.15}$$

We used the correspondence Eq. (0.58) between partial derivatives $\partial/\partial g$ and functional derivatives $\delta/\delta g$.

With Δx small but finite, the integral in (11.15) must be zero. As at Eq. (3.20), we demand that the zero be attained at each value $\mathbf{x}$ of the integrand:

$$\sum_{ijkl} G_{ijkl}(\mathbf{x}) \frac{\delta^2 \psi}{\delta g_{ij}(\mathbf{x}) \delta g_{kl}(\mathbf{x})} = 0. \tag{11.16}$$

This is the Wheeler–DeWitt equation in the absence of sources and in negligibly curved space (see Kuchar (1973) for details).

11.2.9 Klein–Gordon type solution

The answer Eq. (11.14) or Eq. (11.16) has the mathematical form of a Klein–Gordon equation for the scalar wave ψ in a curved space, i.e., Riem space $g_{ij}(\mathbf{x})$. The mixed partial derivatives allow for the presence of such curvature. By comparison, the Klein–Gordon equation was found, at Eq. (4.28), to be obeyed by a particle of zero spin in $\mathbf{x}$-space. If we further specialize that scenario to one of a massless and charge-free particle, then Eq. (4.28) resembles Eqs. (11.14) and (11.16) even more closely. Thus, relativistic quantum mechanics and quantum gravity obey parallel theories even though their spaces are vastly different.

11.2.10 EPI solution for a pure radiation universe

Although mass–energy sources are still assumed to be absent, we now allow for a non-negligible curvature of space. This means the presence of a generally non-zero Ricci curvature scalar, $^3R = {}^3R(g_{11}, \ldots, g_{33})$. Such a situation defines a 'pure radiation universe'. This explicit dependence upon the Fisher coordinates g_{ij} will now lead to a non-zero bound information J. As before, we develop the theory in analogy to that of the quantum development in Sec. 4.1.

11.2.11 Momentum conjugate space

For convenience, we stay with the case of the continuum of points $\mathbf{x}$. In this limit, by definition (0.58) the information expression (11.9) becomes

$$I = 4 \int d\mathbf{x} \int Dg \sum_{ijkl} G_{ijkl} \frac{\delta \psi^*}{\delta g_{ij}(\mathbf{x})} \frac{\delta \psi}{\delta g_{kl}(\mathbf{x})},$$

(11.17)

$$Dg \equiv \lim_{N \to \infty} \prod_{ij} \prod_{n=1}^{N} dg_{ij}(\mathbf{x}_n).$$

(An irrelevant constant multiplier $\Delta\mathbf{x}$ has been ignored; see Eq. (11.15)). The partial derivatives become functional derivatives. The inner integral is called a 'functional integral' (Ryder, 1987, pp. 162–4, 176–9). In effect, it is integrated over all possible functions g_{ij} at all possible argument values $\mathbf{x}$. For fixed i, j this represents a kind of a metric 'path' through $\mathbf{x}$ space, analogous to the paths through temporal space of the Feynman path integral.

We now define a *conjugate momentum space* that is analogous with that in Chap. 4 describing conventional quantum mechanics. The Fisher coordinates for the problem are $g_{ij}(\mathbf{x})$ for all i, j. Denote these as $\mathbf{g}(\mathbf{x})$. The new momentum space will connect with the space of coordinates $\mathbf{g}(\mathbf{x})$ by a unitary transformation, in particular, a Fourier transformation. This unitary transformation validates the EPI approach for this problem [see Secs. 3.8.5, 3.8.7].

Our unknown functional $\psi[\mathbf{g}(\mathbf{x})]$ represents the probability amplitude for all metrics evaluated at all points of four-space. Then likewise define a conjugate function $\phi[\boldsymbol{\mu}(\mathbf{x})]$ representing the probability amplitude for all *momentum values* μ^{ij} as evaluated over all points of four-space. In analogy with conventional quantum mechanics, a Fourier relation is to connect the two,

$$\psi[\mathbf{g}(\mathbf{x})] = \int D\mu \, \phi[\boldsymbol{\mu}(\mathbf{x})] \, e^{-i \int dx \mathbf{g} \cdot \boldsymbol{\mu}}, \quad \text{where}$$

(11.18)

$$D\mu \equiv \lim_{N \to \infty} \prod_{ij} \prod_{n=1}^{N} (2\pi)^{-1} \, d\mu^{ij}(\mathbf{x}_n), \quad \mathbf{g} \cdot \boldsymbol{\mu} \equiv \sum_{ij} g_{ij}(\mathbf{x}) \mu^{ij}(\mathbf{x}).$$

(Compare with Eq. (4.4).) This is a *functional* Fourier relation. It is mathematically similar in form to the well known path integrals used by Feynman to reformulate quantum mechanics (Ryder, 1987, p. 178). Although Eq. (11.18) seems a reasonable relation on these grounds, and will lead to the desired end Eq. (11.32), we have no other evidence for its validity at this time.

Regarding units in the Fourier Eq. (11.18), the exponent as it stands has units of *length⁴–momentum* (note that the metric g_{ij} is unitless). But mathematically it must be unitless. This implies that it should be divided by a constant that carries the units of *length⁴–momentum*. It seems reasonable to use for this purpose $\hbar L^3$ where L is the Planck distance given by Eq. (6.64). Hence, momentum $\boldsymbol{\mu}$ is assumed to be expressed in units of $\hbar L^3$.

11.2.12 Finding bound information J

As in Sec. 4.1.11 we find J by representing the Fisher information (11.17) in momentum space. This entails use of a Parseval's theorem, as follows.

A functional derivative (see Eq. (0.59)) of Eq. (11.18) gives

$$\frac{\delta\psi}{\delta g_{kl}(\mathbf{x})} = -i \int D\mu \phi[\boldsymbol{\mu}(\mathbf{x})] \mu^{kl}(\mathbf{x}) e^{-i \int d\mathbf{x} \mathbf{g}\cdot\boldsymbol{\mu}}. \qquad (11.19)$$

A similar expression holds for $\delta\psi^*/\delta g_{ij}$. We will also have need for a *functional* representation of the Dirac delta function:

$$\int Dg \, e^{i \int d\mathbf{x}\mathbf{g}\cdot(\boldsymbol{\mu}'-\boldsymbol{\mu})} = (2\pi)^{6N}\delta[\boldsymbol{\mu}'(\mathbf{x}) - \boldsymbol{\mu}(\mathbf{x})]. \qquad (11.20)$$

(Compare with Eq. (0.70).) This expression is easily derived by evaluating the $6N$ separated integrals of the left-hand side that follow the use of the identity (11.17) for Dg.

Substituting Eq. (11.19) and its complex conjugate into Eq. (11.17), switching orders of integration, and using the Dirac delta function representation (11.20) gives

$$I = 4(2\pi)^{-6N} \int d\mathbf{x} \sum_{ijkl} G_{ijkl} \int D\mu \phi(\boldsymbol{\mu}) \mu^{kl} \int D\boldsymbol{\mu}' \phi^*(\boldsymbol{\mu}') \mu'^{ij} (2\pi)^{6N} \delta(\boldsymbol{\mu}' - \boldsymbol{\mu}).$$

$$(11.21)$$

In analogy with quantum mechanical Eq. (4.9), this contracts to

$$I = 4 \int d\mathbf{x} \sum_{ijkl} G_{ijkl} \int D\mu \phi[\boldsymbol{\mu}(\mathbf{x})] \phi^*[\boldsymbol{\mu}(\mathbf{x})] \mu^{ij} \mu^{kl} \qquad (11.22)$$

after use of the 'sifting' property (Eq. (0.69)) of the Dirac delta function. The right-hand sides of Eqs. (11.17) and (11.22) are equal, since they both equal I, which shows that the transformation Eq. (11.18) is unitary (see definition Eq. (4.5)). Then EPI holds identically for this measurement scenario (Secs. 3.8.5, 3.8.7).

Since $\phi(\boldsymbol{\mu})$ is, by definition, a probability amplitude, its squared amplitude $\phi\phi^*$ is a probability density. Then the inner integral in Eq. (11.22) becomes an expectation $\langle \, \rangle$ over all momenta at a given $\mathbf{x}$, and

$$I = 4 \int d\mathbf{x} \sum_{ijkl} G_{ijkl} \langle \mu^{ij} \mu^{kl} \rangle. \qquad (11.23)$$

As a check on this calculation, we note that in quantum mechanics (Eqs. (4.12), (4.13)) the Fisher information I was, as here, proportional to the squared momentum. Hence, quantum mechanics and quantum gravity continue to develop along parallel lines. Also, the formulation of quantum gravity by DeWitt (1967, p. 1118) is through a Hamiltonian that, likewise, contains squared momentum terms. A Lagrangian approach (as here) would, then, also contain squared momentum terms.

By Eq. (11.4), Eq. (11.23) further collapses to

$$I = 4 \int d\mathbf{x} \langle |\boldsymbol{\mu}|^2 \rangle, \qquad (11.24)$$

the mean-squared momentum over all space.

As is usual in quantum gravity, the squared momentum is associated with the kinetic energy. We are assuming a pure radiation universe, i.e., no matter–energy inputs to the system, and zero cosmological constant (Sec. 6.3.23). The total energy of such a universe should be zero (Hawking, 1988), so that the kinetic energy equals the potential energy. The latter is associated with the quantity $g^{1/2}\,{}^3R$ (DeWitt, 1967, p. 1117), a quantity that Wheeler (1962), pp. 41, 60) calls the 'intrinsic curvature invariant' of energy. Thus, Eq. (11.24) becomes

$$I = 4 \int d\mathbf{x} \langle g^{1/2}\,{}^3R \rangle \equiv J. \qquad (11.25)$$

By Eq. (3.18), the information transfer efficiency is now $\kappa = 1$, exactly as it was for quantum mechanics (Sec. 4.1.11).

All quantities within the $\langle\ \rangle$ signs of Eq. (11.25) are functions of $\mathbf{g}$. Therefore, the averaging is appropriately done in $\mathbf{g}$-space,

$$\langle g^{1/2}\,{}^3R \rangle = \int Dg\, \psi^*(\mathbf{g})\psi(\mathbf{g})g^{1/2}\,{}^3R. \qquad (11.26)$$

Using this in Eq. (11.25) gives

$$J = 4 \int d\mathbf{x} \int Dg\, \psi^*\psi\, g^{1/2}\,{}^3R. \qquad (11.27)$$

11.2.13 Physical information K

Use of Eqs. (11.17) and (11.27) gives

$$K \equiv I - J = 4 \int d\mathbf{x} \int Dg \left(\sum_{ijkl} G_{ijkl} \frac{\delta\psi^*}{\delta g_{ij}} \frac{\delta\psi}{\delta g_{kl}} - \psi^*\psi\, g^{1/2}\,{}^3R \right). \qquad (11.28)$$

11.2.14 Wheeler–DeWitt equation

The Lagrangian in Eq. (11.28) uses functional derivatives. To find its extremum solution we need an Euler–Lagrange equation of the same type,

$$\sum_{ij} \frac{\delta}{\delta g_{ij}} \left(\frac{\delta \mathscr{L}}{\delta \psi_{,ij}^*} \right) = \frac{\delta \mathscr{L}}{\delta \psi^*}, \tag{11.29}$$

$$\psi_{,ij} \equiv \frac{\delta \psi}{\delta g_{ij}}. \tag{11.30}$$

Using as $\mathscr{L}$ the integrand of Eq. (11.28), we have

$$\frac{\delta \mathscr{L}}{\delta \psi_{,ij}^*} = 4 \sum_{kl} G_{ijkl} \frac{\delta \psi}{\delta g_{kl}}, \quad \frac{\delta \mathscr{L}}{\delta \psi^*} = -4\psi g^{1/2} \, ^3R. \tag{11.31}$$

Then the Euler–Lagrange solution Eq. (11.29) is

$$\sum_{ijkl} G_{ijkl} \frac{\delta^2 \psi}{\delta g_{ij} \delta g_{jk}} + g^{1/2} \, ^3R\psi = 0. \tag{11.32}$$

This is the Wheeler–DeWitt equation for a pure radiation universe. It is essentially a Klein–Gordon equation, with the form of Eq. (4.26). Hence, ordinary quantum mechanics and quantum gravity continue to develop along parallel lines. In particular, the unitary transformation is vital to both derivations.

It is interesting to compare this approach with the path-integral approach taken by Hawking (1979) and by Hartle and Hawking (1983). These authors *postulate* the use of a path integral approach, in analogy to the use of path integrals by Feynman and Hibbs (1965) in ordinary quantum mechanics. Moreover, the integrand is of a postulated *form*, involving exponentiation of the classical gravitational action.

By comparison, the EPI use of path integrals appears to arise a little more naturally. There are two path integrals in the EPI approach. The path integral Eq. (11.17) arises as the continuous limit of Fisher information over Riem space. The form of that integral is fixed as the usual Fisher form. The second path integral, Eq. (11.18), arises as the continuous limit over momentum space of a Fourier relation between the metric and its associated momentum. The form of that integral is the usual Fourier one.

11.2.15 Need for the inclusion of spin 1/2 in formalism

In the preceding, the wave function ψ was a scalar quantity. As we saw in Sec. 4.2.1, scalar quantum mechanics describes the kinematics of a spin 0 particle.

Particles with integral values of the spin are called bosons. Correspondingly, the Wheeler–DeWitt equation (11.32) describes the kinematics of a spin 2 particle, the graviton. This is, then, another boson. On the other hand, Wheeler (1962) has emphasized that quantum gravity must allow for the inclusion of spin $1/2$ particles as well, such as the neutrino. Particles having half-integer spin, like the neutrino, are called fermions. How, then, can a gravitational theory of fermions be formulated?

Recall that in Sec. 4.2 we were able to develop the Dirac equation, which describes the quantum mechanics of a spin $1/2$ particle, by using a *vector* wave function ψ. Since quantum mechanics and quantum gravity have, to this point, been developed by analogous steps, it should follow that a quantum gravity for a spin $1/2$ particle can be developed analogously. This entails using the zero-aspect Eq. (3.18) of EPI to form a solution. We have not carried through on such a derivation, but it should be eminently possible. The following steps are proposed.

Factor the integrand of Eq. (11.28), just as the integrand of Eq. (4.32) was factored. See also Matveev and Salle (1991, pp. 26–7). This requires abandonment of the scalar nature of ψ for a vector ψ. Next, matrices analogous to the Dirac matrices Eq. (4.34) need to be found. The results will be replacement of the second-order differential Lagrangian in Eq. (11.28) with a product of two first-order factors. Setting either to zero (via EPI principle (3.18)) will give the vector gravitational wave equation. It will be interesting to compare the answer with other vector wave equations of quantum gravity.

11.2.16 *The original deus ex machina?*

It was mentioned in Sec. 3.10, and confirmed by the behavior of Eq. (10.17), that measurements act, to the object under measurement, as unpredictable, discontinuous, irreversible, instantaneous operations, somewhat like so many *deus ex machina* activities. We now view the current measurement problem from this viewpoint.

This approach to quantum gravity agrees with the approach taken in Chap. 4 to derive the Klein–Gordon equation, except for one important factor: the number of real measurements that are required to be taken. We assume, here, essentially an infinity of measurements of the metric tensor to be made, i.e., throughout all of four-space. By comparison, in Chap. 4 a single measurement was required. In principle, a single measurement should work here as well: if a universal wave function $\psi[g_{ij}(\mathbf{x})]$ exists that truly *connects jointly* the behavior of all metric values over all space, then but a single measurement of the metric would suffice to provide the perturbation $\delta\psi[g_{ij}(\mathbf{x})]$ over all space that EPI needs in order to execute.

On the other hand, we did achieve the 'correct' answer (Eq. (11.32)) by the use of an infinity of measured points, implying that this approach is correct as well. (*Caveat*: this assumes that the Wheeler–DeWitt equation is fully correct. Not everyone agrees with this, however, notably DeWitt (DeWitt, 1997); Wheeler seems to still support the equation (Wheeler, 1994). We will proceed on the assumption that it is at least partially valid. The 'end' that we are working towards, in this section, might justify the 'means'.)

But, if both approaches are correct, this implies that, somehow, a single measurement is equivalent to an infinity of measurements over all space. How could this occur?

An obvious answer is that the single measurement takes place when all of space is squeezed into a single point. This, presumably, is at the Big Bang. The implication of this is that a single observation of the metric occurred at the onset of the Universe, and this generated the Wheeler–DeWitt equation for the pure radiation universe which existed then. In other words, the gravitational structure of the Universe was created out of a single, primordial quest for knowledge.

11.3 Nearly incompressible turbulence

11.3.1 Introduction

Turbulence is one of the last great mysteries of classical physics. It describes the seemingly random, or quasi-random, flow of gases and liquids that occur under normal operating conditions. A good general reference is Bradshaw (1978).

A qualitative description of turbulence is easy to give. Turbulence is a three-dimensional nonstationary motion of the medium. The motion is characterized by a continuous range of fluctuations of velocity, with wavelengths ranging from (i) minimum lengths that are determined by viscous forces, to (ii) maximum lengths that are determined by the boundary conditions of the flow. The turbulent process is initiated at the maximal lengths. An example is a body of warm air just beginning to rise from a parked car on a windless day. The maximal length is the length of the automobile. At this stage, the flow is smooth, laminar and deterministic. But, because of frictional and viscous interaction with the surrounding air, as this body of air rises it breaks down into many smaller lengths of rising air, called 'eddies', with conservation of energy transmitting the internal energy of the original mass into these eddies. The process continues in this manner, to ever smaller eddies. At the smallest eddies the energy is transformed into heat due to the action of the viscous forces, which is then radiated away.

The quantitative picture is, however, less precise. Turbulent flow obeys classical physics, whose laws are well understood. However, the process is what is called mathematically 'ill-posed', in that the slightest imprecision in the knowledge of initial conditions propagates into huge errors in any subsequent physical (deterministic) analysis after but a finite amount of time. This is the so-called 'butterfly effect' that makes weather forecasting so tenuous an activity. Interestingly, there is an intermediary state between the initial, deterministic flow and complete randomness of subsequent small scale states. The situation at this intermediary state is only quasi-random. This state is the subject of 'chaos' theory; see, e.g., Sagdeev *et al.* (1988).

For these reasons, turbulence is often analyzed statistically. See, e.g., Ehlers *et al.* (1972). That is, probability laws are sought to model the joint fluctuations in velocity, pressure, temperature, etc., that define a turbulent process. It is known that the key such PDF is the one on joint velocity at all points in space (Batchelor, 1956). But this PDF is, so far, too complicated to find or to use. It is also known that lower-order joint PDFs or single-point PDFs on velocity are close to Gaussian in form, with any departures arising out of the essentially nonlinear behavior of the turbulence (Bradshaw, 1978, p. 15). The theory below verifies this effect.

Another statistical quantity of interest is the joint PDF on density and velocity in a fluid. This PDF was recently determined by Cocke (1996) using an approach that is very close to that of EPI. We present it next.

11.3.2 Fisher coordinates

All physical quantities are assumed to be evaluated at a given position and time in the fluid. We do not consider the correlation between such quantities at different positions or at different times.

The Fisher coordinates of the problem should comprise a four-vector (see Sec. 3.5). Mass flux

$$w^\mu \equiv \rho v^\mu, \ \mu = 0, 1, 2, 3 \tag{11.33}$$

with ρ the local density and v^μ the four-velocity is one such quantity. (We initially use tensor notation; see Sec. 6.2.1.) Here,

$$v^\mu \equiv \frac{dx^\mu}{d\tau} \tag{11.34}$$

with x the position and τ the proper time. As is usual of a four-vector, w obeys a conservation (of mass) equation

$$\frac{\partial w^\mu}{\partial x^\mu} = 0, \tag{11.35}$$

using summation notation (see Sec. 6.2.2).

We restrict attention to nonrelativistic velocities v^μ. This is on the grounds of simplicity, and also because data which one would need to confirm the theory are lacking for relativistic turbulence. In the nonrelativistic limit, by Eq. (3.33) the proper time $\tau = t$, the laboratory time, so that $dx^0/d\tau = d(ct)/dt = c$. It follows from Eqs. (11.33) and (11.34) that

$$w^\mu = \rho(c, \mathbf{v}), \quad \mathbf{v} \equiv \frac{d\mathbf{x}}{dt} = (v_x, v_y, v_z), \tag{11.36}$$

the ordinary Newtonian fluid velocity. Hence, the Fisher coordinates become

$$\rho c \text{ and } \mathbf{w}, \quad \mathbf{w} \equiv \rho\mathbf{v}. \tag{11.37}$$

The variables $(\rho c, \mathbf{w})$ are fluctuations from ideal (here, mean) values θ_ρ, θ_w that need not be specified. Thus, we seek the PDF $p(\rho c, \mathbf{w})$. As usual, we work with the corresponding amplitude function $q(\rho c, \mathbf{w}) \equiv q(\rho, \mathbf{w})$ in simpler notation.

11.3.3 Fisher information I

Given this choice of coordinates, the information I is, from Eq. (2.19),

$$I = 4 \int d\mathbf{w} \, d\rho \left[\nabla_w q \cdot \nabla_w q + \frac{1}{c^2} \left(\frac{\partial q}{\partial \rho} \right)^2 \right]. \tag{11.38}$$

The notation ∇_w signifies the gradient in three-dimensional $\mathbf{w}$-space.

11.3.4 Energy effects

The kinetic energy density is defined as

$$E_{kin} \equiv \tfrac{1}{2}\rho v^2 = \frac{w^2}{2\rho} \tag{11.39}$$

by Eq. (11.37). The medium also has an internal energy density $\epsilon(\rho)$ which can, it turns out, be left unspecified.

11.3.5 Skewness effects

The PDF on pressure is known to be highly skewed. It is an exponential for negative pressure and nearly a Gaussian for positive pressure (Pumir, 1994). Since pressure and density ρ often vary nearly linearly with one another (see Sec. 11.3.11), one would expect the PDF on ρ to likewise exhibit strong skewness. Given these facts, it is reasonable to build the possibility of skewness into the one input to EPI that brings prior physical knowledge into the theory: the bound information J.

11.3.6 Information J

A turbulent medium may be regarded as a non-ideal gas. We found the PDF $p(\mathbf{v})$ for an ideal gas in Chap. 7. This suggest that we represent the bound information J for a turbulent medium as that for an ideal gas, except for extra terms expressing skewness effects as mentioned above. The tactic turns out to be valid in that it leads to empirically correct results.

From Eqs (7.40), (7.47) and (7.49) the J for an ideal gas was equivalent to a normalization constraint plus a constraint on mean-squared velocity. The latter is equivalent to mean kinetic energy. The turbulent medium is characterized by the internal energy $\epsilon(\rho)$ as well. Hence, we express J here as a sum of constraints on normalization, kinetic energy and internal energy, suitably skewed:

$$J = 4 \int d\mathbf{w}\, d\rho \left[\frac{\lambda_1 w^2}{2\rho} + H(\rho - \rho_1)(\lambda_2 + \lambda_3 \epsilon(\rho)) \right] q^2. \qquad (11.40)$$

Function $H(\rho)$ is the step function $H(\rho) = 1$ for $\rho \geqslant 0$, $H(\rho) = 0$ for $\rho < 0$. The constant ρ_1 allows for the skewness effects mentioned above. Using the step function in Eq. (11.40), these are:

$$\int_{\rho_1}^{\infty} d\mathbf{w}\, d\rho q^2 = const., \quad \int_{\rho_1}^{\infty} d\mathbf{w}\, d\rho \epsilon(\rho) q^2 = const. \qquad (11.41)$$

Constants $\lambda_1, \lambda_2, \lambda_3$ are found by the theory.

It is to be noted that we have, here, departed from the usual procedure for finding J, namely *solving for it* via knowledge of the form of I and of an invariance principle. Hopefully this can be carried through in future research.

11.3.7 Net EPI variational principle

By the use of Eqs. (11.38) and (11.40), the EPI principle Eq. (3.16) becomes
$$K \equiv I - J$$

$$= 4 \int d\mathbf{w}\, d\rho \left[\nabla_w q \cdot \nabla_w q + \frac{1}{c^2} \left(\frac{\partial q}{\partial \rho} \right)^2 - \frac{\lambda_1 w^2}{2\rho} q^2 - H(\rho - \rho_1)(\lambda_2 + \lambda_3 \epsilon(\rho)) q^2 \right]$$

$$= extrem. \qquad (11.42)$$

This is to be solved for the amplitude function $q(\rho, \mathbf{w})$ under the boundary conditions that q^2 be integrable and continuous, with continuous first derivatives, and that $q^2(0, \mathbf{w}) = 0$.

11.3.8 Euler–Lagrange solution

The Euler–Lagrange Eq. (0.34) for this problem is

$$\sum_{i=1}^{3} \frac{d}{dw_i}\left(\frac{\partial \mathscr{L}}{\partial q_{,i}}\right) + \frac{d}{d\rho}\left(\frac{\partial \mathscr{L}}{\partial q_{,\rho}}\right) = \frac{\partial \mathscr{L}}{\partial q},$$

$$q_{,i} \equiv \frac{\partial q}{\partial w_i}, \; i = 1, 2, 3, \; q_{,\rho} \equiv \frac{\partial q}{\partial \rho}. \tag{11.43}$$

The Lagrangian $\mathscr{L}$ is the integrand of Eq. (11.42). Then the Euler–Lagrange solution Eq. (11.43) is

$$\nabla_w^2 q + \frac{1}{c^2}\frac{\partial^2 q}{\partial \rho^2} + \left[\frac{\lambda_1}{2\rho} w^2 + H(\rho - \rho_1)(\lambda_2 + \lambda_3 \epsilon(\rho))\right] q = 0. \tag{11.44}$$

Some simplification arises out of an assumption of isotropy for the turbulence. This allows us to use

$$q(\rho, \mathbf{w}) = q(\rho, w), \text{ and } \nabla_w^2 q = \frac{1}{w^2}\frac{\partial}{\partial w}\left(w^2 \frac{\partial q}{\partial w}\right) \tag{11.45}$$

in Eq. (11.44).

11.3.9 Numerical solutions

The problem (11.44), (11.45) is very difficult to solve, even numerically. However, an approximate solution may be found, for cases of low Mach number $M \equiv v_{rms}/v_0$, where v_{rms} is the root mean-square velocity in the medium and v_0 is the speed of sound. Then any velocity v obeys

$$v^2 \ll v_0^2. \tag{11.46}$$

We will sketch the steps of the derivation. Details are found in Cocke (1996).

Solutions are found for two different regions of density, $\rho < \rho_1$ and $\rho > \rho_1$. These are denoted by subscripts 1 and 2 in the following.

The solution in region 1 is

$$q_1(\rho, w) = R(\rho)W(w), \; R(\rho) = Ae^{-B\rho},$$

$$W(w) = Ce^{-Dw^2}, \; A, B, C, D = const. \tag{11.47}$$

Thus, q (and the PDF p) obey an exponential falloff in density and a Gaussian dependence upon velocity.

It is well known that turbulent velocity does approximately obey Gaussian statistics. However, there are more zeros and more strong gusts than would be the case for a pure Gaussian dependence. The low Mach number approximation Eq. (11.46) may be responsible for the Gaussian result. If so, then relaxing this condition will produce better agreement with the known facts. This is left for future research.

The solution in region 2 is

$$q_2(\rho, w) = S(\rho)W(w), \ S(\rho) \approx Ai(a\rho + b), \ a, b = const. \qquad (11.48)$$

Here it was assumed that the fluctuations in density are small enough that the internal energy function $\epsilon(\rho)$ is well approximated by a linear function of ρ. Function $Ai(\rho)$ is Airy's function. An asymptotic formula is

$$Ai(x) \approx \frac{e^{-2x^{3/2}/3}}{2\sqrt{\pi}x^{1/4}}. \qquad (11.49)$$

The PDF on density is, therefore, proportional to the square of Airy's function.

The two solutions $q_1(\rho)$ and $q_2(\rho)$ are made to be continuous in their values and in their derivatives $\partial q/\partial \rho$ at the connecting point $\rho = \rho_1$. This is accomplished through adjustment of the constants of the problem (the λ_i, a, b, etc.).

11.3.10 *EPI zero solution*

The preceding was the solution to EPI principle Eq. (3.16). This is the extremization half of the principle. But the other half, the zero-condition Eq. (3.18), must also be satisfied. The premise of EPI is that all solutions to *both* the extremization and the zero-condition must have physical significance.

If we take the information efficiency constant $\kappa = 1$ here, then Eq. (11.42) is now to be zeroed. Hence, what is the solution q to this problem? In fact, the solutions we obtained above for extremizing Eq. (11.42) also must zero it! This follows essentially because the constraint terms in Eq. (11.42) are proportional to factor q^2. For a proof, see Frieden (1994).

Hence, the EPI *zero-solution* for this problem does not elicit any additional physical phenomena. This is for a *scalar* amplitude q. By contrast, the use of vector amplitudes **q**, as in Sec. 4.2, might lead to useful new results.

11.3.11 *Discussion*

It was found that, in the limit of low Mach number, velocity fluctuations obey Gaussian statistics. Also, for $\rho < \rho_1$ the density fluctuations obey exponential statistics whereas for $\rho > \rho_1$ the PDF for density fluctuations is the square of an Airy function.

It is possible to compare these density fluctuation results with some of the simulation results of Pumir (1994). Pumir computed PDF curves of *pressure* fluctuation, based upon simulations using the Navier–Stokes equations with a viscosity term. These pressure fluctuation curves may be interpreted as *density* fluctuation curves as well since, in our low-Mach limit, the pressure is a linear function of the density ρ.

One such simulation curve is shown in Fig. 11.1 (dotted line). The

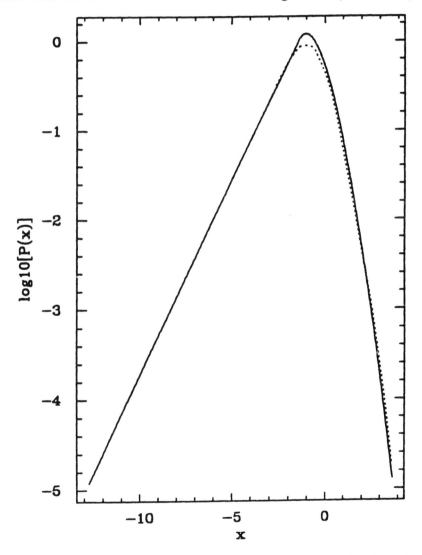

Fig. 11.1. The solid curve is the logarithm of our unnormalized model PDF $P(\rho) \equiv q^2(\rho)$ as a function of the dimensionless variable x, with $x \propto \rho$. Pumir's simulations are represented by the dotted curve. (Reprinted with permission from *Physics of Fluids* **8**, 1609 (1996). Copyright 1996, American Institute of Physics.)

x-coordinate (horizontal) is linear in the density ρ. The peak probability value corresponds to density value ρ_1. To its left is an exponential solution and to its right is a near-Gaussian. These can be considered as 'ground truth', since they are based upon physical simulation of the turbulence.

By comparison, adjustment of the constants in the above EPI approach give the solid curve. The agreement with Pumir's simulation curve is seen to be very good over four decades of density. The only significant disagreement occurs near the maximum, where Pumir's curve lies below the EPI curve. However, Pumir's approach included physical effects which we ignored – the Navier–Stokes equations with a viscosity term. It may be that the inclusion of such effects into the EPI principle would give better agreement in this region. Rigorous use of the EPI principle, in place of the *ad hoc* approach taken here, might improve things as well.

12

Summing up

As we have seen, the principle of extreme physical information (EPI) permits physics to be viewed within a unified framework of measurement. Each phenomenon of physics is elicited by a particular real measurement. The following is a summary of the new physical effects, approaches and concepts that have arisen out of the principle.

Chapter 1

The Fisher information I in a single measurement of a phenomenon obeys Eq. (1.24),

$$I = 4 \int dx \left(\frac{dq(x)}{dx} \right)^2 . \tag{12.1}$$

Here $q(x)$ is the *real* probability amplitude for a fluctuation x in the measurement.

Under certain conditions, information I obeys an 'I-theorem' Eq. (1.30),

$$\frac{dI(t)}{dt} \leq 0. \tag{12.2}$$

This means that I is a monotonic measure of system disorder. The I-theorem probably holds under more general conditions than does the corresponding 'H-theorem' for the Boltzmann entropy H.

In the same manner by which a positive increment in thermodynamic time is defined by an increase in Boltzmann entropy, there is a positive increment in a 'Fisher time' that is defined by a decrease in the information I. (Both times increase as their corresponding measures of disorder increase.) The two times might not always agree, however, as was found by numerical experiments.

Let θ identify the phenomenon that is being measured. A Fisher temperature T_θ corresponding to θ may be defined as

$$\frac{1}{T_\theta} \equiv -k_\theta \frac{\partial I}{\partial \theta}, \ k_\theta = const. \tag{12.3}$$

(Eq. (1.39)). When θ is taken to be the system energy E, then the Fisher temperature has analogous properties to those of the (ordinary) Boltzmann temperature. Among these properties is a perfect gas law (Eq. (1.44))

$$\overline{p}V = k_E T_E I \tag{12.4}$$

where $\overline{p}$ is the pressure. It may be that all of thermodynamics can be formulated by the use of information I in place of the usual concept.

Chapter 2

The I-theorem derived in Chapter 1 for a single parameter measurement may be extended to a multiparameter, multicomponent measurement scenario. A scalar measure of the information for this general scenario is developed. This information, called I, now obeys (Eq. (2.19))

$$I = 4 \int d\mathbf{x} \sum_n \nabla q_n \cdot \nabla q_n \tag{12.5}$$

where $q_n \equiv q_n(\mathbf{x})$ is the nth component probability amplitude for the four-fluctuation $\mathbf{x} = (x_0, \dots, x_3)$. I is called the 'intrinsic' information, since it is a functional of the probability amplitudes that are intrinsic to the measured phenomenon.

This multi-component I still obeys the I-theorem and, when used in the EPI principle, gives rise to the major laws of physics.

In summary, I is at the same time (i) a thermodynamic measure of disorder and (ii) a universal measure of information whose variation gives rise to most (perhaps all) of physics.

Chapter 3

Where does the information in acquired data come from? It must originate in the physical phenomenon (or, system) that is under measurement. Any measurement of physical parameters initiates a *transformation of Fisher information $J \to I$* connecting the phenomenon with the intrinsic data (Sec. 3.3.2). This information transition occurs in the object, or input, space to a measuring instrument (Sec. 3.8.1). The phenomenological, or 'bound', information is denoted as J. The acquired information in the intrinsic data is what is called I. Information J is ultimately identified by an invariance principle that characterizes the measured phenomenon (Sec. 3.4.5).

Suppose that, due to a measurement, the system is perturbed, causing a perturbation δJ in the bound information (Secs. 3.3.2, 3.8.1, 3.11). What will be the effect on the data information I? In analogy to a thermodynamic model Eq. (3.5) of Brillouin and Szilard it must be that $\delta I = \delta J$, there is no loss of the perturbed Fisher information in its relay from the phenomenon to the intrinsic data. This is a new conservation law. It is also axiom 1 of an axiomatic approach for deriving physical laws.

Since $\delta I = \delta J$ necessarily $\delta(I - J) = 0$. Hence, if we define $I - J \equiv K$ as a net 'physical' information, a variational principle

$$K \equiv I - J = extrem. \tag{12.6}$$

(Eq. (3.16)) results. This is called the EPI variational principle.

The EPI zero principle

$$I - \kappa J = 0, \quad 0 \leqslant \kappa \leqslant 1 \tag{12.7}$$

(Eq. (3.18)) follows, as well, on the grounds that generally $K \neq 0$ so that $I \neq J$ or $I = \kappa J$. That $\kappa \leqslant 1$ follows from the I-theorem Eq. (1.30) as applied to the information transition $J \to I$.

Eqs. (12.6) and (12.7) together define the EPI principle. These equations also follow identically, i.e., independent of the axiomatic approach taken, and of the I-theorem, if there is a *unitary transformation* connecting the measurement space with a physically meaningful conjugate space (Secs. 3.8.5, 3.8.7).

The solution $q_n(\mathbf{x})$ to EPI defines the physics of the measured particle or field in the object, or input, space to the measuring instrument (Sec. 3.8.5). (The physics of the *output space* is taken up in Chap. 10.)

It follows that the Lagrangian density for any phenomenon or system is not simply an *ad-hoc* construction for producing a differential equation. The Lagrangian has a definite prior significance. It represents the physical information density $k(\mathbf{x}) \equiv \sum_n k_n(\mathbf{x})$, with components $k_n(\mathbf{x}) \equiv i_n(\mathbf{x}) - j_n(\mathbf{x})$ in terms of data information and bound information components. The integral of $k(\mathbf{x})$ represents the total *physical information K* for the system.

For real coordinates $\mathbf{x}$, the solution to such a variational problem represents the payoff in a mathematical game of information 'hoarding' between the observer and the measured phenomenon (Secs. 3.4.11–3.4.13).

A requirement of invariance of the mean-square error of estimation to reference frame leads to the *Lorentz transformation* equations (Sec. 3.5.5), and to the *requirement of covariance* for the amplitude functions $\mathbf{q}(\mathbf{x})$. As a corollary, $\mathbf{x}$ must be four-vector, whose physical nature depends upon that of the measured parameters $\boldsymbol{\theta}$. Since the squared amplitude functions are PDFs, all PDFs that are derived by EPI must obey *four-dimensional* normalization, including integration over time if $\mathbf{x}$ is a four-position. A measurement of time is regarded *a priori* to be as imprecise and random as that of space (Sec. 3.5.8).

The dynamics and kinematics of the system are direct manifestations of the information densities $i_n(\mathbf{x})$, $j_n(\mathbf{x})$ (Sec. 3.7).

The constancy of the physical constants is an expression of the constancy of the corresponding information quantities J for the scenarios that generate them (Sec. 3.4.14). The constants do not change with time, contrary to Dirac's proposal (1937, 1938).

The Von Neumann measurement–wavefunction separation effect is derived, by the use of a simple coherent optics measuring setup (Sec. 3.8.3). Photons are the carriers of the information from object to image space.

Chapter 4

An attempt to measure the classical four-position of a particle gives, via the EPI principle, both the Klein–Gordon equation (if the particle is a boson), and the Dirac equation (if the particle is a fermion). The Klein–Gordon result follows from Eq. (3.16) of EPI, while the Dirac result follows from Eq. (3.18) of EPI.

The Schroedinger wave equation follows either as the non-relativistic limit of the Klein–Gordon result (Appendix G) or, independently, in a one-dimensional EPI derivation (Appendix D).

The intrinsic information I in the four-position of a free-field particle is proportional to the square of its relativistic energy mc^2 (Eq. (4.18)).

The equivalence Eq. (4.17) of energy, momentum and mass is a consequence of the theory; as is the Compton 'resolution length' Eq. (4.21).

The constancy of the particle rest mass m and Planck's constant $\hbar$ is implied by the constancy of information J.

The well-known problem of a variable normalization integral in ordinary Klein–Gordon theory is solved by using the four-dimensional approach afforded by EPI.

Derivation of the Dirac equation is not by a variational principle, but by satisfying the EPI zero condition Eq. (3.18).

The Heisenberg uncertainty principle Eq. (4.53) is usually regarded as a quantum effect *par excellence*. But in fact, we find that it is simply an expression of the Cramer–Rao inequality Eq. (1.1) of *classical* measurement theory, as applied to a position determination. Furthermore, such use of the Cramer–Rao inequality gives a generalization of the uncertainty principle – it limits the spread in *any function* of the position measurement.

Chapter 5

The vector wave equation of electromagnetic theory arises, via EPI, as the reaction to an attempt to determine a four-position within an electromagnetic field. The four-position is intrinsically uncertain and is characterized by random errors $\mathbf{x}$.

The PDF on these errors varies as the square of the four-potential (Eq. (5.52)). In general, this PDF is unique and normalizeable (Secs. 5.1.25–5.1.28). In the special case of zero sources the PDF becomes simply that for the position of a photon within a coarse-grained field (with 'grain' size the order of the wavelength).

That classical electromagnetic theory should arise out of a statistical approach is, perhaps, surprising. The tie-in is the correspondence Eq. (5.48) between the probability amplitudes for the measurements (statistical quantities) and the electromagnetic four-potential (physical quantities). This correspondence is based on the recognition that the given sources are electromagnetic.

With the vector wave equation so derived, Maxwell's equations follow as a consequence. In this manner, classical electromagnetic theory derives from the EPI principle. Ultimately, it derives from the admission that any measurement of the electromagnetic field must suffer from random errors in position.

Chapter 6

The weak-field equation of classical gravity arises as a reaction to an attempt at four-position measurement within a gravitational field.

The entire EPI derivation follows along analogous lines to that in Chap. 5. Thus, EPI unifies the two classical theories of electromagnetism and gravitation.

The PDF on four-position fluctuation that is implied by the theory is the square of the weak-field metric. In the special case of zero energy-stress tensor this becomes the PDF on position of a graviton. The marginal PDF on the time is consistent with the hypothesis that gravitons can be created and absorbed (Sec. 6.3.25).

The Planck length is derived (Sec. 6.3.22), and the cosmological constant Λ is found to be zero (Sec. 6.3.23).

With the weak-field equation so derived, it is a simple matter to 'riff up' to the general Einstein field equation.

Chapter 7

The Boltzmann energy distribution law and the Maxwell–Boltzmann velocity law are derived by considering a four-dimensional space of energy-momentum fluctuations (iE, $c\mathbf{\mu}$). For the purposes of these derivations, this four-dimensional space replaces the usual six-dimensional phase space ($\mathbf{\mu}$, $\mathbf{x}$) of momentum-coordinate values that is conventionally used.

The Boltzmann and Maxwell–Boltzmann PDFs are the equilibrium laws for this problem. The EPI approach finds, as well, a set of PDF laws on velocity $\mathbf{x} = (x, y, z)$ for which I is stationary but not at its absolute minimum value. These obey Eq. (7.59), a weighted superposition of Hermite–Gauss functions

$$p(\mathbf{x}) =$$

$$Ae^{-|\mathbf{x}|^2/2a_0^2}\left\{1 + \left[\sum_{\substack{nijk \\ i+j+k=n \\ n>0}} a_{nijk} H_i(x/a_0\sqrt{2})H_j(y/a_0\sqrt{2})H_k(z/a_0\sqrt{2})\right]^2\right\}$$

(12.8)

with parameters a_0, A, $a_{nijk} = const$. This solution is confirmed by previous work of Rumer and Ryvkin (1980). The lowest-order Hermite–Gauss function – a simple Gaussian – gives the absolute minimum value for I, and hence the equilibrium solution.

The Second Law of thermodynamics states that the time rate of change dH/dt of entropy must exceed or equal 0. The zero is, then, a lower bound. Is there an upper bound? Eq. (7.87) is

$$\left(\frac{dH(t)}{dt}\right)^2 \leq I(t)\int d\mathbf{r}\,\frac{\mathbf{P}\cdot\mathbf{P}}{p},$$

(12.9)

where $\mathbf{P}$ is the flow vector for the phenomenon. This shows that an upper bound to dH/dt exists, and, is proportional to the square root of the Fisher information for the system. In other words, 'entropy shall increase, but not by too much'.

Like inequalities of this type follow in other fields as well. In application to electromagnetics, Eq. (7.99) gives the upper bound to the entropy rate for charge density $\rho(\mathbf{r}, t)$ in terms of the Fisher information of the charge density distribution. Also, Eq. (7.101) gives a like expression for the entropy of the scalar potential $\phi(\mathbf{r}, t)$.

There is a similar entropy bound for gravitational flow phenomena, as discussed.

An upper bound to the entropy rate dH/dt for electron flow is given by (Eq. (7.114))

$$\left(\frac{1}{c}\frac{\partial H}{\partial t}\right)^2 \leq 3I.$$

(12.10)

The speed of light c plays the fundamental role of the proportionality constant between $\partial H/\partial t$ and $\sqrt{I}$ in this case.

Chapter 8

$1/f$-type power spectral noise is shown to follow as an equilibrium state of a measurement procedure. The measurements are the complex amplitudes of one or more pure tones (in the language of acoustics).

The $1/f$ answer holds over an infinite range of values (κ, N) of the EPI theory. This accounts for the fact that $1/f$ noise characterizes a wide variety of phenomena.

Chapter 9

EPI predicts that the fundamental physical constants should obey a $1/x$ probability law. This is equivalent to uniform randomness in their logarithms (i.e., their powers of 10). The effect is independent of the choice of units, i.e., whether m.k.s., c.g.s., f.p.s., etc. It is also independent of inversion or combination operations on any number of the constants. The currently known constants confirm the $1/x$ law quite well (chi-square $\alpha = 0.95$).

Theoretically, the median value of the physical constants should be 1 (Sec. 9.3.14). Again, this effect is independent of the choice of units. The currently known constants confirm this value fairly well.

An implication of the $1/x$ law is that, *a priori*, small constants are more probable than larger ones. This implies that, given a number N of experimental determinations of a constant, the maximum probable estimate x_{MP} of the constant is *not* merely the arithmetic average $\bar{x}$ of the data; it must be corrected *downward* from the average via (Eq. (9.56))

$$x_{MP} = \tfrac{1}{2}\left(\bar{x} + \sqrt{\bar{x}^2 - \sigma^2/4N}\right). \tag{12.11}$$

Quantity σ^2 is the variance of error in one determination.

Chapter 10

Previous chapters showed how to derive the physical laws that occur at the *input space* to a measuring instrument. Here, we want to modify the approach to acquire, as well, the physical law that occurs at the instrument's output, measurement space. A one-dimensional analysis of this kind was given in Sec. 3.8, but was severely limited in scope. Accordingly, a new, covariant approach to quantum measurement is developed. Consistent with the preceding EPI principle (Eqs. (3.16), (3.18)), this approach treats time on an equal basis with the three space variables, i.e., as a quantity that is only imperfectly (statistically) known.

How are observed data values accomodated by the theory? Given our premise that the observer is part of the phenomenon under measurement, it follows that *observation* of a data value affects the *physical state* of the overall observer–phenomenon system (Sec. 10.10). This suggests that we can accomodate real data values in the overall EPI approach by simply adding Lagrange constraint terms to the EPI variational principle Eq. (3.16), two for each data value (Eq. (10.13)).

On this basis, a Klein–Gordon equation with data constraints is found (Eq. (10.14)) as well as a Schroedinger Eq. (10.17). Solving the latter at the particular time of data collection gives rise to the Von Neumann separation effect, Eq. (10.22).

The analysis defines, as well, the sequence of physical events that define EPI as a physical process (Sec. 10.10). The sequence defines an 'epistemactive' process. That is, a question is asked (what is the value of $\boldsymbol{\theta}$?), and the response is not only a measured value of $\boldsymbol{\theta}$, but also the mechanism for determining the measured value. This mechanism is the probability law from which the measured value is randomly sampled.

Chapter 11

Here we describe some problems that are currently being attacked using EPI. One of these is quantum gravity. We unite quantum mechanical and gravitational effects, through the activity of measuring the metric tensor over all of four-space. The derivation is completely analogous to that of the Klein–Gordon equation of Chap. 4.

Thus, we postulate a momentum space that is the Fourier conjugate (Eq. (11.18)) to measurement space. The existence of such a unitary space implies that EPI rigorously holds for this measurement scenario (see Sec. 11.2.12). The result is the Wheeler–DeWitt equation of quantum gravity (Eq. (11.32)), for a pure radiation universe.

The point of departure of this derivation from that of the Klein–Gordon equation in Chap. 4 is replacement of the ordinary derivatives and integrals of the Klein–Gordon derivation by *functional* derivatives and integrals.

Interestingly, the latter are naturally in the form of Feynman path integrals. Past derivations of the Wheeler–DeWitt equation have had to postulate the Feynman approach. A generalization of the approach is suggested which will result in a *vector* wave equation of gravity.

The turbulence problem is being attacked using a quasi-EPI approach. Probability laws on fluid density and velocity have been found for a situation of low velocity flow of a nearly incompressible medium. The acquired laws agree well with published data arising out of detailed simulations of such turbulence.

The Last Word

After all is said and done, this can only be a repetition of the remarks of Prof. J. A. Wheeler (1990) that were quoted at the beginning of the book. And so we come full circle:

All things physical are information-theoretic in origin and this is a participatory universe ... Observer participancy gives rise to information; and information gives rise to physics.

Acknowledgements

This project could not have progressed nearly as far as it has without the aid of many people. I want to thank the following.

W. J. Cocke, my co-worker on the connection between EPI and classical gravitation, for many engaging and instructive sessions on covariance, gravitation and the universe at large.

B. Ebner, whose probing questions have been a real stimulus.

H. A. Ferwerda, for carefully reading sections of the manuscript and clarifying many issues that arose.

P. W. Hawkes, for continued support and encouragement of the research. It is much appreciated.

R. J. Hughes, my co-worker in research on the $1/f$ law, for many imaginative and stimulating conversations by email on subjects ranging from the fundamental nature of the physical constants to cosmic radiation.

A. S. Marathay, for help in electromagnetic theory and other topics.

B. Nikolov, my co-worker in establishing links between entropy and Fisher information.

R. N. Silver, for encouragement, constructive criticism of the initial theory and insights on basic quantum theory.

B. H. Soffer, my 'spiritual' mentor overall, and co-worker in research on the knowledge acquisition game. His encouragement, perceptive remarks, thoughts and observations have also been of immeasurable help to the overall research. He introduced me to the work of J. A. Wheeler and insisted, ever so gently (but correctly), that EPI had a lot in common with Wheeler's views of measurement theory. Many of his ideas have been of fundamental importance, including the notion that unitarity is fundamental to the overall theory, and the ideas that EPI is intimately connected with the Szilard–Brillouin model of statistical physics.

E. M. Wright, for bringing to my attention many aspects of modern quantum theory, especially its wonderful mysteries and paradoxes, bringing me up to date on the subject of measurement theory, and (hopefully) constraining my enthusiastic forays of theory to stay within physical bounds.

Appendix A

Solutions common to entropy and Fisher
I-extremization

Maximum entropy (ME) and extreme physical information (EPI) are two variational principles for finding an unknown probability amplitude function $q(x)$. Eq. (1.23) then gives the PDF. Suppose that both are subject to the same constraints. For what constraints do the two solutions coincide? And what are the coincident solutions?

A.1 Coincident solutions

The two variational problems are as follows. By Eq. (2.19) and (3.16), the EPI approach for a single unknown amplitude $q(x)$ is

$$4\int dx\, q'^2 + \sum_{k=1}^{K} \lambda_k \int dx\, q^\alpha f_k(x) = extrem., \quad q \equiv q(x). \tag{A1}$$

Information J is the sum. (*Note*: To aid in making comparisons, in this appendix we often refer to the sum as 'constraint terms.')

The constrained ME approach is, from Eqs. (0.20) and (1.23),

$$-\int dx\, q^2 \ln(q^2) + \sum_{k=1}^{K} \mu_k \int dx q^\alpha f_k(x) = extrem. \tag{A2}$$

We want to know what constraint kernels $f_k(x)$ cause the same solution $q(x)$ to both problems (A1), (A2). To obtain solutions, we regard (A1) and (A2) as variational problems, and so use the Euler–Lagrange Eq. (0.13)

$$\frac{d}{dx}\left(\frac{\partial \mathscr{L}}{\partial q'}\right) = \frac{\partial \mathscr{L}}{\partial q} \tag{A3}$$

to define the solution. The Lagrangian $\mathscr{L}$ is the total integrand of the integral to be extremized.

First consider the case $\alpha = 2$ where J is an expectation in Eq. (A1). From the Lagrangian of (A1), solution (A3) to EPI obeys

283

$$q'' - q \sum_k \lambda_k f_k(x) = 0. \tag{A4}$$

We have renamed the λ_k for convenience. (This is permitted since they are, at this point, variables to be fixed by the constraint Eqs. (1.48b).)

Likewise, from the Lagrangian of (A2) the Euler–Lagrange solution to the ME approach obeys

$$q = \exp\left(-\frac{1}{2} + \sum_k \mu_k f_k(x)\right). \tag{A5}$$

The μ_k were renamed for convenience, as in the preceding.

Since we are seeking a common solution $q(x)$, we want to substitute the solution (A5) into solution (A4). This will produce a requirement on the constraint functions $f_k(x)$. Two differentiations of (A5) give

$$q'' = q\left[\sum_k \mu_k f''_k + \left(\sum_k \mu_k f'_k\right)^2\right]. \tag{A6}$$

Substituting this into (A4) gives the requirement on the $f_k(x)$ that

$$\sum_k \mu_k f''_k + \left(\sum_k \mu_k f'_k\right)^2 = \sum_k \lambda_k f_k. \tag{A7}$$

There are two main problem areas to consider.

A.2 Moment constraint information

One possibility is for the constraints to be moments, with

$$f_k(x) = x^k, \; k = 0, 1, \ldots, K. \tag{A8}$$

(Note that $k = 0$ corresponds to normalization.) What values K allow a common solution?

The general approach is to balance equal powers of x on both sides of Eq. (A7) after substitution (A8). Doing this, we find the following.

For $K = 0$, $\lambda_0 \equiv 0$ with μ_0 any value. This means that there is a normalization constraint for ME, but none for EPI. This violates the premise of the same moments for both approaches and, hence, is a negative result.

For $K = 1$, $\lambda_1 \equiv 0$. Therefore there is a zeroth and first moment constraint for ME, but no first moment for EPI. This is again a negative result, since it violates the premise of the same moments in both approaches. It also indicates that the solution to ME with up to a first-moment constraint is the same as that for EPI with just the zeroth-moment constraint. The common solution is an exponential or Boltzmann law; see Sec. 7.3.4.

For $K = 2$, we obtain the requirements

$$x^2: \lambda_2 = 4\mu_2^2$$

$$x^1: \lambda_1 = 4\mu_1\mu_2 \tag{A9}$$

$$x^0: \lambda_0 = \mu_1^2 + 2\mu_2$$

resulting from the indicated powers of x. This is a *positive* result since the relations may be satisfied by generally nonzero values of the λ_i and the μ_i. There is now no requirement that a moment of EPI be deleted. Hence, both ME and EPI now allow the three moments as constraints, and with the above relations (A9) satisfied they will give the same answer for $q(x)$.

For cases $K > 2$ the result is that coefficients $\mu_k \equiv 0$, $k > 1 + K/2$. Once again the constraints cannot be the same in the two approaches. What this is saying is that only the above case $K = 2$ gives the same solution for the same moment constraints.

A.3 Maxwell–Boltzmann case

We now observe that the common solution $q(x)$ to the $K = 2$ problem is, by Eq. (A5), a normal law. Therefore its square, the PDF $p(x)$, is normal as well. In statistical mechanics, the PDF on the velocity of particles subject to conditions of zero mean velocity and fixed mean-square velocity (kinetic energy) due to a fixed temperature is a normal law with mean zero, the Gaussian law. These physical 'conditions' correspond mathematically to the power-law constraints for the case $K = 2$. Hence, solving for the unknown PDF on velocity by *either* ME or EPI subject to these constraints would produce the same, correct answer.

A.4 Constraints imposed by information J

In EPI, the constraints are physically effected by a functional J (Eq. (3.4)) defining the information that is intrinsic, or 'bound', to the phenomenon under measurement (see Secs. 3.1 and 3.3). In all physical applications of EPI, with the exception of the preceding Maxwell–Boltzmann one, J is thereby defined as a *single* term. Hence, we now compare EPI and ME approaches (A1), (A2) for a case $K = 1$, seeking the form of a common single constraint kernel $f(x)$ that leads to the same solution by the two approaches. This is done as follows.

Eq. (A7) in the case $K = 1$ is

$$\mu f'' + \mu^2 f'^2 = \lambda f. \tag{A10}$$

This differential equation has the unique solution (Kamke, 1948)

$$f'^2 = Ce^{-2\mu f} - \frac{\lambda}{2\mu^3}(1 - 2\mu f), \quad C = const. \tag{A11}$$

This cannot be directly solved for $f(x)$, but may be solved for $x(f)$ by taking the square root of both sides, solving for dx, and then integrating (if need be, numerically). By (A1), this solution corresponds to a 'bound' information case

$$J = \int dx q^2(x) f(x), \tag{A12}$$

where f is given by the solution to Eq. (A11). We do not know of a scenario that is describable by this specific J.

A.5 Constraint case $\alpha = 1$

From Eq. (A4) onwards, a constraint exponent case $\alpha = 2$ was assumed. This corresponds to expectation constraints. However, in certain scenarios (Chaps. 5 and 6) the constraint term J is actually *linear* in q, i.e., $\alpha = 1$. We now consider this case. Again address the possibility of obtaining common solutions to the two approaches. The same analysis as the preceding was followed. This shows that (a) regarding the possibility of moment constraint (A8) solutions in common, there are now none; and (b) for a single general constraint kernel $f(x)$ a very specialized integral condition must be satisfied, and this probably does not define a physically realizable J functional.

In summary, there are only two possible scenarios for which EPI and ME give the same solution: where the constraints are moments of $p(x)$ up to the quadratic one, or where J obeys Eqs. (A11, A12). The former case corresponds to the Maxwell–Boltzmann velocity dispersion law while the latter corresponds to an, as yet, unknown scenario. This scenario is probably unphysical.

The Maxwell–Boltzmann answer is an equilibrium distribution. Hence, equilibrium statistical mechanics is the common meeting ground of the EPI and ME approaches to estimating PDFs. Next, consider the more general circumstance of non-equilibrium statistics. The solution must obey the Boltzmann transport differential equation, and the solution to this is a general superposition of Hermite–Gauss functions (Rumer and Ryvkin, 1980). In fact, EPI generates these solutions as subsidiary minima in I, with the absolute minimum attained by the Maxwell–Boltzmann solution (see Sec. 7.4.8). This is under the constraints of normalization and mean energy.

However, under the same constraint inputs ME only gives the equilibrium, Maxwell–Boltzmann answer (Jaynes, 1957a, b); it fails to produce any of the higher-order Hermite–Gauss solutions. Basically, this is because multiple solutions follow from differential equations, which ME cannot produce; see Sec. 1.3.1. Hence, EPI and ME only coincide at the most elemental level of statistical mechanics, that of equilibrium statistics. Beyond this level ME does not apply.

Appendix B

Cramer–Rao inequalities for vector data

Here we consider the case of vector data

$$\mathbf{y} = \boldsymbol{\theta} + \mathbf{x}, \tag{B1}$$

where each vector has N components. Each estimator $\hat{\theta}_i(\mathbf{y})$ is allowed to be a general function of all the data $\mathbf{y}$. Also, it is given to be unbiased,

$$\int d\mathbf{y}\hat{\theta}_i(\mathbf{y})p \equiv \theta_i, \; i = 1, \ldots, N$$

$$p \equiv p\,(\mathbf{y}|\boldsymbol{\theta}). \tag{B2}$$

We want to find a relation that defines the mean-square error

$$e_i^2 \equiv \int d\mathbf{y}[\hat{\theta}_i(\mathbf{y}) - \theta_i]^2 p, \; i = 1, \ldots, N \tag{B3}$$

in each estimate in terms of an information quantity of some kind. The derivation is entirely different from that of the scalar result (1.1), but equally simple (Van Trees, 1968).

B.1 Fisher information matrix

First we establish the relation for e_1^2. Form an auxiliary vector

$$\mathbf{v} \equiv [(\hat{\theta}_1(\mathbf{y}) - \theta_1)\, (\partial \ln p/\partial\theta_1) \ldots (\partial \ln p/\partial\theta_N)]^T, \tag{B4}$$

where T denotes the transpose. Next, form the matrix

$$[M] \equiv \langle \mathbf{v}\mathbf{v}^T \rangle, \tag{B5}$$

where $\langle\ \rangle$ denotes an expectation over all data $\mathbf{y}$. By the use of (B4) we get

287

$$[M] = \begin{bmatrix} e_1^2 & 1 & 0 & \cdots & 0 \\ 1 & F_{11} & F_{12} & \cdots & F_{1N} \\ 0 & F_{21} & F_{22} & \cdots & F_{2N} \\ \vdots & \vdots & \vdots & & \vdots \\ 0 & F_{N1} & F_{N2} & \cdots & F_{NN} \end{bmatrix}. \tag{B6}$$

Elements F_{ij} obey

$$F_{ij} \equiv \int d\mathbf{y} \frac{\partial \ln p}{\partial \theta_i} \frac{\partial \ln p}{\partial \theta_j} p. \tag{B7}$$

With i and j ranging independently from 1 to N, the F_{ij} form a matrix [F] called the 'Fisher information matrix'. Notice that [F] is a straightforward generalization of the scalar information quantity (1.9).

In (B6), the elements of [F] derive in straightforward fashion. Also, the (1, 1) element is obviously e_1^2 as given. We next derive the elements that are 0 or 1. Consider element

$$M_{12} \equiv \int d\mathbf{y}[\hat{\theta}_1(\mathbf{y}) - \theta_1] p \partial \ln p / \partial \theta_1. \tag{B8}$$

This is directly

$$M_{12} = M_1 - M_2, \quad M_1 \equiv \int d\mathbf{y}\hat{\theta}_1 p \partial \ln p / \partial \theta_1,$$

$$M_2 \equiv \theta_1 \int d\mathbf{y} p \partial \ln p / \partial \theta_1. \tag{B9}$$

Using the formula for the derivative of a logarithm gives

$$M_1 = \int d\mathbf{y}\hat{\theta}_1 p \frac{1}{p} \frac{\partial p}{\partial \theta_1} = \frac{\partial}{\partial \theta_1} \int d\mathbf{y}\hat{\theta}_1 p = \frac{\partial \theta_1}{\partial \theta_1} = 1, \tag{B10}$$

by unbiasedness condition (B2). Regarding M_2,

$$\int d\mathbf{y} p \frac{\partial \ln p}{\partial \theta_1} = \int d\mathbf{y} p \frac{1}{p} \frac{\partial p}{\partial \theta_1} = \frac{\partial}{\partial \theta_1} \int d\mathbf{y} p = \frac{\partial}{\partial \theta_1} 1 = 0. \tag{B11}$$

Hence $M_2 = 0$. Therefore by the first Eq. (B9), $M_{12} = 1$, as was required to show.

B.2 Exercise

By the same steps, show that $M_{13} = 0$. The proof is easily generalized to $M_{1i} = 0$, $i = 4, \ldots, N$. The top row of (B6) is now accounted for; as is the first column, by the symmetry in [M].

B.3 [F] *is diagonal for independent data*

Next, consider the special case of independent data **y**. Then by definition

$$p(\mathbf{y}|\boldsymbol{\theta}) = \prod_i p_i(y_i|\boldsymbol{\theta}) = \prod_i p_i(y_i|\theta_i), \tag{B12}$$

the latter by the fact that, by the form of (B1), for independent data **y** a change in θ_j does not influence in any way y_i for $i \neq j$. Then taking the logarithm and differentiating gives

$$\frac{\partial \ln p}{\partial \theta_i} = \frac{\partial \ln p_i}{\partial \theta_i}. \tag{B13}$$

Then by Eqs. (B12), (B13) and definition (B7),

$$F_{ij} = \int \prod_k dy_k \, p_k(y_k|\theta_k) \frac{\partial \ln p_i}{\partial \theta_i} \frac{\partial \ln p_j}{\partial \theta_j}. \tag{B14}$$

This evaluates differently according to whether $i \neq j$ or $i = j$. In the former case we get

$$F_{ij} = F_i F_j, \quad F_i \equiv \int dy_i \, p_i \frac{\partial \ln p_i}{\partial \theta_i}, \quad p_i \equiv p_i(y_i|\theta_i), \tag{B15}$$

since all other integrals dy_k integrate out to unity by normalization. But then by (B15)

$$F_i = \int dy_i \frac{\partial p_i}{\partial \theta_i} = \frac{\partial}{\partial \theta_i} \int dy_i \, p_i = \frac{\partial}{\partial \theta_i} 1 = 0. \tag{B16}$$

Then by the first Eq. (B15) $F_{ij} = 0$, $i \neq j$. This proves the supposition that [F] is diagonal in this case.

The diagonal elements are evaluated by taking $i = j$ in Eq. (B14). This gives

$$F_{ii} = \int dy_i \, p_i \left(\frac{\partial \ln p_i}{\partial \theta_i} \right)^2 \tag{B17}$$

since the remaining integrals dy_k integrate out to unity, by normalization. The information (B17) is precisely of the form (1.9) for one-dimensional data.

B.4 *Cramer–Rao inequalities for general data*

We now use the fact that matrix [M], by its definition (B5), is positive-definite. Hence its determinant is greater than or equal to zero. Expanding form (B6) for [M] by cofactors along its top row, we get

$$\det[M] = e_1^2 \det[F] - 1 \cdot cof[F_{11}] \geq 0. \tag{B18}$$

Therefore

$$e_1^2 \geqslant \frac{cof\,[F_{11}]}{det\,[F]} = [F]_{11}^{-1}, \tag{B19}$$

the $(1,\,1)$ element of the inverse matrix to [F]. This is the Cramer–Rao inequality obeyed by error e_1^2.

The Cramer–Rao inequality for the general error component e_i^2, $i = 1, \ldots, N$ may be derived entirely analogously to the preceding steps (B4)–(B19). The result is

$$e_i^2 \geqslant [F]_{ii}^{-1}, \tag{B20}$$

the ith element along the diagonal of the inverse matrix to [F].

B.5 Cramer–Rao inequalities for independent data

We found in Sec. B.3 that when the data are independent the Fisher information matrix [F] is diagonal. Now, the inverse of a diagonal matrix [F] is a diagonal matrix whose elements are the reciprocals of the respective diagonal elements of [F]. Then the general Cramer–Rao inequality (B20) becomes

$$e_i^2 \geqslant 1/F_{ii}, \; i = 1, \ldots, N, \tag{B21}$$

where F_{ii} is given by Eq. (B17). This verifies Eqs. (2.6), (2.7).

Appendix C

Cramer–Rao inequality for an imaginary parameter

Our EPI approach is covariant, so that one or more parameters are always purely imaginary. This has the ramification that in the Euler–Lagrange solution to a given physical problem there is differentiation with respect to the imaginary quantity as well as with respect to the other, real quantities. This is beneficial to the theory, as this common treatment of real and imaginary parameters leads to correct answers. However, there is a small complication.

As it turns out, the Fisher information I for an imaginary parameter is *negative* (see below). This sets it apart from real parameters, for which I is positive (by the form of Eq. (1.9)). Furthermore, the mean-square error e^2 is also negative. However, the result is that the product $e^2 I$ is positive, so that a Cramer–Rao inequality (1.1) is obeyed once again. These effects are shown next.

C.1 Analysis

Suppose that we have an estimation problem where all coordinates are imaginary: the unknown parameter $\theta \equiv ia$, the vector of data values $\mathbf{y} \equiv i\boldsymbol{\xi}$ and the scalar estimator $\hat{\theta} \equiv \hat{\theta}(\mathbf{y}) \equiv i\hat{a}(\boldsymbol{\xi})$, where the quantities a, $\boldsymbol{\xi}$, $\hat{a}(\boldsymbol{\xi})$ are purely real. Parameter θ is one component of a general vector $\boldsymbol{\theta}$, as in Chap. 2. Can we define appropriate expressions for the information I and the error e^2 in the imaginary coordinates?

To find out, we first find the Cramer–Rao inequality for the real parts (amplitudes) of the imaginary coordinates above. Start out as at Eq. (1.3),

$$\int d\boldsymbol{\xi} \, p(\boldsymbol{\xi}|a)[\hat{a}(\boldsymbol{\xi}) - a] = 0. \tag{C1}$$

This is simply a statement that the (real part of the) estimator is unbiased. Next, we follow the same procedure as beyond (1.3), differentiating $\partial/\partial a$, etc. Corresponding to Eq. (1.8) we get

$$\left[\int d\boldsymbol{\xi} \left(\frac{\partial \ln p}{\partial a} \right)^2 p \right] \left[\int d\boldsymbol{\xi} (\hat{a} - a)^2 p \right] \geq 1. \tag{C2}$$

This is the Cramer–Rao relation for the real parts of the coordinates of the problem.

Intuitively, one might expect it to be obeyed by the full (imaginary) coordinates as well. This is, in fact, verified below.

The definition Eq. (1.9) of Fisher information may be more simply stated as

$$I \equiv \left\langle \left(\frac{\partial}{\partial \theta} \ln p \right)^2 \right\rangle, \quad p = p(\mathbf{y}|\theta) = p(\xi|a). \tag{C3}$$

The notation $\langle\ \rangle$ denotes an expectation. Parameter θ is ordinarily a real quantity. However, our aim here is to define I for an imaginary parameter. Therefore, we extend definition (C3) to describe an imaginary parameter θ as well. In other words, definition (C3) is generalized to represent an *analytical function* of the parameter θ. Treating both real and imaginary parameters the same way is appropriate on the basis of simple consistency as well.

The last equality in Eq. (C3) follows because the PDF (probability density function) $p(z)$ for a complex event $z = x + iy$ does not exist *per se*. Instead one uses a PDF $p(x, y)$ on the real arguments (x, y) (Frieden, 1991).

Now since $\theta = ia$ we have $da/d\theta = 1/i$, so that

$$\frac{\partial \ln p(\mathbf{y}|\theta)}{\partial \theta} = \frac{\partial \ln p(\xi|a)}{\partial a} \frac{da}{d\theta} = \frac{1}{i} \frac{\partial \ln p(\xi|a)}{\partial a}. \tag{C4}$$

The second Eq. (C3) was also used. Combining Eqs. (C3) and (C4) gives

$$I = -\left\langle \left(\frac{\partial \ln p(\xi|a)}{\partial a} \right)^2 \right\rangle = -\int d\xi \left(\frac{\partial \ln p}{\partial a} \right)^2 p. \tag{C5}$$

The Fisher information for an imaginary parameter is negative.

Also, the mean-square error e^2 obeys, from Eq. (1.10),

$$e^2 \equiv \langle [\hat{\theta}(\mathbf{y}) - \theta]^2 \rangle. \tag{C6}$$

Again using $\theta = ia$ and $\hat{\theta} = i\hat{a}$, we get

$$e^2 = -\langle [\hat{a}(\xi) - a]^2 \rangle = -\int d\xi [\hat{a}(\xi) - a]^2 p(\xi|a). \tag{C7}$$

The mean-square error is negative.

Hence, the Fisher information (C5) and the mean-square error (C7) for an imaginary parameter are both negative. This is not surprising: by their definitions, both I and e^2 are squared distance measures, i.e., L2 norms. Squared imaginary distances are negative. But, do the negative I and e^2 obey Cramer–Rao theory? The basic inequality (C2) may be rewritten as

$$\left[-\int d\xi \left(\frac{\partial \ln p}{\partial a} \right)^2 p \right] \left[-\int d\xi (\hat{a} - a)^2 p \right] \geqslant 1. \tag{C8}$$

Comparison with Eqs. (C5) and (C7) verifies that the Cramer–Rao inequality is still obeyed by these negative quantities.

Also, at efficiency we then still have

$$e^2_{\text{eff}} = 1/I. \tag{C9}$$

This was exactly the relation needed, at Eq. (2.9), to attain the indicated equality (for this component θ of $\boldsymbol{\theta}$ in the indicated sum). The result is an added (negative) term in Eq. (2.10), our basic information form. The negativity of this term is exactly what is needed in many of the derived wave equations. It supplies the minus sign for the time derivative term in the d'Alembertian (see Chaps. 4–6).

Appendix D

Simplified derivation of the Schroedinger wave equation

Here, a one-dimensional analysis is given. The derivation runs parallel to the fully covariant EPI derivation in Chap. 4 of the Klein–Gordon equation. We point out corresponding results as they occur.

The position θ of a particle of mass m is measured as a value $y = \theta + x$ (see Eq. (2.1)), where x is a random excursion whose probability amplitude law $q(x)$ is sought. Since the time t is ignored, we are in effect seeking a stationary solution to the problem. Notice that the one-dimensional nature of x violates the premise of covariant coordinates as made in Sec. 4.1.2.

Since the approach is no longer covariant, it is being used improperly. However, a benefit of EPI is that it has a kind of 'robustness' to improper use. It gives approximate answers when used in an approximate sense. The approximate answer will be the non-relativistic, Schroedinger wave equation.

Assume that the particle is moving in a conservative field of scalar potential $V(x)$. Then the total energy W is conserved.

Again using definition (2.24) to define new, complex wave functions $\psi_n(x)$, the information expression (2.19) now becomes Eq. (3.38), which in one dimension is

$$I = 4N \sum_{n=1}^{N/2} \int dx \left| \frac{d\psi_n(x)}{dx} \right|^2. \tag{D1}$$

As at Eq. (4.4) we define a Fourier transform space consisting of functions $\psi_n(\mu)$ of momentum μ obeying

$$\psi_n(x) = \frac{1}{\sqrt{2\pi\hbar}} \int d\mu \, \phi_n(\mu) \exp(-i\mu x/\hbar). \tag{D2}$$

The unitary nature of this transformation guarantees the validity of the EPI variational procedure (Sec. 3.8.7).

Since ψ and ϕ are Fourier mates, we may use Parseval's theorem to get

$$I = \frac{4N}{\hbar^2} \int d\mu \mu^2 \sum_n |\phi_n(\mu)|^2 \equiv J. \tag{D3}$$

This corresponds to Eq. (4.9), and is the invariance principle for the given measurement problem.

The x-coordinate expressions analogous to Eqs. (4.10) and (4.11) show that the sum in (D3) is a probability density $P(\mu)$ on μ. Then (D3) is actually an expectation

$$J = \frac{4N}{\hbar^2} \langle \mu^2 \rangle. \tag{D4}$$

Now we use the specifically non-relativistic approximation that the kinetic energy E_{kin} of the particle is $\mu^2/(2m)$. Then

$$J = \frac{8Nm}{\mu^2} \langle E_{kin} \rangle = \frac{8Nm}{\hbar^2} \langle [W - V(x)] \rangle. \tag{D5}$$

This may be re-expressed in x-coordinate space as

$$J = \frac{8Nm}{\hbar^2} \int dx [W - V(x)] \sum_n |\psi_n(x)|^2 \tag{D6}$$

since the latter sum is the PDF $p(x)$ by Eq. (2.25). Eq. (D6) is the bound information functional $J[\mathbf{q}] \equiv J[\psi]$ for the problem. By Eqs. (D3) and (3.18), parameter $\kappa = 1$ for this problem.

Use of Eqs. (D1) and (D6) in definition (3.16) of K gives a variational problem

$$K = N \sum_{n=1}^{N/2} \int dx \left[4 \left| \frac{d\psi_n(x)}{dx} \right|^2 - \frac{8m}{\hbar^2} [W - V(x)] |\psi_n(x)|^2 \right] = Extrem. \tag{D7}$$

The Euler-Lagrange equation for the problem is

$$\frac{d}{dx} \left(\frac{\partial \mathscr{L}}{\partial \psi_{nx}^*} \right) = \frac{\partial \mathscr{L}}{\partial \psi_n^*}, \quad n = 1, \ldots, N/2, \ \psi_{nx}^* \equiv \partial \psi_n^*/\partial x. \tag{D8}$$

Using as Lagrangian $\mathscr{L}$ the integrand of (D7), (D8) gives a solution

$$\psi_n''(x) + \frac{2m}{\hbar^2} [W - V(x)] \psi_n(x) = 0, \ n = 1, \ldots, N/2. \tag{D9}$$

This is the Schroedinger wave equation without the time.

We see that the form of Eq. (D9) is the same for each index value n. Therefore, the scenario admits of $N = 2$ degrees of freedom $q_n(x)$ or, by Eq. (2.24), one complex degree of freedom $\psi(x)$, for its description. This is the usual result that the SWE defines a single complex wave function.

A noteworthy aspect of this derivation is that it works with a purely real Fisher coordinate x. This implies that the information transfer game of Sec. 3.4.12 is played here. The payoff of the game is the Schroedinger wave equation.

Appendix E

Factorization of the Klein–Gordon information

The aims here are (a) to establish result (4.39), the factorization property of the helper vectors $\mathbf{v}_1$, $\mathbf{v}_2$; and (b) to show that $S_4 + S_5 = 0$ for $\mathbf{v}_1 = 0$ or $\mathbf{v}_2 = 0$. These require that we evaluate all terms in the dot product indicated in Eq. (4.40). Commutation rules for the matrices $[\alpha]$, $[\beta]$ will result.

Accordingly, using definitions (4.37), we form

$$
\mathbf{v}_1 \cdot \mathbf{v}_2 = \sum_{n=1}^{N/2} \left(i \sum_{m=1}^{3} \sum_{j=1}^{N/2} \alpha_{mnj} \psi_{mj} - \sum_{j=1}^{N/2} \beta_{nj} \eta \psi_j + i\lambda \psi_{4n} \right)
$$

$$
\times \left(i \sum_{l=1}^{3} \sum_{k=1}^{N/2} \alpha_{lkn} \psi_{lk}^* + \sum_{k=1}^{N/2} \beta_{kn} \eta \psi_k^* - i\lambda \psi_{4n}^* \right). \tag{E1}
$$

The notation is as follows. Quantity α_{lmn} denotes the (m, n) element of matrix $[\alpha_l]$, $l = 1, 2, 3$ corresponds to components x, y, z respectively, and

$$
\partial \psi_j / \partial x \equiv \psi_{1j}, \ \partial \psi_j / \partial y \equiv \psi_{2j}, \ \partial \psi_j / \partial z \equiv \psi_{3j}, \ \partial \psi_j / \partial t \equiv \psi_{4j}. \tag{E2}
$$

We also assumed that the matrices are Hermitian,

$$
\alpha_{lnk}^* \equiv \alpha_{lkn}, \qquad \beta_{nk}^* \equiv \beta_{kn}. \tag{E3}
$$

We proceed to evaluate all individual terms in Eq. (E1).

Product of first terms in (E1)

This is

$$
S_1 \equiv - \sum_{n=1}^{N/2} \sum_{l,m=1}^{3} \sum_{j,k=1}^{N/2} \alpha_{lkn} \alpha_{mnj} \psi_{mj} \psi_{lk}^*,
$$

$$
= - \sum_{\substack{j,k,l,m,n \\ m<l}} \alpha_{lkn} \alpha_{mnj} \psi_{mj} \psi_{lk}^* - \sum_{\substack{j,k,l,m,n \\ m\leqslant l}} \alpha_{mkn} \alpha_{lnj} \psi_{lj} \psi_{mk}^* \tag{E4}
$$

296

after renaming $l \to m$, $m \to l$ in the second sum. Rearranging orders of summation and combining the two sums in (E4) gives

$$S_1 = - \sum_{\substack{j,k,l,m \\ m < l}} \left(\psi_{mj} \psi_{lk}^* \sum_n \alpha_{lkn} \alpha_{mnj} + \psi_{lj} \psi_{mk}^* \sum_n \alpha_{mkn} \alpha_{lnj} \right)$$

$$- \sum_{j,k,m} \psi_{mj} \psi_{mk}^* \sum_n \alpha_{mkn} \alpha_{mnj} \tag{E5}$$

where the last sum is for $m = l$. Now suppose that

$$\sum_n \alpha_{lkn} \alpha_{mnj} = - \sum_n \alpha_{mkn} \alpha_{lnj}, \text{ all } j, k, l, m, l \neq m. \tag{E6}$$

In matrix notation this means that

$$[\alpha_l][\alpha_m] = -[\alpha_m][\alpha_l], \ l \neq m. \tag{E7}$$

The matrices $[\alpha_x]$, $[\alpha_y]$, $[\alpha_z]$ anticommute. The effect on Eq. (E5) is directly

$$S_1 = - \sum_{\substack{j,k,l,m \\ m \leqslant l}} \sum_n \alpha_{mkn} \alpha_{lnj} (\psi_{lj} \psi_{mk}^* - \psi_{mj} \psi_{lk}^*) - \sum_{j,k,m} \psi_{mj} \psi_{mk}^* \sum_n \alpha_{mkn} \alpha_{mnj}. \tag{E8}$$

When S_1 is integrated $d\mathbf{r}dt$, the first double sum contributes zero. This is shown in Appendix F, Eqs. (F1)–(F5). Hence, the first double sum effectively contributes zero to S_1 of (E4).

Regarding the second double sum in (E8), suppose that matrices $[\alpha_m]$ obey

$$\sum_n \alpha_{mkn} \alpha_{mnj} = \delta_{jk}.$$

This is equivalent to the matrix statement

$$[\alpha_m]^2 = [1], \ m = 1, 2, 3, \tag{E9}$$

where [1] denotes the diagonal unit matrix. Then directly the second double sum becomes

$$- \sum_{j,k,m} \psi_{mj} \psi_{mk}^* \delta_{jk} = - \sum_{m,j} \psi_{mj} \psi_{mj}^* = - \sum_{j=1}^{N/2} \nabla \psi_j \cdot \nabla \psi_j^*. \tag{E10}$$

The latter is by use of the notation (E2). Equation (E10) gives the first right-hand term in the required Eq. (4.39), i.e.,

$$\iint d\mathbf{r} \, dt S_1 = - \iint d\mathbf{r} \, dt \sum_j \nabla \psi_j^* \cdot \nabla \psi_j. \tag{E11}$$

Product of second terms in (E1)

This is

$$S_2 \equiv -\eta^2 \sum_{j,k,n} \beta_{kn} \beta_{nj} \psi_j \psi_k^*. \tag{E12}$$

Suppose that matrix $[\beta]$ obeys

$$\sum_n \beta_{kn} \beta_{nj} = \delta_{jk}.$$

This is equivalent to a matrix requirement

$$[\beta]^2 = [1].$$ (E13)

Then S_2 becomes directly

$$S_2 = -\eta^2 \sum_{j,k} \psi_j \psi_k^* \delta_{jk} = -\eta^2 \sum_{j=1}^{N/2} \psi_j \psi_j^*.$$ (E14)

This has the form of the third right-hand term in form (4.39), as required.

Cross products of first and second terms

These are

$$S_3 \equiv i\eta \sum_{j,k,l} \psi_k^* \psi_{lj} \sum_n \beta_{kn} \alpha_{lnj} - i\eta \sum_{j,k,l} \psi_j \psi_{lk}^* \sum_n \alpha_{lkn} \beta_{nj}.$$ (E15)

We renamed dummy summation index m to l in the first sum. Suppose that

$$\sum_n \alpha_{lkn} \beta_{nj} = -\sum_n \beta_{kn} \alpha_{lnj}.$$ (E16)

In matrix form this states that

$$[\alpha_l][\beta] = -[\beta][\alpha_l].$$ (E17)

That is, matrix $[\beta]$ *anticommutes with matrices* $[\alpha_x]$, $[\alpha_y]$, $[\alpha_z]$.

Using identity (E16) in Eq. (E15) gives

$$S_3 = -i\eta \sum_{j,k,l,n} \alpha_{lkn} \beta_{nj} (\psi_k^* \psi_{lj} + \psi_j \psi_{lk}^*).$$ (E18)

It is shown in Appendix F that the integral $d\mathbf{r}\, dt$ of this sum vanishes,

$$\iint d\mathbf{r}\, dt S_3 = 0.$$ (E19)

(In particular, see Eqs. (F6)–(F9).)

Cross products of second and third terms

The product of these terms is a sum

$$S_4 \equiv i\eta\lambda \sum_{k,n} (\psi_k^* \psi_{4n} \beta_{kn} + \psi_k \psi_{4k}^* \beta_{nk}),$$ (E20)

after renaming the dummy index j to k. Replacing β_{kn} by β_{nk}^* (hermiticity property (E3)) gives

$$S_4 = i\eta\lambda \sum_{k,n} (\psi_k^* \psi_{4n} \beta_{nk}^* + \psi_k \psi_{4n}^* \beta_{nk}).$$ (E21)

This is a purely imaginary term. It will cancel another imaginary term at the solution for $\mathbf{v}_1$, $\mathbf{v}_2$. That term is found next.

Cross products of first and third terms

This is a sum

$$S_5 \equiv -\lambda \left(\sum_{k,l,n} \alpha^*_{lnk} \psi^*_{lk} \psi_{4n} - \sum_{k,l,n} \alpha_{lnk} \psi_{lk} \psi^*_{4n} \right) \tag{E22}$$

after renaming dummy indices j, m to k, l, respectively, and replacing α_{lkn} by α^*_{lnk}, hermiticity property (E3). The first sum in (E22) is the complex conjugate of the second. Therefore, the difference is pure imaginary. Hence, S_5 is pure imaginary.

Product of last terms

These two terms multiply each other to give a contribution

$$S_6 \equiv \lambda^2 \sum_n \psi_{4n} \psi^*_{4n},$$

or

$$S_6 = \lambda^2 \sum_n \frac{\partial \psi_n}{\partial t} \frac{\partial \psi^*_n}{\partial t} \tag{E23}$$

by notation (E2).

Total sum of products

Using sum results (E11), (E14), (E19), (E21), (E22) and (E23) in Eq. (E1), we have

$$\iint d\mathbf{r}\, dt \mathbf{v}_1 \cdot \mathbf{v}_2 \equiv \iint d\mathbf{r}\, dt \sum_{j=1}^{6} S_j$$

$$= \iint d\mathbf{r}\, dt \sum_{n=1}^{N/2} \left[-(\nabla \psi)^* \cdot \nabla \psi_n + \lambda^2 \left(\frac{\partial \psi_n}{\partial t} \right)^* \frac{\partial \psi_n}{\partial t} - \eta^2 |\psi_n|^2 \right]$$

$$+ i \iint d\mathbf{r}\, dt (S_4 + S_5). \tag{E24}$$

This accomplishes the required goal (a) of the appendix.

Evaluation at $\mathbf{v}_1 = 0$

We can rearrange the terms of Eqs. (E21) and (E22) to give

$$S_4 + S_5 = \lambda \sum_n \psi_{4n} \left(i\eta \sum_k \psi^*_k \beta^*_{nk} - \sum_{k,l} \psi^*_{lk} \alpha^*_{lnk} \right)$$

$$+ \lambda \sum_n \psi^*_{4n} \left(i\eta \sum_k \psi_k \beta_{nk} + \sum_{k,l} \psi_{lk} \alpha_{lnk} \right). \tag{E25}$$

By Eqs. (4.36) and (4.37), we can express the fact that the nth component $v_{1n} = 0$ as

$$v_{1n} \equiv i \sum_k \left[\alpha_{1nk} \frac{\partial \psi_k}{\partial x} + \alpha_{2nk} \frac{\partial \psi_k}{\partial y} + \alpha_{3nk} \frac{\partial \psi_k}{\partial z} \right] - \eta \sum_k \beta_{nk} \psi_k + i\lambda \frac{\partial \psi_n}{\partial t} \equiv 0. \quad (E26)$$

Using the derivative notation (E2) gives

$$v_{1n} = i \sum_{l,k=1}^{3} \alpha_{lnk} \psi_{lk} - \eta \sum_k \beta_{nk} \psi_k + i\lambda \psi_{4n} = 0. \tag{E27}$$

Multiplying this by i gives

$$i\eta \sum_k \beta_{nk} \psi_k + \sum_{l,k=1}^{3} \alpha_{lnk} \psi_{lk} = -\lambda \psi_{4n}. \tag{E28}$$

The left-hand side is the last factor in Eq. (E25). Moreover, its complex conjugate is the first factor. The result is that (E25) becomes

$$S_4 + S_5 = \lambda \sum_n \psi_{4n} \lambda \psi_{4n}^* + \lambda \sum_n \psi_{4n}^* (-\lambda \psi_{4n}) = 0 \tag{E29}$$

identically.

Evaluation at **$v_2 = 0$**

Because of relation (4.38) between $\mathbf{v}_1$ and $\mathbf{v}_2$, result (E29) must hold if the alternative solution $\mathbf{v}_2 = 0$ is used. We leave this as an exercise for the reader.

Hence, goal (b) of this appendix has been accomplished.

Appendix F

Evaluation of certain integrals

The aims are to show that (a) the integral of the first sum in Eq. (E8) is zero; and (b) the integral of $S_3 = 0$.

To show (a), arbitrarily consider the case $l = 3$, $m = 1$. The integrals in question are then (for arbitrary j, k)

$$T = \iint d\boldsymbol{r}\, dt(\psi_{3j}\psi_{1k}^* - \psi_{1j}\psi_{3k}^*) \equiv T_{zx} - T_{xz}. \tag{F1}$$

It is convenient to work in frequency space. Represent each $\psi_j(\boldsymbol{r}, t)$ by its Fourier transform, Eq. (4.4). Then by Parseval's theorem

$$T_{zx} \equiv \iint d\boldsymbol{r}\, dt\, \frac{\partial \psi_j}{\partial z}\frac{\partial \psi_k^*}{\partial x} = \iint d\boldsymbol{\mu}\, dE \phi_j(\boldsymbol{\mu}, E)\phi_k^*(\boldsymbol{\mu}, E)\mu_3\mu_1. \tag{F2}$$

Likewise we get

$$T_{xz} \equiv \iint d\boldsymbol{r}\, dt\, \frac{\partial \psi_j}{\partial x}\frac{\partial \psi_k^*}{\partial z} = \iint d\boldsymbol{\mu}\, dE \phi_j(\boldsymbol{\mu}, E)\phi_k^*(\boldsymbol{\mu}, E)\mu_1\mu_3. \tag{F3}$$

Comparing Eqs. (F2) and (F3) shows that

$$T_{xz} = T_{zx}. \tag{F4}$$

Then by (F1)

$$T = 0. \tag{F5}$$

But the integral of the first double sum in Eq. (E8) is a weighted sum of integrals of the form (F1). Hence, its value is zero. This was goal (a) of the appendix.

By Eq. (E18), goal (b) will be accomplished if we can show that

$$\iint d\boldsymbol{r}\, dt(\psi_j\psi_{lk}^* + \psi_k^*\psi_{lj}) = 0, \quad \text{all } j, k, l. \tag{F6}$$

First consider the case $l = 3$, with j, k arbitrary. Then the integral is

301

$$\iint d\boldsymbol{r}\,dt(\psi_j\psi^*_{3k} + \psi^*_k\psi_{3j}). \tag{F7}$$

Again working in frequency space, we have by Parseval's theorem

$$\iint d\boldsymbol{r}\,dt\ \psi_j\psi^*_{3k} \equiv \iint d\boldsymbol{r}\,dt\ \psi_j\frac{\partial\psi^*_k}{\partial z} = \frac{i}{\hbar}\iint d\boldsymbol{\mu}\,dE\ \phi_j(\boldsymbol{\mu})\phi^*_k(\boldsymbol{\mu})\mu_3. \tag{F8}$$

Likewise

$$\iint d\boldsymbol{r}\,dt\ \psi^*_k\psi_{3j} \equiv \iint d\boldsymbol{r}\,dt\ \psi^*_k\frac{\partial\psi_j}{\partial z} = -\frac{i}{\hbar}\iint d\boldsymbol{\mu}\,dE\ \phi^*_k(\boldsymbol{\mu})\phi_j(\boldsymbol{\mu})\mu_3, \tag{F9}$$

the negative of (F8). Therefore, the sum of (F8) and (F9) is zero, for this case *l*. But obviously the same derivation can be carried through for any chosen *l*. Therefore, task (b) is accomplished.

Appendix G

Schroedinger wave equation as a non-relativistic limit

It is straightforward to derive the equation as the non-relativistic limit of the Klein–Gordon equation (Schiff, 1955). Multiply out the factors and dot-products in the Klein–Gordon Eq. (4.28), bringing all time-derivative terms to the left and all space-derivative terms to the right:

$$\left(\hbar^2 \frac{\partial^2}{\partial t^2} + 2i\hbar e\phi \frac{\partial}{\partial t} - e^2\phi^2 + i\hbar e \frac{\partial\phi}{\partial t} \right)\psi$$

$$= [c^2\hbar^2\nabla^2 - 2iec\hbar(\boldsymbol{A}\cdot\nabla) - iec\hbar(\nabla\cdot\boldsymbol{A}) - e^2|\boldsymbol{A}|^2 - m^2c^4]\psi. \quad (G1)$$

We used the previously found scalar nature of the wave function, $\psi_n = \psi$, and the easily established identity

$$\nabla\cdot(\boldsymbol{A}\psi) = \psi\nabla\cdot\boldsymbol{A} + \boldsymbol{A}\cdot\nabla\psi. \quad (G2)$$

Use, now, the non-relativistic representation for ψ of

$$\psi(\boldsymbol{r},\, t) = \psi'(\boldsymbol{r},\, t)e^{-imc^2 t/\hbar}. \quad (G3)$$

Effectively, the rest energy mc^2 has been separated out of ψ to form a new wave function ψ'. Note that this step violates relativistic covariance by separating out a time t component $\exp(-imc^2 t/\hbar)$ from the rest of the wave function. This is part of the non-relativistic limit taken.

The overall aim is to substitute the form Eq. (G3) into Eq. (G1). The left-hand side of the latter requires time derivatives of ψ. Differentiating Eq. (G3) once, and then twice, gives

$$\frac{\partial\psi}{\partial t} = \left(\frac{\partial\psi'}{\partial t} - \frac{imc^2}{\hbar}\psi' \right)e^{-imc^2 t/\hbar} \quad (G4)$$

and

$$\frac{\partial^2\psi}{\partial t^2} = \left(\frac{\partial^2\psi'}{\partial t^2} - 2\frac{imc^2}{\hbar}\frac{\partial\psi'}{\partial t} - \frac{m^2c^4}{\hbar^2}\psi' \right)e^{-imc^2 t/\hbar}. \quad (G5)$$

303

The first right-hand terms in these equations can be neglected. Then, ignoring as well the last two terms on the left-hand side (LHS) of Eq. (G1), substitute the resulting Eqs. (G4), (G5) into the LHS of Eq. (G1). The result is

$$\text{LHS} = -\left(2imc^2\hbar \frac{\partial \psi'}{\partial t} + m^2 c^4 \psi' \right) e^{-imc^2 t/\hbar} - 2i\hbar e\phi(-imc^2/\hbar)\psi' e^{-imc^2 t/\hbar}. \quad (G6)$$

We now substitute this for the LHS of Eq. (G1). A term $-m^2 c^4 \psi' \exp(-imc^2 t/\hbar)$ common to both the LHS and RHS cancels. Multiplying both sides of the result by $-1/2mc^2$ and renaming $\psi' = \psi$ gives a result

$$i\hbar \frac{\partial \psi}{\partial t} - e\phi\psi = -\frac{\hbar^2}{2m}\nabla^2\psi + \frac{ie\hbar}{mc} A\cdot\nabla\psi + \frac{ie\hbar}{2mc}(\nabla\cdot A)\psi + \frac{e^2|A|^2}{2mc^2}\psi. \quad (G7)$$

This is the Schroedinger wave equation (SWE) in the presence of general electromagnetic fields A, ϕ. In the usual case of a purely scalar field ϕ, it becomes the standard expression

$$i\hbar \frac{\partial \psi}{\partial t} = -\frac{\hbar^2}{2m}\nabla^2\psi + e\phi\psi. \quad (G8)$$

This is the SWE with the time.

Appendix H

Non-uniqueness of potential **A** for finite boundaries

The proof is due to Ferwerda (1998). Any one component $g = g(\boldsymbol{r}, t)$ of Eq. (5.51) is of the form

$$\Box g = h, \tag{H1}$$

where $h = h(\boldsymbol{r}, t)$ is an arbitrary source function. The question we address is whether g is unique, given that it obeys either Dirichlet or Neumann conditions on (now) a closed *four*-dimensional surface M_0 that encloses space $(\boldsymbol{r}, t)$. *Note*: The Dirichlet condition is that $g = 0$ on M_0; the Neumann condition is that $\boldsymbol{\nabla} g = 0$ on M_0.

Assume, contrarywise, that g is not unique, so that Eq. (H1) has two different solutions g_1 and g_2. The difference $g_{12} \equiv g_1 - g_2$ then satisfies, by Eq. (H1),

$$\Box g_{12} = 0. \tag{H2}$$

We would like to show that the only solution to this problem is that $g_{12} = 0$ within the bounding four-dimensional surface M_0. However, this will not be generally forthcoming.

Start from Green's first identity (Jackson, 1975) in four-space $(\boldsymbol{r}, ct)$,

$$c \int_V d\boldsymbol{r}\, dt (\boldsymbol{\nabla}\phi \cdot \boldsymbol{\nabla}\psi + \phi \Box \psi) = \int_{M_0} d\boldsymbol{\sigma} \cdot \boldsymbol{\nabla}\psi\, \phi \tag{H3}$$

$$\text{where } \boldsymbol{\nabla} \equiv \left(\boldsymbol{\nabla}, \frac{1}{ic}\frac{\partial}{\partial t}\right), \; d\boldsymbol{\sigma} \equiv (d\boldsymbol{\sigma}_3, icdt).$$

Quantity $d\boldsymbol{\sigma}$ is a four-component 'surface' element with space part $d\boldsymbol{\sigma}_3$, and V is the four-volume enclosed by surface M_0. Functions ϕ and ψ are arbitrary but differentiable.

Now take $\phi = \psi = g_{12}$ in Eq. (H3). Then using Eq. (H2) as well, we obtain

$$c \int_V d\boldsymbol{r}\, dt \boldsymbol{\nabla} g_{12} \cdot \boldsymbol{\nabla} g_{12} = \int_{M_0} d\boldsymbol{\sigma} \cdot \boldsymbol{\nabla} g_{12}\, g_{12}. \tag{H4}$$

But then by the Dirichlet or Neumann conditions the right-hand side is zero. Since volume V in the left-hand side integral is arbitrary, its integrand must be zero,

$$\nabla g_{12} \cdot \nabla g_{12} = 0, \text{ i.e., } (\nabla g_{12})^2 - \left(\frac{1}{c}\frac{\partial g_{12}}{\partial t}\right)^2 = 0. \tag{H5}$$

This equation is satisfied by any constant or any function $g_{12}(x \pm ct)$, $g_{12}(y \pm ct)$, $g_{12}(z \pm ct)$ or $g_{12}(r \pm ct)$, as substitution will show. These represent travelling waves of arbitrary shape. Hence, Eq. (H5) is not satisfied only by $g_{12} = 0$, as we hoped.

In conclusion, Eq. (H1), and hence, Eq. (5.51), does not have a unique solution in the case of a finite space (r, t) and Dirichlet or Neumann conditions on the boundary.

References

Abramowitz, A. and Stegun, I., *Handbook of Mathematical Functions*, National Bureau of Standards, Washington, D.C. (1965)

Akhiezer, A. I. and Berestetskii, V. B., *Quantum Electrodynamics*, Wiley, New York (1965)

Allen, C. W., *Astrophysical Quantities*, 3rd edn, Athlone Press, London (1973)

Amari, S., *Differential-Geometrical Methods in Statistics, Lecture Notes in Statistics, vol. 28*, Springer-Verlag, New York (1985)

Arthurs, E. and Goodman, M., *Phys. Rev. Lett.* **60**, 2447 (1988)

Barnes, J. A. and Allan, D. W., *Proc. IEEE* **54**, 176 (1966)

Barrow, J. D., *Theories of Everything*, Clarendon Press, Oxford (1991), p. 101

Batchelor, G. K., *The Theory of Homogeneous Turbulence*, Cambridge University Press, Cambridge (1956)

Beckner, W., *Ann. Math.* **102**, 159 (1975)

Bell, D. A., *J. Phys.* C**13**, 4425 (1980)

Bialynicki-Birula, I., *Quantum Electrodynamics*, Pergamon Press, Oxford (1975)

Bialynicki-Birula, I., *Prog. in Optics, vol. XXXVI*, ed. E. Wolf, Elsevier, Amsterdam (1996)

Blachman, N. M., *IEEE Trans. Inf. Theory*, IT-**11**, 267 (1965)

Bohm, D., *Quantum Theory*, Prentice-Hall, Englewood Cliffs, New Jersey (1951), pp. 591–8

Bohm, D., *Phys. Rev.* **85**, 180 (1952); first 3 pages of article

Born, M., *Zeits. f. Physik* **37**, 863 (1926); *Nature* **119**, 354 (1927)

Bracewell, R., *The Fourier Transform and its Applications*, McGraw-Hill, New York (1965)

Bradshaw, P., ed., *Turbulence*, Topics in Applied Physics, vol. 12, Springer-Verlag, New York (1978)

Braunstein, S. L. and Caves, C. M., *Phys. Rev. Lett.* **72**, 3439 (1994)

Brillouin, L., *Science and Information Theory*, Academic, New York (1956)

Caianiello, E. R., *Rivista del Nuovo Cimento* **15**, 7 (1992)

Campbell, M. J. and Jones, B. W., *Science* **177**, 889 (1972)

Carter, B., *Confrontation of Cosmological Theories with Observational Data*, ed. M. S. Longair, IAU Symposium (1974), p. 291

Caves, C. M., *Phys. Rev.* D**33**, 1643 (1986)

Caves, C. M. and Milburn, G. J., *Phys. Rev.* A**36**, 5543 (1987)

Cocke, W. J., *Phys. of Fluids* **8**, 1609 (1996)

Cocke, W. J. and Frieden, B. R., *Found. Physics* **243**, 1397 (1997)

Compton, A. H., *Phys. Rev.* **21**, 715; **22**, 409 (1923)

Cook, R. J., *Phys. Rev.* A**26**, 2754 (1982)

Cover, T. and Thomas, J., *Elements of Information Theory*, Wiley, New York (1991)

Cvitanovic, P., Percival, I. and Wirzba, A., eds., *Quantum Chaos – Quantum Measurement*, Kluwer Academic, Dordrecht (1992)

DeGroot, S. R. and Suttorp, L. G., *Foundations of Electrodynamics*, North-Holland, Amsterdam (1972)

Deutsch, I. H. and Garrison, J. C., *Phys. Rev.* A**43**, 2498 (1991)

DeWitt, B. S., *Phys. Rev.* **160**, 1113 (1967)

DeWitt, B. S., 'The quantum and gravity: the Wheeler–DeWitt equation', *Proc. of 8th Marcel Grossman Conf.*, Jerusalem, eds. R. Ruffini and T. Piran, World Scientific (1997)

Dicke, R. H., *Nature Letters* **192**, 440 (1961)

Dirac, P. A. M., *Proc. R. Soc. Lond.* A**117**, 610 (1928); also *Principles of Quantum Mechanics*, 3rd edn, Oxford, New York (1947)

Dirac, P. A. M., *Nature* **139**, 23 (1937); also *Proc. R. Soc. Lond.* A**165**, 199 (1938)

Ehlers, J., Hepp, K. and Weidenmuller, H. A., eds., *Statistical Models and Turbulence*, Lecture Notes in Physics, vol. 12, Springer-Verlag, New York (1972)

Einstein, A., *Ann. Physik* **17**, 891 (1905); also *The Theory of Relativity*, 6th edn, Methuen, London (1956)

Eisberg, R. M., *Fundamentals of Modern Physics*, Wiley, New York (1961)

Ferwerda, H. A., personal communication (1998)

Feynman, R. P., *Rev. Mod. Phys.* **20**, 367 (1948)

Feynman, R. P. and Hibbs, A. R., *Quantum Mechanics and Path Integrals*, McGraw-Hill, New York (1965), p. 26

Feynman, R. P. and Weinberg, S., *Elementary Particles and the Laws of Physics*, Cambridge University Press, Cambridge (1993)

Fisher, R. A., *Phil. Trans. R. Soc. Lond.* **222**, 309 (1922)

Fisher, R. A., *Ann. Eugenics* **12**, 1 (1943)

Fisher, R. A., *Statistical Methods and Scientific Inference*, 2nd edn, Oliver and Boyd, London (1959)

Fisher Box, J., *R. A. Fisher, the Life of a Scientist*, Wiley, New York (1978); the author is his daughter

Frieden, B. R., *Founds. Physics* **16**, 883 (1986)

Frieden, B. R., *J. Modern Optics* **35**, 1297 (1988)

Frieden, B. R., *Am. J. Phys.* **57**, 1004 (1989)

Frieden, B. R., *Phys. Rev.* A**41**, 4265 (1990)

Frieden, B. R., *Probability, Statistical Optics and Data Testing*, 2nd edn, Springer-Verlag, Berlin (1991)

Frieden, B. R., in *Advances in Imaging and Electron Physics*, ed. P. W. Hawkes, Academic, Orlando (1994)

Frieden, B. R. and Cocke, W. J., *Phys. Rev.* E**54**, 257 (1996)

Frieden, B. R. and Hughes, R. J., *Phys. Rev.* E**49**, 2644 (1994)

Frieden, B. R. and Soffer, B. H., *Phys. Rev.* E**52**, 2274 (1995)

Gardiner, C. W., *Handbook of Stochastic Methods*, 2nd edn, Springer-Verlag, Berlin (1985)

Gardiner, C. W., *Quantum Noise*, Springer-Verlag, Berlin (1991)

Goldstein, H., *Classical Mechanics*, Addison-Wesley, Cambridge, Massachusetts (1950)

Good, I. J., in *Foundations of Probability Theory, Statistical Inference and Statistical Theories of Science, Vol. II*, eds. W. L. Harper and C. A. Hooker, Reidel, Boston (1976)

Gottfried, K., *Quantum Mechanics, vol. I: Fundamentals*, Benjamin, New York (1966), pp. 170–1

Granger, C. W. J., *Econometrics* **34**, 150 (1966)

Halliwell, J. J., Perez-Mercader, J. and Zurek, W. H., eds., *Physical Origins of Time Asymmetry*, Cambridge University Press, Cambridge (1994)

Handel, P. H., *Phys. Rev.* **A3**, 2066 (1971)

Hartle, J. B. and Hawking, S. W., *Phys. Rev.* **D28**, 2960 (1983)

Hawking, S. W., in *General Relativity: an Einstein Centenary Survey*, ed. W. Israel, Cambridge University Press, Cambridge (1979)

Hawking, S. W., *A Brief History of Time*, Bantam Books, Toronto (1988), p. 129

Heitler, W., *The Quantum Theory of Radiation*, Dover, New York (1984); reprint of 3rd edn, Oxford University Press (1954)

Hirschman, I. I., *Am. J. Math.* **79**, 152 (1957)

Hooge, F. N., *Physica* **8B**, 14 (1976)

Holden, A. V., *Models of the Stochastic Activity of Neurons*, Springer-Verlag, Berlin (1976)

Huber, P. J., *Robust Statistics*, Wiley, New York (1981), pp. 77–86

Hughes, R. J., suggested by discussions with (1994)

Jackson, J. D., *Classical Electrodynamics*, 2nd edn, Wiley, New York (1975), pp. 536–541, 549

Jauch, J. M. and Rohrlich, F., *Theory of Photons and Electrons*, Addison-Wesley, Cambridge, Massachusetts (1955), p. 23

Jaynes, E. T., *Phys. Rev.* **106**, 620 (1957a); **108**, 171 (1957b)

Jaynes, E. T., in *Maximum Entropy and Bayesian Methods in Inverse Problems*, eds. C. R. Smith and W. T. Grandy, Reidel, Dordrecht (1985)

Jensen, H. J., *Phys. Scr.* **43**, 593 (1991)

Johnson, J. B., *Phys Rev.* **26**, 71 (1925)

Kamke, E., *Differentialgleichungen Losungsmethoden und Losungen*, Chelsea, New York (1948), p. 551

Kelvin, Lord, *Popular Lectures and Addresses, Vol. I*, Epigraph to Chap. 37, MacMillan, London (1889). We quote: 'When you cannot measure [what you are speaking about], your knowledge is of a meager and unsatisfactory kind: . . . you have scarcely . . . advanced to the stage of science.'

Keshner, M. S., *Proc. IEEE* **70**, 212 (1982)

Korn, G. A. and Korn, T. M., *Mathematical Handbook for Scientists and Engineers*, 2nd edn, McGraw-Hill, New York (1968)

Kuchar, K., in *Relativity, Astrophysics and Cosmology*, ed. W. Israel, Reidel, Dordrecht, Holland (1973), p. 268

Kullback, S., *Information Theory and Statistics*, Wiley, New York (1959)

Lagrange, J. L., *Mecanique Analytique* (1788); republished in series *Oeuvres de Lagrange*, Gauthier-Villars, Paris (1889)

Landau, L. D. and Lifshitz, E. M., *Classical Theory of Fields*, Addison-Wesley, Cambridge, Massachusetts (1951), p. 258

Landauer, R., *IBM J. Research and Development* **5**, 183 (1961)

Lawrie, I. D., *A Unified Grand Tour of Theoretical Physics*, Adam Hilger, Bristol, England (1990)

Lindblad, G., *Non-equilibrium Entropy and Irreversibility*, Reidel, Boston (1983)

Lindsay, R. B., *Concepts and Methods of Theoretical Physics*, Van Nostrand, New York (1951)

Louisell, W. H., in *Quantum Optics*, eds. S. M. Kay and A. Maitland, Academic Press, New York (1970)

Lowen, S. B. and Teich, M. C., *Electron. Lett.* **25**, 1072 (1989)

Mandelbrot, B. B., *The Fractal Geometry of Nature*, Freeman, New York (1983)

Mandelbrot, B. B. and Wallis, J. R., *Water Resour. Res.* **5**, 321 (1969)

Marathay, A. S., *Elements of Optical Coherence Theory*, Wiley, New York (1982)

Martens, H. and de Muynck, W. M., *Phys. Lett.* A**157**, 441 (1991)

Matheron, G., *Adv. Appl. Probab.* **5**, 439 (1973)

Matveev, V. B. and Salle, M. A., *Darboux Transformations and Solitons*, Springer-Verlag, Berlin (1991)

Mensky, M. B., *Phys. Rev.* D**20**, 384 (1979)

Mensky, M. B., *Continuous Quantum Measurements and Path Integrals*, Institute of Physics Publishing, Bristol, England (1993)

Merzbacher, E., *Quantum Mechanics*, 2nd edn, Wiley, New York (1970)

Misner, C. W., Thorne, K. S. and Wheeler, J. A., *Gravitation*, Freeman, New York (1973)

Morgenstern, O. and von Neumann, J., *Theory of Games and Economic Behavior*, Princeton Univ. Press, Princeton (1947)

Morse, P. M. and Feshbach, H., *Methods of Theoretical Physics, Part I*, McGraw-Hill, New York (1953)

Motchenbacher, C. D. and Connelly, J. A., *Low-Noise Electronic System Design*, Wiley, New York (1993)

Musha, T. and Higuchi, H., *Jpn. J. Appl. Phys.* **15**, 1271 (1976)

Nikolov, B., personal communication (1992)

Nikolov, B. and Frieden, B. R., *Phys. Rev.* E**49**, 4815 (1994)

Panofsky, W. K. H. and Phillips, M., *Classical Electricity and Magnetism*, Addison-Wesley, Reading, Massachusetts (1955)

Papoulis, A., *Probability, Random Variables and Stochastic Processes*, McGraw-Hill, New York (1965)

Pauli, W., *Zeits. f. Physik* **43**, 601 (1927)

Pike, E. R. and Sarkar, S., eds., *Quantum Measurement and Chaos*, Plenum, New York (1987)

Plastino, A. R. and Plastino, A., *Phys. Rev.* E**52**, 4580 (1995)

Plastino, A. R. and Plastino, A., *Phys. Rev.* E**54**, 4423 (1996)

Pumir, A., *Phys. Fluids* **6**, 2071 (1994)

Reif, F., *Fundamentals of Statistical and Thermal Physics*, McGraw-Hill, New York (1965)

Reza, F. M., *An Introduction to Information Theory*, McGraw-Hill, New York (1961)

Risken, H., *The Fokker–Planck Equation*, Springer-Verlag, Berlin (1984)

Roman, P., *Theory of Elementary Particles*, North-Holland, Amsterdam (1960), p. 377

Rumer, I. B. and Ryvkin, M. Sh., *Thermodynamics, Statistical Physics and Kinetics*, Mir Publishers, Moscow (1980), p. 526

Ryder, L. H., *Quantum Field Theory*, Cambridge University Press, Cambridge (1987)

Sagdeev, R. Z., Usikov, D. A. and Zaslavskii, G. M., *Nonlinear Physics: From the Pendulum to Turbulence and Chaos*, Harwood Academic, New York (1988)

Schiff, L. I., *Quantum Mechanics*, 2nd edn, McGraw-Hill, New York (1955)

Schroedinger, E., *Ann. Phys.* **79**, 361 (1926)

Shannon, C. E., *Bell Syst. Tech. J.* **27**, 379; 623 (1948)

Shirai, H., Reinterpretation of quantum mechanics based on a statistical interpretation, *Found. Phys.*, to be published (1998)

Sipe, J. E., *Phys. Rev.* A**52**, 1875 (1995)

Soffer, B. H., personal communication (1993)

Soffer, B. H., personal communication (1994)

Solo, V., *SIAM J. Appl. Math.* **52**, 270 (1992)

Stam, A. J., *Information and Control* **2**, 101 (1959)

Stein, E. M. and Weiss, G., *Fourier Analysis on Euclidean Spaces*, Princeton University Press, Princeton, New Jersey (1971)

Szilard, L., *Z. Physik* **53**, 840 (1929)

Takayasu, H., *J. Phys. Soc. Jpn.* **56**, 1257 (1987)

Van Trees, H. L., *Detection, Estimation, and Modulation Theory, Part I*, Wiley, New York (1968)

Vilenkin, A., *Phys. Lett.* **117B**, 25 (1982); *Phys. Rev.* D**27**, 2848 (1983)

Von Mises, R., *Wahrscheinlichkeit, Statistik und Wahrheit*, Springer-Verlag OHG, Vienna (1936)

Von Neumann, J., *Mathematical Foundations of Quantum Mechanics* (English transl.), Princeton University Press, Princeton, New Jersey (1955)

Voss, R. F. and Clarke, J., *J. Acoust. Soc. Am.* **63**, 258 (1978)

Vstovsky, G. V., *Phys. Rev.* E**51**, 975 (1995)

Wheeler, J. A., *Geometrodynamics*, Academic, New York (1962)

Wheeler, J. A., in *Proceedings of the 3rd International Symposium on Foundations of Quantum Mechanics, Tokyo, 1989*, eds. S. Kobayashi, H. Ezawa, Y. Murayama and S. Nomura, Physical Society of Japan, Tokyo (1990), p. 354

Wheeler, J. A., in *Physical Origins of Time Asymmetry*, eds. J. J. Halliwell, J. Perez-Mercader and W. H. Zurek, Cambridge University Press (1994), pp. 1–29

Woodroofe, M. and Hill, B., *J. Appl. Prob.* **12**, 425 (1975)

Wootters, W. K., *Phys. Rev.* D**23**, 357 (1981)

Wyss, W., *J. Math. Phys.* **37**, 2782 (1986)

Zeh, H.-D., *Physical Basis of the Direction of Time*, Springer-Verlag, Berlin (1992)

Index

312